An Introduction to Mathematical Thinking

Algebra and Number Systems

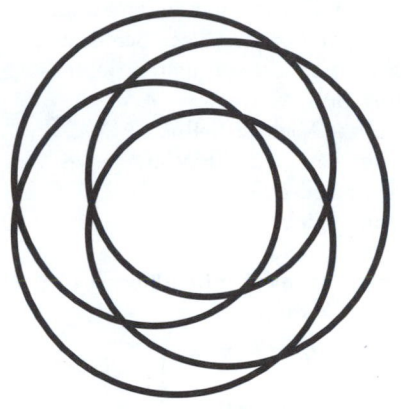

William J. Gilbert
Scott A. Vanstone

University of Waterloo

PEARSON

Prentice
Hall

Upper Saddle River, New Jersey 07458

Library of Congress Cataloging-in-Publication Data

Gilbert, William J.
   An introduction to mathematical thinking: algebra and number systems / William J.
   Gilbert, Scott Vanstone.
      p. cm.
   Includes bibliographical references and index.
   ISBN 0-13-184868-2
   1. Algebraic logic. 2. Number concept. 3. Mathematics–Philosophy. I. Vanstone, Scott
   A. II. Title.
   QA10.G55 2005
   510dc22   2004044797

Editor-in-Chief: Sally Yagan
Acquisition Editor: George Lobell
Vice President/Director of Production and Manufacturing: David W. Riccardi
Executive Managing Editor: Kathleen Schiaparelli
Senior Managing Editor: Linda Mihatov Behrens
Production Editor: Bob Walters
Manufacturing Buyer: Michael Bell
Manufacturing Manager: Trudy Pisciotti
Marketing Manager: Halee Dinsey
Marketing Assistant: Vince Jansen
Director of Marketing: John Tweeddale
Editorial Assistant: Melanie VanBenthuysen
Art Director: Jayne Conte
Cover Image: William J. Gilbert

©2005 Pearson Education, Inc.
Pearson Prentice Hall
Pearson Education, Inc.
Upper Saddle River, New Jersey 07458

Printed in the United States of America

10   9   8

ISBN: 0-13-184868-2

Pearson Education LTD., *London*
Pearson Education Australia PTY, Limited, *Sydney*
Pearson Education, Singapore, Pte. Ltd.
Pearson Education North Asia Ltd., *Hong Kong*
Pearson Education Canada, Ltd., *Toronto*
Pearson Educación de Mexico S.A. de C.V.
Pearson Education – Japan, *Tokyo*
Pearson Education Malaysia, Pte. Ltd.

Contents

Preface

Objectives

This book is designed for a first course in abstract mathematics for university students who wish to major in mathematics or computer science. At the same time as teaching algebra that will be useful in later studies, the book provides an introduction to mathematical thinking and to the art of writing proofs. The idea behind this book is that students will be better motivated to understand proofs, and the mathematical way of thinking, if they are absorbed in interesting topics rather than analyzing the idea of a proof in the abstract. To this end, the first chapter on mathematical language and proofs is reasonably brief.

The algebra in the book is centered around the number systems, from the integers to the complex numbers, and the solution of polynomial equations in these systems. Even though the mathematics in the book is classical, we include a very up-to-date application to cryptography that appeals to students as being very relevant.

This book will prepare the student for more advanced courses in linear and modern algebra as well as analysis.

Material Covered

The first chapter on logic and proofs discusses the ideas of propositional logic that are essential in computer science and mathematics. It also describes the various standard strategies for writing mathematical proofs.

The next two chapters, on integers and modular arithmetic, introduce students to the new ideas of the Euclidean Algorithm and congruences, even though they are based on the integers, with which the students are very familiar. Divisibility, greatest common divisors, and congruences naturally give rise to proofs that are meaningful but not overcomplicated. These provide a good starting point for learning how to write proofs. Besides the more abstract idea of proof, the chapters contain interesting computational methods, such as the complete solution of linear Diophantine equations, which should provide a sense of accomplishment to the student.

Chapter 4 discusses induction and recursion, which are important ideas in mathematics and computer science.

Chapter 5 gives a brief introduction to rational and real numbers from an algebraic point of view.

Functions are introduced in Chapter 6 so that we can talk about function inverses and bijections. Examples of inverses of some familiar real-valued functions are given in the brief sections on log, exponential, and inverse trigonometric functions; this may overlap with a calculus course, but the emphasis here is on the domain and range of the function and its inverse.

Chapter 7 gives a modern application of algebra to cryptography. The RSA scheme for encrypting and decrypting messages relies on congruences and prime factorizations and is well within the reach of students. This topic ties together several of the ideas from the chapters on the integers.

The remaining two chapters introduce complex numbers and the solution to polynomial equations over various number fields.

It is assumed that the reader has some knowledge of trigonometry. A brief summary is provided in the Appendix, along with the basics of manipulating inequalities, which tends not to be taught formally in the curriculum. While a knowledge of calculus is not necessary for this book, many students learn a little in high school, so we occasionally use the derivative of a polynomial to help us sketch a graph. However, differentiation of polynomials is needed in the optional Section 9.7 on Multiple Roots.

The numbering scheme in the book is as follows: 1.23 means Chapter 1, Section 1.2, and the 3rd numbered item in that section. All the items, whether Theorems, Propositions, Examples, or whatever, are numbered together. Hopefully this will provide an easy way of finding any reference in the text.

Exercises and Problems

There are ample questions for the student to tackle at the end of each chapter. They are divided into two types; the Exercises are routine applications of the material in the chapter, while the Problems usually require more ingenuity and range from easy to nearly impossible. Brief answers to all the odd-numbered questions are given at the back of the book.

To the Instructor

Chapter 1 on Logic and Proofs is fairly concise, and instructors should refer the student back to this chapter when a particular point in logic or type of proof is used in practice in the remainder of the book.

Some instructors may wish to start with Chapter 2 and integrate the results of Chapter 1 with the material as they come across proof techniques and logical ideas in the remainder of the book.

Induction is introduced formally after the bulk of the material on the integers, even though it is used implicitly earlier. This was done for psychological and pedagogical reasons, so that the abstract idea of induction comes after some of the more practical and computational ideas, such as solving integer equations and congruences.

The theoretical Section 3.3, on Equivalence Relations, could be omitted without jeopardizing later sections, since we will only be concerned with the relatively concrete example of congruence. However, it has been included here since the notion of equivalence relation is such an important general idea in later mathematics, and congruence is a prime example.

Instructors can choose how many, if any, of the sections to cover in Chapter 5, on Rational and Real Numbers; Chapter 6, on Functions and Bijections; and Chapter 9, on Polynomial Equations, to fit in with their time schedule and the topics of other courses at their institution.

Other less important sections that could be omitted are Section 2.4, on writing integers in different bases; Section 3.7, on the Euler-Fermat Theorem; and Section 8.8, on the Fundamental Theorem of Algebra except, of course, to mention it. The Euler phi function and the Euler-Fermat Theorem of Section 3.7 provide motivation for the RSA cryptographic scheme in Chapter 7, but the proofs in that chapter are written so as not to explicitly refer to the results of Section 3.7.

Chapter 7, on Cryptography, is not essential for the logical development of the book. However, it provides a concrete example that the students find very interesting since it is a rather unexpected application of algebra, and Web security is becoming very relevant to student's lives. Cryptography could be done immediately after Chapter 3, on Congruences, so long as the idea of encryption being a bijection, with decryption as its inverse, is not emphasized.

Instructors may obtain complete solutions to all the Exercises and Problems from the publisher.

Acknowledgments

We are grateful for the assistance we received from our colleagues and students in the first-year algebra course at the University of Waterloo, which used earlier versions of this book. Above all we would like to thank James Geelen, Alfred Menezes, Stephen New, and Paul Schellenberg for their ideas and suggestions. We would like to thank the following reviewers of the manuscript for their valuable comments: Edward Azoff, University of Georgia, Christopher Byrne, Pennsylvania State University, Carl FitzGerald, University of California, San Diego, John Gimbel, University of Alaska, Fairbanks, Sergei Ivanov, University of Illinois, Urbana, and Shlomo Libeskind, University of Oregon. We are especially indebted to Alejandro Morales for his excellent work in writing out the solutions for all the exercises and problems and in proofreading.

Finally, we would like to thank Andrea Vanstone and Ruth Gilbert for typing earlier versions of the book. This edition was typeset in LaTeX, using TeXShop on an Apple Macintosh G4, and most of the figures were produced with OmniGraffle.

William J. Gilbert
Scott A. Vanstone

C H A P T E R   1

Logic and Proofs

1.1 THE LANGUAGE OF MATHEMATICS

Mathematics makes precise use of language in stating and proving its results. Mathematicians use the English (or other) language to express their ideas and arguments, though they give some common English words a more precise meaning, so as to make them unambiguous.

Your aim in writing mathematics should be to convince the reader (and yourself) that your ideas are correct. Good mathematical writing consists of complete sentences, allowing for the fact that symbols stand for words. For example, "$A = B$." is a sentence in which the subject is "A", the verb is *equals* and the object is "B".

Symbols are supposed to make the mathematics easier to comprehend, not to confuse the reader, so do not go overboard in using a multitude of nonstandard symbols. The standard symbols that we use in this book are listed in the List of Symbols on page 293. Nonmathematical readers may be baffled by pages of mathematics full of symbols, but writing it out without symbols would not make it any clearer to them.

Mathematicians use a variety of terms to label their results. We shall use the terms **theorem**, **proposition**, and **lemma** to describe results, in decreasing order of importance. These will be general statements that usually apply to a large number of cases. A *theorem* will be a major landmark in the mathematical theory, a *proposition* a lesser result, while a *lemma* will usually be a result that is needed to prove a theorem or proposition but is not very interesting on its own. A **corollary** is a result that follows almost immediately from a theorem. An **example** is not normally a general result, but often a particular case of a theorem or proposition. An **algorithm** is an explicit procedure for solving a problem in a finite number of steps.

A **definition** gives a precise meaning to a mathematical term so that the reader knows what the author intends. A given definition may not be the one you would use, but you have to accept it to make sense of the following mathematics. For example, some people would define the natural numbers as the positive integers 1, 2, 3, ..., while others think that zero is natural, and would define the natural numbers as the nonnegative integers 0, 1, 2, 3, ...

A **proof** is a mathematical argument intended to convince us that a result is correct. A proof of a theorem, or other result, is a series of logical deductions, using the assumptions of the theorem, the definitions of the terms involved, and previous results that have been proven. You have to use your judgment when writing a proof on how much detail to include, and this will depend on your audience. If you put in too much detail, the overall argument will be cluttered. On the other hand, if you do not put in enough detail, your readers may not understand all your argument and, more important, you may omit some technicality that invalidates your proof. In this chapter, we introduce the more common types of mathematical reasoning used in proofs and the standard strategies for attacking proofs.

In mathematics, we tend to use more complicated, and more compound expressions than we do in everyday language, so the next section explains some methods for dealing with these expressions.

1.2 LOGIC

Logic is the study of correct reasoning. The rules of logic give precise meaning to mathematical statements and allow us to make correct arguments about these statements.

Definition.   In mathematics, a **statement** or **proposition** is a sentence that is either TRUE or FALSE.

For example,

- The year 2000 is a leap year.
- $3 + 2 = 6$.
- $\pi^2 < 10$.
- The decimal expansion of π contains one hundred consecutive 3's.

They are sentences that are either TRUE or FALSE, though you most probably do not know the truth value of the last statement.

However, the following sentences are not statements.

- Let $x = 4$.
- Find the nearest integer to $\sqrt{5}^{13}$.
- Is it Friday today?

Questions are never statements. If we changed the last sentence to "It is Friday today" it would be a statement.

Propositional logic is the part of logic that deals with combining statements using connectives such as AND, OR, NOT, or implies. We use these connectives in everyday language, but in mathematics and computer science we tend to use them in more complex combinations. We (and the computers) need to know precisely what these combinations mean.

We can always combine any two statements using AND or OR to form a third statement. For example,

- Ottawa is the capital of Canada and New York is the capital of the United States.

If P denotes the statement "Ottawa is the capital of Canada" and Q denotes the statement "New York is the capital of the United States" then the preceding statement could be denoted by P AND Q.

Every statement has a negation. For example, if P is the statement "$3+2 = 6$" then its negation is "It is not true that $3 + 2 = 6$" or more simply "$3 + 2 \neq 6$."

Definition. Let P and Q be statements. The statement P **AND** Q is called the **conjunction** of P and Q. The statement P AND Q will be TRUE if P is TRUE and Q is TRUE but will be FALSE otherwise.

The statement P **OR** Q is called the **disjunction** of P and Q. The statement P OR Q will be TRUE if either P is TRUE, or Q is TRUE, or both are TRUE.

The **negation** of the statement P is denoted by **NOT** P. Another very common notation is $\sim P$.

We can define these connectives by the following truth tables, in which T stands for TRUE and F stands for FALSE.

P	Q	P AND Q
T	T	T
T	F	F
F	T	F
F	F	F

P	Q	P OR Q
T	T	T
T	F	T
F	T	T
F	F	F

P	NOT P
T	F
F	T

A *truth table* for a statement containing unknowns lists the truth values of the statement for all possible truth values of the unknowns. Since each statement has two possible values, T or F, the number of possibilities for n independent unknown statements is 2^n, and this will be the number of rows (not counting the headings) in the truth table of a statement with n unknowns. The statement P AND Q has the two unknowns P and Q, so the truth table will have four rows, while the truth table for NOT P has two rows.

In everyday language, the word *or* can be used in two different ways. Consider the meaning of *or* in the following two sentences.

- The prerequisite for this course is algebra or trigonometry.

- You will be served tea or coffee.

In the first sentence, the *or* is used in an inclusive way, which means that you can still take the course if you have both algebra and trigonometry. However, in the second sentence, *or* is used in an exclusive way; you should not expect tea *and* coffee. Mathematics always uses the *inclusive* OR, which means that P OR Q is true even if both P and Q are true.

Example 1.21. Show that the statement NOT (P AND Q) has the same truth table as the statement (NOT P) OR (NOT Q).

Solution.

P	Q	P AND Q	NOT (P AND Q)
T	T	T	F
T	F	F	T
F	T	F	T
F	F	F	T

P	Q	NOT P	NOT Q	(NOT P) OR (NOT Q)
T	T	F	F	F
T	F	F	T	T
F	T	T	F	T
F	F	T	T	T

The two final columns are the same, so the two statements have the same truth table. □

The symbol □ denotes the end of a proof or solution.

The negation of a conjunction in the previous example can be illustrated in everyday language. Consider the statement "It is not true that it is raining and windy." This means that "Either it is not raining or it is not windy." Notice that we are using the mathematical *inclusive or* here; it could be sunny and calm.

In a similar way, you are asked to show in Exercise 22 that NOT (P OR Q) and (NOT P) AND (NOT Q) have the same truth tables.

In mathematics, we often use statements of the form "If P, then Q" such as "If an angle bisector of a triangle is also a median, then the triangle is isosceles." This is called a conditional statement or implication, where P is the *hypothesis* and Q is the *conclusion*.

If P is a true statement, then clearly the truth value of "If P then Q" should be the same as the truth value of Q. However, if P is false, it is not obvious what the truth value of the statement should be. For example, if pigs can fly then I will eat my hat. If P is false, then the statement "If P then Q" normally imparts no meaning in everyday language, though you would certainly not say that it is false. Since we would like "If P then Q" to be a mathematical statement that is either true or false, we define it to be true, whenever P is false. This yields the following definition.

Definition.   Given two statements P and Q, the **conditional statement** "If P, then Q" is denoted by $P \Longrightarrow Q$ (pronounced "P implies Q") and is defined by the following truth table.

P	Q	$P \Longrightarrow Q$
T	T	T
T	F	F
F	T	T
F	F	T

The conditional in mathematics is an extension of everyday usage, but it has some unusual consequences because $P \implies Q$ is always true if P is false. For example, "If $5 > 7$, then $2+2 = 3$" is a true statement. When we use the conditional "If P, then Q" in everyday language we are usually suggesting that the truth of P *causes* Q to be true. In mathematics, there does not have to be any relation whatsoever between P and Q. However, this leads to logically correct but pretty useless statements, such as "If $2 < 4$, then $2+2 = 4$." One reason for not considering causality in the definition is that it would be very hard to make precise. "If you take this pill, you will get better." Is there a cause and effect here? Maybe, maybe not.

An implication in which the hypothesis is false is sometimes called **vacuously true**, because there is nothing to check. For example, "If x is an integer between 2.2 and 2.8, then x is even" is vacuously true, as there are no integers in that range.

The following are all alternative ways of expressing a conditional statement.

- $P \implies Q$.
- P implies Q.
- If P, then Q.
- If P, Q.
- Q if P.
- P only if Q.
- P is sufficient for Q.
- Q is necessary for P.

The **converse** of the conditional statement "If P, then Q" is "If Q, then P." Of course, even if a statement is true, its converse does not have to be true. For example, "If a quadrilateral has its four sides equal, then it is a parallelogram" is true, but the converse "If a quadrilateral is a parallelogram, then it has its four sides equal" is false.

If the conditional statement "If P, then Q" and its converse "If Q, then P" are both true, then we say that "P if and only if Q." For example, if $xy = 0$ then $x = 0$ or $y = 0$. Also, if $x = 0$ or $y = 0$, then $xy = 0$. Hence

$$xy = 0 \quad \text{if and only if} \quad x = 0 \text{ or } y = 0.$$

The statement "$xy = 0$" is true if and only if the statement "$x = 0$ or $y = 0$" is true. The statements express the same idea in different words.

Example 1.22.   Find the truth table for the statement

$$(P \implies Q) \text{ AND } (Q \implies P).$$

Solution.

P	Q	$P \Rightarrow Q$	$Q \Rightarrow P$	$(P \Rightarrow Q) \text{ AND } (Q \Rightarrow P)$
T	T	T	T	T
T	F	F	T	F
F	T	T	F	F
F	F	T	T	T

We shall now use this for the truth table for the statement "P if and only if Q." □

Definition.   Given two statements P and Q, we denote the statement "P **if and only if** Q" by $P \Longleftrightarrow Q$, and define it by $(P \Rightarrow Q)$ AND $(Q \Rightarrow P)$. The truth table is given below.

P	Q	$P \Longleftrightarrow Q$
T	T	T
T	F	F
F	T	F
F	F	T

The statement $P \Longleftrightarrow Q$ is true precisely when P and Q have the same truth values, in which case we say that P and Q are **equivalent statements**. Hence equivalent statements have the same truth tables.

The expression "if and only if" is used so often in mathematics that it is often abbreviated as *iff*. The following are alternative ways of expressing an "if and only if" statement.

- $P \Longleftrightarrow Q$.
- P if and only if Q.
- P iff Q.
- P is equivalent to Q.
- P is necessary and sufficient for Q.

Example 1.23.   Show that the statement $P \Longrightarrow Q$ is equivalent to the statement Q OR NOT P.

Solution.   Compare their truth tables.

P	Q	$P \Rightarrow Q$	NOT P	Q OR NOT P
T	T	T	F	T
T	F	F	F	F
F	T	T	T	T
F	F	T	T	T

Since the statements $P \Longrightarrow Q$ and Q OR NOT P have the same truth tables, they are equivalent.

As an example, the statement "If I am late, then I am running" is equivalent to "Either I am running or I am not late." □

Example 1.24.   Is the statement $R \Longrightarrow (P$ OR $Q)$ equivalent to the statement P OR $(Q$ AND NOT $R)$?

Solution.   Compare their truth tables.

P	Q	R	P OR Q	$R \Longrightarrow (P$ OR $Q)$
T	T	T	T	T
T	T	F	T	T
T	F	T	T	T
T	F	F	T	T
F	T	T	T	T
F	T	F	T	T
F	F	T	F	F
F	F	F	F	T

P	Q	R	Q AND NOT R	P OR $(Q$ AND NOT $R)$
T	T	T	F	T
T	T	F	T	T
T	F	T	F	T
T	F	F	F	T
F	T	T	F	F
F	T	F	T	T
F	F	T	F	F
F	F	F	F	F

Since the truth tables differ in two positions, the statements are not equivalent. They differ when P is false and both Q and R are true and also when all the variables are false. $\qquad\square$

Alternative Notations for Connectives		
Connective	Propositional Logic	C and Java syntax
AND	$\wedge$	&&
OR	$\vee$	||
NOT	$\neg$ or $\sim$	!
$\Longrightarrow$	$\longrightarrow$	
$\Longleftrightarrow$	$\longleftrightarrow$	

1.3   SETS

Mathematics not only deals with individual objects, such as the integer 2007 or the number $\sqrt{3}$, but also with collections of objects, such as all the real numbers, or all the solutions to an equation. In mathematics, such a collection is called a set. We shall not give a rigorous definition of set, but we shall describe it.

A *set* is any well-defined collection of objects; the objects are called the *elements* or *members* of the set. If x is an element of the set S, we say x belongs to S and write

$$x \in S.$$

If y is not an element of S, we write $y \notin S$. The set of *real numbers* is denoted by the blackboard bold symbol $\mathbb{R}$, so "$\sqrt{3} \in \mathbb{R}$" is a true statement.

There are two basic ways of describing a set. One method is to list all its elements. For example, a set S whose elements are 2, 4, 6 and 8 can be written as

$$S \ = \ \{2, 4, 6, 8\}.$$

In mathematics, a set is considered as an unordered collection, so this same set S could be written as $\{4, 8, 6, 2\}$, or even $\{2, 2, 8, 6, 4\}$. We might write the set of letters of the alphabet as $\{A, B, C, \ldots, X, Y, Z\}$, where the three dots indicate that all the letters between C and X are also to be included. By abusing this notation, we write the set of *positive integers* (or *natural numbers*) as

$$\mathbb{P} = \{1, 2, 3, 4, \ldots\}$$

and also the set of all *integers* as

$$\mathbb{Z} = \{\ldots, -2, -1, 0, 1, 2, 3, \ldots\}.$$

The set $\mathbb{Z}$ consists of the positive integers, zero, and the negative integers.

The other basic method of describing a set is by means of a rule. For example,

$$S \ = \ \{x \in \mathbb{R} \mid 1 < x < 2\}$$

is read as "S is the set of all real numbers x, such that x is greater than 1 and less than 2"; in other words, S is the set of real numbers lying between 1 and 2. Other variants of this notation for the same set are $\{x \in \mathbb{R} \ : \ 1 < x < 2\}$ and $\{x \mid x \in \mathbb{R} \text{ AND } 1 < x < 2\}$.

The set with no elements is called the *empty set* or *null set* and is denoted by the symbol $\emptyset$. There is only one empty set; for example, the set $\{x \in \mathbb{R} \mid x^2 < 0\}$ and the empty set of oranges are the same set.

If S and T are sets such that every element of S is also an element of T, then we say that S is *contained* in T or that S is a *subset* of T and write $S \subseteq T$ or $T \supseteq S$. For example, $\{2, 7, 5\} \subseteq \mathbb{P}$ but $\{-1, 5, 8\}$ is not a subset of $\mathbb{P}$ because $-1 \notin \mathbb{P}$. Two sets S and T are *equal* if $S \subseteq T$ and $T \subseteq S$.

The *intersection* of two sets S and T is the set $S \cap T$, consisting of all elements that are in both S and T; hence

$$S \cap T \ = \ \{x \mid x \in S \text{ AND } x \in T\}.$$

If $S \cap T = \emptyset$, then S and T are said to be *disjoint*. The *union* of the sets S and T is the set $S \cup T$, of all elements that are in either S or T (or both S and T). Hence

$$S \cup T \ = \ \{x \mid x \in S \text{ OR } x \in T\}.$$

The *Cartesian product*, or just the *product*, of two sets S and T is the set $S \times T$, of all ordered pairs (x, y), where $x \in S$ and $y \in T$; hence

$$S \times T \ = \ \{(x, y) \mid x \in S \text{ AND } y \in T\}.$$

For example, $\{a, b, c\} \times \{1, 2\}$ is the six-element set

$$\{(a, 1), \ (b, 1), \ (c, 1), \ (a, 2), \ (b, 2), \ (c, 2)\}.$$

The product $\mathbb{R} \times \mathbb{R} = \{(x, y) \mid x, y \in \mathbb{R}\}$ consists of all the points in the plane.

1.4   QUANTIFIERS

In mathematics we constantly use sentences involving variables, such as "$x > 5$." However, until the value of x is specified, this sentence has no truth value. If we let $P(x)$ denote the sentence "$x > 5$" then $P(7)$ is a true statement while $P(0)$ is a false statement.

If $P(x)$ is a sentence depending on the variable x, we often want to say that $P(x)$ is true for all values of x or that $P(x)$ is true for at least one value of x. This can be done by adding quantifiers, which convert the sentence $P(x)$ into a statement that is either true or false.

Definition.   The *universal quantification* of $P(x)$ is the statement

- $P(x)$ is true for all values of x

and is denoted by

- $\forall x, \quad P(x)$

where the symbol $\forall$ is called the **universal quantifier** and is pronounced "for all."

This statement $\forall x, \ P(x)$ can also be expressed in any of the following ways.

- For all x, $P(x)$.
- For every x, $P(x)$.
- For each x, $P(x)$.
- $P(x)$, for all x.

The values of x are assumed to lie in a particular set called the *universe of discourse*. For example, the universe of discourse may be the integers, or the real numbers, or the set of all people.

If we are dealing with the real numbers, the statement "$\forall x, x^2 > 0$" means that "For all real numbers x, $x^2 > 0$" which is a false statement. However the statement "$\forall x, x^2 \geq 0$" is true.

Definition.   The *existential quantification* of $P(x)$ is the statement

- There exists an x for which $P(x)$ is true

and is denoted by

- $\exists x, \quad P(x)$

where the symbol $\exists$ is called the **existential quantifier** and is pronounced "there exists."

Again this is interpreted to mean that "There exists an x in the universe of discourse for which $P(x)$ is true." This statement $\exists x, \ P(x)$ could also be expressed in any of the following ways.

- There is an x for which $P(x)$.
- For some x, $P(x)$.
- $P(x)$, for some x.

For example, if the universe of discourse is the set of real numbers, then the factorization $x^3 - 1 = (x-1)(x^2 + x + 1)$ is true for all x. So

$$\forall x, \quad x^3 - 1 \;=\; (x-1)(x^2 + x + 1)$$

is a true statement. However the equation $x^2 + x - 6 = 0$ is true for only certain values of x, namely $x = 2$ and $x = -3$. Hence $\forall x, \ x^2 + x - 6 = 0$ is false, but $\exists x, \ x^2 + x - 6 = 0$ is true.

Example 1.41. Express the statement "Every real number has a real square root" as a logical expression using quantifiers.

Solution. If we assume that the universe of discourse is the set of real numbers, we can express this statement as

$$\forall a \ \exists x, \quad x^2 \;=\; a.$$

Note that this is just a statement. It does not have to be true; in fact it is not. $\square$

Example 1.42. Express the statement "There is a real number between any two real numbers" as a logical expression using quantifiers.

Solution. Assume that the universe of discourse is the set of real numbers. Before we start to convert the statement to a logical expression, we have to decide how to interpret the English. Does *between* mean *strictly between* and does *two real numbers* mean *two distinct real numbers*?

If we assume that *between* means *strictly between* and we require the statement to be true, then we have to take distinct real numbers. We could write it in either of the following ways.

$$\forall y \ \forall z, \quad (y \neq z \Longrightarrow (\exists x, \ (y < x < z \text{ OR } z < x < y))).$$
$$\forall y \ \forall z, \quad (y < z \Longrightarrow (\exists x, \ y < x < z)).$$

If we assume that *between* could include equals, then we could write

$$\forall y \ \forall z, \quad \exists x, \ (y \leq x \leq z \text{ OR } z \leq x \leq y). \qquad \square$$

Proposition 1.43. *If S is any set and $\emptyset$ denotes the empty set, then $S \subseteq S$ and $\emptyset \subseteq S$. If A and B are sets, the inclusion relation $A \subseteq B$ can be expressed using a quantifier as*

$$\forall x, \quad (x \in A \Longrightarrow x \in B).$$

Proof.   The statement $S \subseteq S$ is equivalent to $\forall x, (x \in S \implies x \in S)$. If $P(x)$ is the statement $x \in S$, then $P(x) \implies P(x)$ is true for all x. Hence $S \subseteq S$ is true.

The statement $\emptyset \subseteq S$ is equivalent to $\forall x, (x \in \emptyset \implies x \in S)$. Now $x \in \emptyset$ is false for all x. However, $P \implies Q$ is always true if P is false, so that the statement $\emptyset \subseteq S$ is true. $\square$

Example 1.44.   The *divisibility* relation $a|b$, which will be discussed in detail in Chapter 2, can be defined symbolically as

$$\exists q, \quad b = qa,$$

where the universe of discourse is the set of integers. Using this definition, determine whether (i) $0|3$ and (ii) $0|0$.

Solution.   **(i)** $0|3$ is equivalent to $\exists q, \ 3 = q0$; that is, $\exists q, \ 3 = 0$. Since $3 = 0$ is always false, $0|3$ is not true.

(ii) $0|0$ is equivalent to $\exists q \ 0 = q0$; that is, $\exists q, \ 0 = 0$. Since $0 = 0$ is always true, we can choose q as any integer, and so $0|0$ is true. $\square$

How do we negate quantifiers? For example, the statement "All Canadians speak French" is not true. However, we do not have to show that "All Canadians do not speak French" to show that the statement is false. We only have to show that "There exists a Canadian who does not speak French." Also, the statement "There exists a real solution to the equation $x^2 = -1$." is false. However, to show this, we have to show that "For all real x, $x^2 \neq -1$."

Consider the negation of the statement "Everyone has a calculator." It is "Not everyone has a calculator," which has the same meaning as "Someone does not have a calculator." On the other hand, the negation of the statement "Someone has a calculator" is "No one has a calculator," which has the same meaning as "Everyone does not have a calculator." These are examples of the following rules for negating quantifiers.

Quantifier Negation Rules 1.45.

NOT $(\forall x, \ P(x))$ *is equivalent to* $(\exists x, \ \text{NOT } P(x))$.

NOT $(\exists x, \ P(x))$ *is equivalent to* $(\forall x, \ \text{NOT } P(x))$.

In general, two statements involving quantifiers will be *equivalent* if they have the same meaning. We cannot always use truth tables to check for equivalence or implications involving quantifiers, so at this stage we have to reason informally to check the equivalence or implication.

Example 1.46. If the universe of discourse is the integers, what does the following statement mean in English? How would you prove it true or false?

$$\exists x \ \forall y, \ (x \geq y).$$

Solution. The statement says that there is an integer that is greater than or equal to all integers. That is, the statement says that there is a largest integer.

This is false. To show that it is false, we have to prove that

$$\text{NOT} \ (\exists x \ \forall y, \ (x \geq y))$$

is true. This is equivalent to the following statements.

$$\forall x, \ \text{NOT} \ (\forall y, \ (x \geq y)).$$
$$\forall x \ \exists y, \quad \text{NOT} \ (x \geq y).$$
$$\forall x \ \exists y, \ (x < y).$$

This last statement is true because, for every x, we can take $y = x + 1$. $\qquad\square$

Mathematicians would not normally write out the proof in the above example this way, using logical symbols. They would write it out as in the example following the Proof by Contradiction Method 1.55 in the next section, where it is proved that there is no largest integer. This illustrates the fact that mathematics is normally easier to understand if words are used for quantifiers rather than symbols. Compare the following two ways of saying the same thing.

- $\exists x \ \forall y, \ (x \geq y).$
- There is an integer that is greater than or equal to all integers.

The symbolic form is certainly more concise, but that does not necessarily make it easier to understand. Therefore, in later chapters, we shall not normally use quantifier symbols, but we shall write them out in English. However, some complicated manipulations of logical statements, such as finding the negation, may be easier to do using symbols.

Example 1.47. Determine whether each pair of statements are equivalent.

(i) $\forall x, \ (P(x) \ \text{AND} \ Q(x)).$ $(\forall x, \ P(x)) \ \text{AND} \ (\forall x, \ Q(x)).$

(ii) $\forall x, \ (P(x) \ \text{OR} \ Q(x)).$ $(\forall x, \ P(x)) \ \text{OR} \ (\forall x, \ Q(x)).$

Solution. (i) These statements are equivalent. Suppose $\forall x, \ (P(x) \ \text{AND} \ Q(x))$ is true. Hence $\forall x, \ P(x)$ is true and $Q(x)$ is true. In particular, $(\forall x, \ P(x))$ is true. Similarly, $(\forall x, \ Q(x))$ is true. Therefore, $(\forall x, \ P(x)) \ \text{AND} \ (\forall x, \ Q(x))$ is true.

Now suppose $(\forall x, \ P(x)) \ \text{AND} \ (\forall x, \ Q(x))$ is true. If x is any element in the universe of discourse, then $P(x)$ is true and $Q(x)$ is true. Hence the statement $\forall x, \ (P(x) \ \text{AND} \ Q(x))$ is true. We have shown that whenever one of the statements is true, then the other one is also true.

(ii)   These statements are not always equivalent. We shall give a particular example in which they do not have the same meaning.

Let the universe of discourse be the set of real numbers. Let $P(x)$ be the expression $x \geq 0$ and $Q(x)$ be the expression $x < 0$. Then, for all real numbers x, $(P(x) \text{ OR } Q(x))$ is true. However, $(\forall x, \; P(x))$ is not true, and $(\forall x, \; Q(x))$ is not true, so $(\forall x, \; P(x)) \text{ OR } (\forall x, \; Q(x))$ is not true. $\qquad\square$

1.5   PROOFS

Why are mathematicians so fussy about proving their results? Besides wanting to be correct, mathematicians want to be able to rely on a result, so that they can build on it and use it for further theorems. It is not good enough if the result is correct 99.9% of the time. Mathematics is most probably the most cumulative of subjects; later work nearly always relies on previous theorems. If an earlier theorem was found to be incorrect, it may put any subsequent work in jeopardy.

There are many methods for proving theorems, propositions, and lemmas, but there is no procedure that will apply to all proofs. It is extremely difficult to get a computer to write a good proof. Proof writing is an art that requires much practice. There is a delicate balance between writing down too many details and leaving out logical steps that cannot easily be filled by the reader. Remember that a proof is designed to be read and understood by a human!

There are some standard strategies for attacking proofs, and we now introduce the most important of these. These methods of proof are not only important in mathematics but also in computer science, where for example they are used in software specification for verifying programs.

Many mathematical theorems can be expressed symbolically in the form

$$P \implies Q.$$

The statement P is called the *assumption* or *hypothesis* of the theorem, and the statement Q is the *conclusion*. The assumption will consist of one or more statements, normally involving some variables. The theorem says that if the assumption is true, then the conclusion is true.

How do you go about thinking up ways to prove a result? In general, it takes time and practice before you are comfortable in being able to write out a proof. You should start out with simple proofs and build up to multistage proofs. If you cannot immediately see how to prove a result, here are some steps you should go though in tackling a proof.

- **Understand the definitions.** You should know the technical terms involved in the result; this may mean looking up the definition of some terminology with which you are not familiar.

- **Try examples.** Look at various concrete examples that satisfy the hypothesis in order to get a feel for the problem. These examples should convince you that the result is true and may suggest a method of attack for the proof. If the problem involves sets, like the first proposition below, you might draw an appropriate Venn diagram, and if the problem involves functions, you might sketch some suitable graphs.

If you find an example that satisfies the hypothesis but does not satisfy the conclusion, then you have found a counterexample, and the result is false. We discuss counterexamples in the next section.

- **Try standard proof methods.** Try the various techniques in this section that are appropriate for the result you are trying to prove.

Direct Proof Method 1.51. *Proving* $P \Longrightarrow Q$

The direct method of proving $P \Longrightarrow Q$ is to assume that the hypothesis P is true, and use this to prove that the conclusion Q is true.

> PROPOSITION. *If $S \cap T = S$, then $S \subseteq T$.*
>
> *Proof.* Suppose that $S \cap T = S$. To prove that S is a subset of T, we need to prove that if $x \in S$, then $x \in T$.
>
> Let $x \in S$, so that $x \in S \cap T$, since $S \cap T = S$. It follows from the definition of the intersection of sets that $x \in T$. It now follows from the definition of inclusion that $S \subseteq T$. $\square$

If and Only If Proof Method 1.52. *Proving* $P \Longleftrightarrow Q$

This type of result can usually be recognized by the phrase "if and only if" or the phrase "necessary and sufficient." The result "P if and only if Q" can be split up into the two cases, the "only if" part $P \Longrightarrow Q$, and the "if" part $Q \Longrightarrow P$, and then each case can be proved separately.

> PROPOSITION. $S \cap T = S \cup T$ *if and only if $S = T$.*
>
> *Proof.* To prove $(S \cap T = S \cup T) \Longrightarrow (S = T)$, suppose $S \cap T = S \cup T$. If $x \in S$ then $x \in S \cup T$. Since $S \cap T = S \cup T$, $x \in S \cap T$, and hence $x \in T$. This proves that $S \subseteq T$. The problem is symmetric in S and T, since interchanging S and T leaves the problem unchanged. Hence a similar proof, with S and T interchanged, will show that $T \subseteq S$. Combining $S \subseteq T$ with $T \subseteq S$ shows that $S = T$.
>
> The proof in the other direction of $(S = T) \Longrightarrow (S \cap T = S \cup T)$ is very easy. Is is often the case that one direction of an "if and only if" proof is simple. Suppose that $S = T$. Then $S \cap T = S \cap S = S$ and $S \cup T = S \cup S = S$, so $S \cap T = S \cup T$. $\square$

The **contrapositive** of the general implication "If P, then Q" is the statement "If not Q, then not P." The Contrapositive Law will show that these statements are equivalent. For example, the contrapositive of the statement "If it rains, then I get wet." is the statement "If I am not wet, then it is not raining." These statements mean the same thing.

Contrapositive Law 1.53. *$P \Longrightarrow Q$ is equivalent to* NOT $Q \Longrightarrow$ NOT P.

Proof. Consider the truth tables.

P	Q	$P \Rightarrow Q$	NOT Q	NOT P	NOT $Q \Rightarrow$ NOT P
T	T	T	F	F	T
T	F	F	T	F	F
F	T	T	F	T	T
F	F	T	T	T	T

We see from the truth table that $P \Longrightarrow Q$ is equivalent to NOT $Q \Longrightarrow$ NOT P. $\square$

Contrapositive Proof Method 1.54. *Proving $P \Longrightarrow Q$*

In this method, we prove the statement $P \Longrightarrow Q$ by proving its contrapositive NOT $Q \Longrightarrow$ NOT P.

> PROPOSITION.   *If x is a real number such that $x^3 + 7x^2 < 9$, then $x < 1.1$.*
>
> *Proof.*   The contrapositive of the statement that we have to prove is "If $x \geq 1.1$, then $x^3 + 7x^2 \geq 9$." Hence suppose that $x \geq 1.1$. In particular, x is positive, and so
>
> $$x^3 + 7x^2 \;\geq\; 1.1^3 + 7(1.1)^2 \;=\; 1.331 + 8.47 \;=\; 9.801.$$
>
> Therefore, by the Contrapositive Proof Method, the original result must be true. $\square$

Proof by Contradiction Method 1.55.

In the proof technique called proof by contradiction we assume that the statement we want to prove is false and then show that this implies a contradiction.

For example, suppose we wanted to prove the statement Q. If we can show that NOT Q leads to a contradiction, then NOT Q must be false; that is, Q must be true.

> PROPOSITION.   *There is no largest integer.*
>
> *Proof.*   Suppose that n is the largest integer. Then $n + 1$ is also an integer, and it is larger than n. This contradicts our assumption that n was the largest integer. Hence there is no largest integer. $\square$

PROPOSITION.   *There is no real solution to $x^2 - 6x + 10 = 0$.*

Proof.   Assume that the result is false; that is, assume that there is a real number x with $x^2 - 6x + 10 = 0$. Then, by completing the square, we can write this as

$$(x - 3)^2 + 1 \;=\; 0.$$

However $(x - 3)^2 \geq 0$ for any real number x, so the left side of this equation is greater than or equal to 1. This gives a contradiction. Hence the original statement is true. $\square$

We could prove, in a similar way, that there is no real square root of -1. These are examples of nonexistence results, which are normally proved by contradiction.

Other good examples of proof by contradiction are Euclid's Theorem 2.52 on the existence of an infinite number of primes, and Theorem 5.21 on the irrationality of $\sqrt{2}$.

Proof Method 1.56. *Proving* $P \implies (Q \text{ OR } R)$

The statement Q is either true or false. If Q is true then $(Q \text{ OR } R)$ is true, and $P \implies (Q \text{ OR } R)$ is always true, regardless of the truth values of P and R. We therefore only have to prove the result when Q is false. The method of proof therefore consists of assuming that P is true and NOT Q is true and using these to prove that R is true. The statement $P \implies (Q \text{ OR } R)$ is therefore equivalent to the statement

$$(P \text{ AND NOT } Q) \implies R.$$

You are asked to verify this equivalence using truth tables in Problem 76.

PROPOSITION.   *Let m and n be integers. If $m^3 + n^3$ is odd, then m is odd or n is odd.*

Proof.   Suppose that $m^3 + n^3$ is odd and that m is not odd. Therefore, m is even and so m^3 will also be even. Hence $m^3 + n^3 - m^3 = n^3$ will be odd. The contrapositive of the true statement "if n is even, then n^3 is even" is the true statement "if n^3 is odd, then n is odd." Hence we have shown

$$(m^3 + n^3 \text{ is odd }) \text{ AND NOT } (m \text{ is odd }) \implies (n \text{ is odd }).$$

This is equivalent to the statement that was to be proved, namely $(m^3 + n^3 \text{ is odd }) \implies (m \text{ is odd }) \text{ OR } (n \text{ is odd }).$ $\square$

Another good example of this type of proof, which we shall meet later, is in Theorem 2.53, which states that whenever p is a prime number

$$p \text{ divides } ab \implies (p \text{ divides } a) \text{ OR } (p \text{ divides } b).$$

Proof Method 1.57. *Proving* $(P \text{ OR } Q) \implies R$

We have to assume that $P \text{ OR } Q$ is true and then prove R. Whenever P is true, then $(P \text{ OR } Q)$ is true, and so we have to prove $P \implies R$. Similarly, we have to prove $Q \implies R$. These two results are sufficient, and this proof method is equivalent to proving

$$(P \implies R) \text{ AND } (Q \implies R).$$

Therefore, we have to prove both of the separate statements $P \implies R$ and $Q \implies R$. Problem 77 asks you to verify this equivalence using truth tables.

> THEOREM.   *If x is a real number, then $(x - a)(x - b) = 0$ if and only if $x = a$ or $x = b$.*
>
> *Proof.*   We shall first prove $(x - a)(x - b) = 0 \implies (x = a) \text{ OR } (x = b)$, using Proof Method 1.56.
>
> Suppose that $x \neq a$ and that $(x - a)(x - b) = 0$. Then $(x - a) \neq 0$ and we can divide by $(x - a)$ to obtain $(x - b) = 0$ and $x = b$.
>
> Hence $(x - a)(x - b) = 0$ implies $x = a$ or $x = b$.
>
> We shall now prove $(x = a) \text{ OR } (x = b) \implies (x - a)(x - b) = 0$, using Proof Method 1.57.
>
> Suppose $x = a$. Then $x - a = 0$ and so $(x - a)(x - b) = 0$. Similarly, if $x = b$ then $(x - a)(x - b) = 0$.
>
> Hence $x = a$ or $x = b$ implies $(x - a)(x - b) = 0$. $\square$

Proof Method 1.58.   *Proving* $P \implies (Q \text{ AND } R)$

The result can be split up into the two cases, $P \implies Q$, and $P \implies R$, and then each case can be proved separately.

You are asked to verify this common sense result using truth tables in Problem 78.

Proof Method 1.59. *Proving* $(P \text{ AND } Q) \implies R$

In this case we assume that P and Q are true and use any of the previous techniques to prove R, such as the Direct Proof Method 1.51, the Contrapositive Method 1.54, or Proof by Contradiction 1.55.

> PROPOSITION.   *Let x be a real number. Then $x^2 + x < 0$ if and only if $-1 < x < 0$.*
>
> Note that the statement "$-1 < x < 0$" means "$-1 < x \text{ AND } x < 0$."

> *Proof.*   We first prove $x^2 + x < 0 \implies -1 < x \text{ AND } x < 0$.
> Assume that $x^2 + x < 0$ so
>
> $$x(x + 1) \; < \; 0.$$

If the product of two numbers is negative, then one number is positive and the other is negative. There are two cases to consider.

Case (i)   If $x > 0$ and $x + 1 < 0$, then $x > 0$ and $x < -1$. This case is impossible, since no real number is both positive and negative.

Case (ii)   If $x < 0$ and $x + 1 > 0$, then $x < 0$ and $x > -1$. Hence $-1 < x < 0$.

We now prove $-1 < x$ AND $x < 0 \Longrightarrow x^2 + x < 0$.

If $-1 < x$ and $x < 0$, then $x + 1 > 0$ and $x < 0$. Since the product of a positive number and a negative number is negative, it follows that $x(x + 1) < 0$. That is, $x^2 + x < 0$.   ☐

You may find it useful to refer back to the different types of proof methods in this section when you encounter specific examples of proofs in the remainder of the book.

1.6   COUNTEREXAMPLES

How do mathematicians think up their theorems? It is usually a combination of looking at many examples, mimicking a result in a related area, a hunch, and trial and error. However, the methods they use in formulating theorems are almost never disclosed or published; only the finished proof is presented. Before a result can be called a theorem or proposition, it has to be proved. A result that is thought to be true but has not been proven is called a *conjecture*.

Sometimes a conjectured result in mathematics is not true. In that case, we would not be able to prove it. However, we could try to disprove it; that is, try to prove that its negation is true. If the conjectured result is of the form

$$\forall x, \; P(x),$$

then its negation is NOT $(\forall x, \; P(x))$, which by the Quantifier Negation Rules 1.45, is equivalent to

$$\exists x, \; \text{NOT } P(x).$$

Hence to disprove the statement $\forall x, \; P(x)$ we only have to find *one value* of x, say c, such that $P(c)$ is false. This value c is called a **counterexample** to the conjecture $\forall x, \; P(x)$.

If the conjectured result is of the form

$$\forall x, \; P(x) \Longrightarrow Q(x),$$

then its negation is

$$\exists x, \; \text{NOT } (P(x) \Longrightarrow Q(x)),$$

which, by Example 1.23 and Exercise 22, is equivalent to the statement

$$\exists x, \; (P(x) \text{ AND } \text{NOT } Q(x)).$$

Hence $x = c$ is a counterexample to the conjecture if $P(c)$ is true, while $Q(c)$ is false.

A mathematician, when first tackling a proof of a conjecture, is not sure whether it is true and must always be on the lookout for counterexamples. You will be put in this situation when answering a question that asks whether a certain result is true or not.

Often the first attempt at stating a theorem is basically correct but fails in certain cases. If this happens, the hypotheses might be able to be changed to eliminate the bad cases and obtain a true result. When writing proofs, you should get in the habit of making sure that you have used all the hypotheses. If you have not used them all and your proof is correct, then you have proved a more general result, as the unused hypotheses could be removed from the statement of the theorem. However, at this stage in your mathematical development, it usually means that your proof is faulty, as most of the results you will be asked to prove have had any unnecessary hypotheses removed.

EXAMPLE.   Let x be a real number. Disprove the statement

$$\text{If } x^2 > 9 \text{ then, } x > 3.$$

Solution.   One counterexample to the statement is obtained by taking $x = c = -4$, since $c^2 = 16 > 9$ and $c \leq 3$. This counterexample disproves the statement. $\square$

EXAMPLE.   Let m and n be integers. If m or n is odd, is it necessarily true that $m^3 + n^3$ is odd?

Solution.   This is a question about the converse of a result proved on page 16.

The answer to the question is no, since we can easily find a counterexample in which m or n is odd, and $m^3 + n^3$ is even. One such counterexample is $m = 1$ and $n = 1$.

Notice that the mathematical inclusive OR is necessary here. Of course, there is an infinite number of counterexamples in this case; they occur when m and n are both odd. However, one counterexample is enough to disprove the result. $\square$

If we wished to disprove an existence statement such as $\exists x,\ P(x)$, then its negation is NOT $(\exists x,\ P(x))$, which is equivalent to $\forall x,\ \text{NOT } P(x)$. In this case we cannot use a counterexample because we have to show that $P(x)$ is false *for all* values of x.

Exercise Set 1

1–6. Determine which of the following sentences are statements. What are the truth values of those that are statements?

1. $7 > 5$

2. $5 > 7$

3. Is $5 > 7$?

4. $\sqrt{2}$ is an integer.

5. Show that $\sqrt{2}$ is not an integer.

6. If 5 is even then $6 = 7$.

7–12. Write the truth tables for each expression.

7. NOT(NOT P)

8. NOT(P OR Q)

9. $P \Longrightarrow (Q$ OR $R)$

10. $(P$ AND $Q) \Longrightarrow R$

11. $(P$ OR NOT $Q) \Longrightarrow R$

12. NOT $P \Longrightarrow (Q \Longleftrightarrow R)$

13. P UNLESS Q is defined as (NOT $Q) \Longrightarrow P$. Show that this statement has the same truth table as P OR Q. Give an example in common English showing the equivalence of P UNLESS Q and P OR Q.

14. Write down the truth table for the *exclusive or* connective XOR, where the statement P XOR Q means $(P$ OR $Q)$ AND NOT $(P$ AND $Q)$. Show that this is equivalent to NOT$(P \Longleftrightarrow Q)$.

15. Write down the truth table for the *not or* connective NOR, where the statement P NOR Q means NOT$(P$ OR $Q)$.

16. Write down the truth table for the *not and* connective NAND, where the statement P NAND Q means NOT$(P$ AND $Q)$.
[Electronic circuits often use NOR, NAND, and XOR gates.]

17–21. Write each statement using P, Q, and connectives.

17. P whenever Q.

18. P is necessary for Q.

19. P is sufficient for Q.

20. P only if Q.

21. P is necessary and sufficient for Q.

22. Show that the statements NOT $(P$ OR $Q)$ and (NOT P) AND (NOT Q) have the same truth tables, and give an example of the equivalence of these statements in everyday language.

23. Show that the statements P AND $(Q$ AND $R)$ and $(P$ AND $Q)$ AND R have the same truth tables. This is the *associative law* for AND.

24. Show that P AND $(Q$ OR $R)$ and $(P$ AND $Q)$ OR $(P$ AND $R)$ are statements with the same truth tables. This is a *distributive law*.

25. Is $(P$ AND $Q) \Longrightarrow R$ equivalent to $P \Longrightarrow (Q \Longrightarrow R)$? Give reasons.

26–28. Let P be the statement "It is snowing" and let Q be the statement "It is freezing." Write each statement using P, Q, and connectives.

26. If it is snowing, then it is freezing.

27. It is freezing but not snowing.

28. When it is not freezing, it is not snowing.

29–32. *Let P be the statement "I can walk," Q be the statement "I have broken my leg," and R be the statement "I take the bus." Express each statement as an English sentence.*

29. $Q \Longrightarrow$ NOT P

30. $P \Longleftrightarrow$ NOT Q

31. $R \Longrightarrow (Q$ OR NOT $P)$

32. $R \Longrightarrow (Q \Longleftrightarrow$ NOT $P)$

33–39. *Express each statement as a logical expression using quantifiers. State the universe of discourse.*

33. There is a smallest positive integer.

34. There is no smallest positive real number.

35. Every integer is the product of two integers.

36. Every pair of integers has a common divisor.

37. There is a real number x such that, for every real number y, $x^3 + x = y$.

38. For every real number y, there is a real number x such that $x^3 + x = y$.

39. The equation $x^2 - 2y^2 = 3$ has an integer solution.

40. Express the following quote due to Abraham Lincoln as a logical expression using quantifiers: "You can fool some of the people all of the time, and all of the people some of the time, but you cannot fool all of the people all of the time."

41–44. *Negate each expression, and simplify your answer.*

41. $\forall x, (P(x)$ OR $Q(x))$

42. $\forall x, ((P(x)$ AND $Q(x)) \Longrightarrow R(x))$

43. $\exists x, (P(x) \Longrightarrow Q(x))$

44. $\exists x, \forall y (P(x)$ AND $Q(y))$

45–50. *If the universe of discourse is the real numbers, what does each statement mean in English? Are they true or false?*

45. $\forall x \, \forall y, (x \geq y)$

46. $\exists x \, \exists y, (x \geq y)$

47. $\exists y \, \forall x, (x \geq y)$

48. $\forall x \, \exists y, (x \geq y)$

49. $\forall x \, \exists y, (x^2 + y^2 = 1)$

50. $\exists y \, \forall x, (x^2 + y^2 = 1)$

51–54. *Determine whether each pair of statements is equivalent. Give reasons.*

51. $\exists x, (P(x)$ OR $Q(x))$ $\qquad\qquad$ $(\exists x, \, P(x))$ OR $(\exists x, \, Q(x))$

52. $\exists x, (P(x)$ AND $Q(x))$ $\qquad\qquad$ $(\exists x, \, P(x))$ AND $(\exists x, \, Q(x))$

53. $\forall x, (P(x) \Longrightarrow Q(x))$ $\qquad\qquad$ $(\forall x, \, P(x)) \Longrightarrow (\forall x, \, Q(x))$

54. $\forall x, (P(x)$ OR $Q(y))$ $\qquad\qquad$ $(\forall x, \, P(x))$ OR $Q(y)$

55–61. *Write the contrapositive and the converse of each statement.*

55. If Tom goes to the party, then I will go to the party.

56. If I do my assignments, then I get a good mark in the course.

57. If $x > 3$, then $x^2 > 9$.

58. If $x < -3$, then $x^2 > 9$.

59. If an integer is divisible by 2, then it is not prime.

60. If $x \geq 0$ and $y \geq 0$, then $xy \geq 0$.

61. If $x^2 + y^2 = 9$, then $-3 \leq x \leq 3$.

62. Let S and T be sets. Prove that if $x \notin S \cap T$, then $x \notin S$ or $x \notin T$.

63. Let a and b be real numbers. Prove that if $ab = 0$, then $a = 0$ or $b = 0$.

64. Use the Contrapositive Proof Method to prove that

$$(S \cap T = \emptyset) \text{ AND } (S \cup T = T) \Longrightarrow S = \emptyset.$$

***65–70.** Prove or give a counterexample to each statement.*

65. $\forall x \in \mathbb{R}, \ (x^2 + 5x + 7 > 0)$

66. If m and n are integers with mn odd, then m and n are odd.

67. If x and y are real numbers, then $\forall x, \exists y \ (x^2 > y^2)$.

68. $(S \cap T) \cup U = S \cap (T \cup U)$, for any sets S, T, and U.

69. $S \cup T = T \Longleftrightarrow S \subseteq T$

70. If x is a real number such that $x^4 + 2x^2 - 2x < 0$, then $0 < x < 1$.

71. Prove the *distributive law* $A \cap (B \cup C) = (A \cap B) \cup (A \cap C)$.

72. Prove the *distributive law* $A \cup (B \cap C) = (A \cup B) \cap (A \cup C)$.

Problem Set 1

73. If S, T, and U are sets, the statement $S \cap T \subseteq U$ can be expressed as

$$\forall x, \ ((x \in S \text{ AND } x \in T) \Longrightarrow x \in U).$$

Express and simplify the negation of this expression, namely $S \cap T \nsubseteq U$, in terms of quantifiers.

74. If S and T are sets, the statement $S = T$ can be expressed as

$$\forall x, \ (x \in S \Longleftrightarrow x \in T).$$

What does $S \neq T$ mean? How would you go about showing that two sets are not the same?

75. The definition of the limit of a function, $\lim_{x \to a} f(x) = L$, can be expressed using quantifiers as

$$\forall \epsilon > 0 \ \exists \delta > 0 \ \forall x, \ (0 < |x - a| < \delta \Longrightarrow |f(x) - L| < \epsilon).$$

Use quantifiers to express the negation of this statement, which would be a definition of $\lim_{x \to a} f(x) \neq L$.

76. Use truth tables to show that the statement $P \Longrightarrow (Q \text{ OR } R)$ is equivalent to the statement $(P \text{ AND NOT } Q) \Longrightarrow R$.
[This explains the Proof Method 1.56 for $P \Longrightarrow (Q \text{ OR } R)$.]

77. Use truth tables to show that the statement $(P \text{ OR } Q) \implies R$ is equivalent to the statement $(P \implies R) \text{ AND } (Q \implies R)$.
    [This explains the Proof Method 1.57 for $(P \text{ OR } Q) \implies R$.]

78. Use truth tables to show that the statement $P \implies (Q \text{ AND } R)$ is equivalent to the statement $(P \implies Q) \text{ AND } (P \implies R)$.
    [This explains the Proof Method 1.58 for $P \implies (Q \text{ AND } R)$.]

79. Is the statement $(P \text{ AND } Q) \implies R$ equivalent to $(P \implies R) \text{ OR } (Q \implies R)$? Give reasons.

80. Is the statement $P \implies (Q \implies R)$ equivalent to $(P \implies Q) \implies R$? Give reasons.

81. Show that the statement $P \text{ OR } Q \text{ OR } R$ is equivalent to the statement $(\text{ NOT } P \text{ AND } \text{ NOT } Q) \implies R$.

82–83. For each truth table, find a statement involving P and Q and the connectives, AND, OR, *and* NOT *that yields that truth table.*

82.

P	Q	???
T	T	T
T	F	T
F	T	F
F	F	T

83.

P	Q	???
T	T	F
T	F	T
F	T	F
F	F	F

84. (a) How many nonequivalent statements are there involving P and Q?

  (b) How many nonequivalent statements are there involving $P_1, P_2, \ldots, P_n$?

CHAPTER 2

Integers and Diophantine Equations

The next three chapters will study the properties of the integers. The basic tool in this chapter is the Division Algorithm. We use it to demonstrate the Euclidean Algorithm, which in turn enables us to solve linear Diophantine equations, which are linear equations that just use integers. The Division Algorithm is also used to represent numbers in different bases. The Unique Factorization Theorem, which is also called the Fundamental Theorem of Arithmetic, states that any integer can be factored into primes in a unique way, and the proof of this result depends on the Division Algorithm.

2.1 THE DIVISION ALGORITHM

The set of integers, $\mathbb{Z}$, has the property that any two elements can be added, subtracted, and multiplied together, and the result will still be an integer. However, division of one integer by another is not always possible if the quotient is also to be an integer.

Definition. If a and b are integers, we say that a **divides** b and write this symbolically as $a|b$, if there exists an integer q such that $b = qa$.

Alternative ways to express this are to say that a is a *factor* of b, or that b is a *multiple* of a. If no integer q exists such that $b = qa$, we say that a *does not divide* b, and write $a \nmid b$.

Hence $(-3)|12$ because $12 = (-4)(-3)$. Also, $5|10$, $7|7$, $7|(-7)$, and $4|0$. It should be noted that with this definition $0|0$, because $0 = q0$, where q can be chosen to be any integer. However, $0 \nmid 4$ because there is no integer q such that $4 = q0$. We now prove some elementary properties of this divisibility relation.

Proposition 2.11. *Let a, b, and c be integers.*

(i) *If $a|b$ and $b|c$, then $a|c$.*

(ii) *If $a|b$ and $a|c$, then $a|(bx + cy)$ for any $x, y \in \mathbb{Z}$. In particular, $a|(b + c)$ and $a|(b - c)$.*

(iii) *If $a|b$ and $b|a$, then $a = \pm b$.*

(iv) *If $a|b$, and b is nonzero, then $|a| \leq |b|$.*

Proof. **(i)**   If $a|b$ and $b|c$, then there exist $q, r \in \mathbb{Z}$ such that $b = qa$ and $c = rb$. Hence $c = rqa$ and, because $rq \in \mathbb{Z}$, it follows that $a|c$.

(ii)   If $a|b$ and $a|c$, then there exist $q, r \in \mathbb{Z}$ such that $b = qa$ and $c = ra$. Now $bx + cy = (qx + ry)a$ and, since $qx + ry \in \mathbb{Z}$ for any $x, y \in \mathbb{Z}$, it follows that $a|(bx + cy)$.

(iii)   If $a|b$ and $b|a$, then there exist $q, r \in \mathbb{Z}$ such that $b = qa$ and $a = rb$. Hence $a = rqa$ and $0 = a(rq - 1)$. If $a \neq 0$ then rq must be 1, and $r = q = \pm 1$. Therefore, $a = \pm b$. On the other hand, if $a = 0$, then $b = q0 = 0$ and $a = b$.

(iv)   If $a|b$ then there exists $q \in \mathbb{Z}$ such that $b = qa$. If b is nonzero, then so is q. In particular $|q| \geq 1$ and hence $|b| = |q| \cdot |a| \geq |a|$. $\qquad\qquad\square$

Even if the integer b cannot be exactly divided by the integer a, we can try to divide b by a, and obtain a remainder. This familiar process is known as the Division Algorithm, even though the following statement is not really an algorithm, as it does not explicitly give the construction of a remainder.

Division Algorithm 2.12.   *If a and b are integers, with b positive, then there exist unique integers q and r such that*

$$a \;=\; qb + r, \quad \text{where} \quad 0 \leq r < b.$$

The integer q is called the *quotient*, and the integer r is called the *remainder*, when a is divided by b. The important part of the result is the fact that the remainder is always a positive or zero integer less than b and that this remainder is unique. It is easily seen that $b|a$ if and only if the remainder is zero.

The following are examples of the Division Algorithm when $b = 6$.

$$
\begin{aligned}
15 &= 2 \cdot 6 + 3 \\
30 &= 5 \cdot 6 + 0 \\
-10 &= (-2) \cdot 6 + 2 \\
-2 &= (-1) \cdot 6 + 4
\end{aligned}
$$

Note that even if a is negative, there is a positive or zero remainder.

Proof.   We wish to find a remainder $r = a - qb$, which is as small as possible while not being negative. The idea of the proof is to start with a and then subtract (or add) integer multiples of b until the desired remainder is obtained.

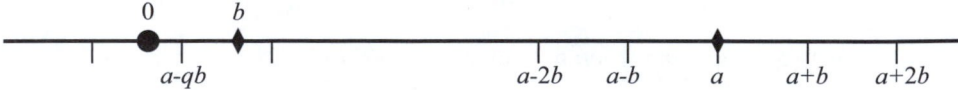

More formally, consider the infinite arithmetic progression of integers

$$S \;=\; \{\ldots, a - 2b, a - b, a, a + b, a + 2b, \ldots\}.$$

This set must contain some nonnegative numbers. This is clear if $a \geq 0$. If $a < 0$, then $a - ab = (-a)(b - 1) \geq 0$.

One of the properties of the integers is that every nonempty set of nonnegative numbers contains a smallest element. Let $r \geq 0$ be the smallest nonnegative integer in S, and let q be the integer such that $a - qb = r$. Now $r < b$, otherwise $r - b = a - (q + 1)b$ would be a nonnegative integer in S that is smaller than r. We have therefore shown the existence of integers q and r with $a = qb + r$ and $0 \leq r < b$.

To show that these integers are unique, suppose that $a = q_1 b + r_1 = q_2 b + r_2$, where $0 \leq r_1 < b$ and $0 \leq r_2 < b$. Suppose $r_1 \neq r_2$, and say $r_1 < r_2$. Then $0 < r_2 - r_1 < b$ and $(q_1 - q_2)b = r_2 - r_1$. Hence $b | r_2 - r_1$ and, by Proposition 2.11 (iv), $b \leq r_2 - r_1$ which contradicts the fact that $r_2 - r_1 < b$. Therefore, the supposition that $r_1 \neq r_2$ is false, and r_1 must equal r_2. Furthermore, $(q_1 - q_2)b = r_2 - r_1 = 0$ and, since $b \neq 0$, $q_1 = q_2$. This proves that the integers q and r are unique. $\qquad \square$

The above proof uses a property of the integers called the *well-ordering principle*, which implies that every nonempty set of positive integers has a smallest element.

The Division Algorithm is equivalent to the fact that each rational number a/b lies in an interval between two integers

$$q \;\leq\; \frac{a}{b} \;<\; q + 1 \quad \text{for some } q \in \mathbb{Z}.$$

If a and b are positive integers, this gives us a way of computing the quotient and remainder using long division, or a calculator. The quotient q is the integer part of a/b.

For example, if $a = 536$ and $b = 19$, then

$$\frac{a}{b} \;=\; \frac{536}{19} \;=\; \mathbf{28\tfrac{4}{19}} \qquad \text{and so} \qquad 536 = 28 \cdot 19 + 4.$$

We can use a calculator to evaluate $536 \div 19 \approx 28.2105$. The quotient is the integer part to the left of the decimal point, namely $q = 28$. The remainder can be calculated from $r = a - qb$, which in this case gives $r = 536 - 28 \cdot 19 = 4$.

For any real number x, the *integer part function* or *floor function* $\lfloor x \rfloor$ is the largest integer less than or equal to x. Hence the quotient is $q = \lfloor a/b \rfloor$.

If a is negative, use long division on the positive numbers, and then compensate for the signs. For example, if $a = -536$ and $b = 19$, then

$$
\begin{aligned}
536 \;&=\; 28 \cdot 19 \;+\; 4; \\
-536 \;&=\; (-28)19 \;-\; 4 \\
&=\; (-29)19 \;+\; 15.
\end{aligned}
$$

In the Division Algorithm it was assumed that b was positive. However, even if b is negative, there are still unique integers q and r such that

$$a \;=\; qb + r \qquad \text{where}, \quad 0 \le r < |b|.$$

For example, if $a = 536$ and $b = -19$, then

$$
\begin{aligned}
536 &= & 28 \cdot 19 & + & 4 \\
&= & (-28)(-19) & + & 4.
\end{aligned}
$$

2.2 THE EUCLIDEAN ALGORITHM

A *common divisor* of two integers is any integer that divides both. For example, the common divisors of 12 and 18 are ± 1, ± 2, ± 3, ± 6. The largest positive one is 6, and this is called the greatest common divisor. This greatest common divisor is used, for example, when reducing the fraction $12/18$ to its lowest terms.

Definition. The **greatest common divisor** of two integers a and b, not both zero, is the largest positive integer dividing both of a and b. It will be denoted by

$$\gcd(a, b).$$

The greatest common divisor is also called the *highest common factor*.

Every integer divides zero, since $0 = 0 \cdot e$ for all $e \in \mathbb{Z}$. Hence, if a is not zero, $\gcd(a, 0) = |a|$. If $a = 0$ and $b = 0$, then all integers are common divisors, and we define $\gcd(0, 0) = 0$. Note that this is not the largest common divisor, but we now have the relationship $\gcd(a, 0) = |a|$ for all $a \in \mathbb{Z}$.

In more advanced number theory books, $\gcd(a, b)$ is usually denoted by just (a, b). However that notation would be too confusing for us.

There is always at least one common divisor, since 1 divides every number. If a is nonzero, it follows from Proposition 2.11 (iv) that there are only a finite number of divisors of a. Hence, if a and b are not both zero, there are only a finite number of common divisors, and $\gcd(a, b)$ exists.

The crude way of finding the greatest common divisor of two integers is to look at all the common divisors of each of them. This method becomes impractical for larger numbers, say 7663 and 8927, because of the difficulty of factoring the numbers. We will now give an algorithm for finding the greatest common divisor in general. If $a \ne 0$, we can use the Division Algorithm to write $b = qa + r$, where $0 \le r < |a|$. The next result will show that the problem of calculating $\gcd(a, b)$ is reduced to the easier problem of calculating $\gcd(b, r)$; it is easier because the numbers involved are smaller.

Proposition 2.21. *If $a = qb + r$, then*

$$\gcd(a, b) = \gcd(b, r).$$

Proof.   Now $a = b = 0$ if and only if $b = r = 0$, and the result is true in this case. Otherwise let $d = \gcd(a, b)$. Since d divides both a and b, it follows from Proposition 2.11 (ii) that $d|(a - qb)$. Therefore $d|r$, and so d is a common divisor of b and r.

Suppose that c is any divisor of b and r. By Proposition 2.11 (ii) again, $c|(qb + r)$; that is, $c|a$. Hence c is a common divisor of a and b. Since d is the greatest common divisor of a and b, and they are not both 0, it follows that $c \le d$. This shows that d is the greatest common divisor of b and r.   □

By repeated use of this result, we obtain the following algorithm for finding the greatest common divisor of any two integers.

Euclidean Algorithm 2.22.   *Let a be an integer and b be a nonzero integer that does not divide a. Then $\gcd(a, b)$ is the last nonzero remainder, r_n, in the following list of equations obtained from the Division Algorithm. If b does divide a, then $\gcd(a, b) = |b|$.*

$$
\begin{aligned}
a &= q_1 b + r_1, & \text{where} && 0 < r_1 < |b| \\
b &= q_2 r_1 + r_2, & \text{where} && 0 < r_2 < r_1 \\
r_1 &= q_3 r_2 + r_3, & \text{where} && 0 < r_3 < r_2 \\
&\;\;\vdots \\
r_{n-2} &= q_n r_{n-1} + r_n, & \text{where} && 0 < r_n < r_{n-1} \\
r_{n-1} &= q_{n+1} r_n + 0.
\end{aligned}
$$

Proof.   The remainders in the above list form a strictly decreasing set of non-negative integers and hence must eventually reach zero. Therefore, the algorithm terminates after a finite number of steps.

If $b \nmid a$, then $r_1 \ne 0$, and r_n is a positive integer. By repeated use of Proposition 2.21 we have

$$
\begin{aligned}
\gcd(a, b) &= \gcd(b, r_1) & = \gcd(r_1, r_2) & = \gcd(r_2, r_3) & = \cdots \\
\cdots &= \gcd(r_{n-2}, r_{n-1}) & = \gcd(r_{n-1}, r_n) & = \gcd(r_n, 0) & = r_n.
\end{aligned}
$$

If $b|a$, then $a = q_1 b + 0$, and there is no last nonzero remainder. In this case, any divisor of b is also a divisor of a. Hence $\gcd(a, b) = |b|$.   □

This algorithm is surprisingly efficient, even for very large numbers.

Example 2.23.   Find $\gcd(1053, 481)$.

Solution.   Putting $a = 1053$ and $b = 481$ in the Euclidean Algorithm, we have

$$
\begin{aligned}
1053 &= 2 \cdot 481 + 91 \\
481 &= 5 \cdot 91 + 26 \\
91 &= 3 \cdot 26 + 13 \\
26 &= 2 \cdot 13 + 0.
\end{aligned}
$$

Hence $\gcd(1053, 481) = 13$.

Check.   Since the Euclidean Algorithm usually involves many operations, it is good practice to check the answer for arithmetical mistakes. We can easily check whether 13 divides 1053 and 481. We have $1053 = 81 \cdot 13$ and $481 = 37 \cdot 13$, so 13 is certainly a common divisor. $\qquad\square$

In this example, it was easy to check that the answer was a common divisor, and with a little more work you could check that it is the largest. However, if the numbers were bigger, it would be more difficult to check that the common divisor was the greatest. The following theorem gives us a good characterization of the greatest common divisor; that is, if you can somehow find integers d, x, and y satisfying the theorem, then you have a guarantee that $d = \gcd(a, b)$. These integers are called a *certificate* for the greatest common divisor of a and b.

GCD Characterization Theorem 2.24.   *If d is a nonnegative common divisor of the integers a and b and there exist integers x and y such that $ax + by = d$, then $d = \gcd(a, b)$.*

Proof.   Let c be any common divisor of a and b. By Proposition 2.11 (ii), $c \mid ax + by$ so $c \mid d$ If d is positive then, by Proposition 2.11 (iv), $c \leq d$, and so $d = \gcd(a, b)$. If $d = 0$, then $0 \mid a$ and $0 \mid b$. This implies $a = b = 0$, and so $d = \gcd(a, b) = 0$. $\qquad\square$

It is alway possible to use the Euclidean Algorithm 2.22 to find a pair of integers x and y satisfying the equation $ax + by = \gcd(a, b)$. We showed in Example 2.23 that $\gcd(1053, 481) = 13$. Let us use the Euclidean Algorithm there to find a solution to the equation $1053x + 481y = 13$. Start with the equation next to the end involving the greatest common divisor, namely $91 = 3 \cdot 26 + 13$. Then work up the equations in the Euclidean Algorithm, eliminating successive remainders at each stage, until we have written 13 in terms of 1053 and 481.

$$\begin{aligned} 13 \; &= \; 91 - 3 \cdot 26 \; = \; 91 - 3(481 - 5 \cdot 91) \; = \; 16 \cdot 91 - 3 \cdot 481 \\ &= \; 16(1053 - 2 \cdot 481) - 3 \cdot 481 \; = \; 16 \cdot 1053 - 35 \cdot 481 \end{aligned}$$

Hence $x = 16$, $y = -35$ is one solution to the equation $1053x + 481y = 13$. This pair of integers is by no means the only pair that will satisfy the equation. For example, $x = -58$, $y = 127$ also satisfies the equation.

It can be confusing keeping track of the numbers in this back-substitution, so the following Extended Euclidean Algorithm organizes the computation. It also provides an efficient algorithm that can easily be implemented on a computer or programmable calculator because the program only has to store two consecutive rows in the table. This Extended Euclidean Algorithm provides a converse to the GCD Characterization Theorem 2.24, namely if $d = \gcd(a, b)$, then there exist integers x and y such that $ax + by = d$.

Extended Euclidean Algorithm 2.25.

INITIALIZE: *Construct a table with three columns labeled by x, y, and r. Set the first two rows to be $(1, 0, a)$ and $(0, 1, b)$, where a and b are positive integers, so that $r_1 = a$ and $r_2 = b$.*

ax	$+$ by	$= r$
1	0	a
0	1	b
x_3	y_3	r_3
x_4	y_4	r_4
$\vdots$	$\vdots$	$\vdots$
x_n	y_n	r_n
x_{n+1}	y_{n+1}	0

GENERAL STEP: *Row i is obtained from the previous two rows by calculating*

$$\text{row } i \quad = \quad \text{row } (i - 2) - q_i \cdot \text{row } (i - 1),$$

where $q_i = \lfloor r_{i-2}/r_{i-1} \rfloor$; that is, q_i is the quotient in the Division Algorithm when r_{i-2} is divided by r_{i-1}:

$$r_{i-2} \quad = \quad q_i r_{i-1} + r_i \quad \text{where } 0 \le r_i < r_{i-1}.$$

STOP: *Stop the algorithm when $r_{n+1} = 0$.*

CONCLUSION:

(i) *The last nonzero element in the third column is $\gcd(a, b) = r_n$.*

(ii) *Every row (x_i, y_i, r_i) satisfies the equation $ax_i + by_i = r_i$.*

(iii) *One integer solution to $ax + by = \gcd(a, b)$ is $x = x_n$, $y = y_n$.*

If a or b is negative, it is easier to solve the equation $|a|x + |b|y = \gcd(a, b)$ and then change the sign of the variables. If $b > a$, then $q_3 = 0$, and row 3 is the same as row 1; however, the remainders decrease after r_3 and the algorithm is still valid. If you have met matrices, then you might recognize the general step as row reduction but using only integer multiples of rows.

Example 2.26.   Use the Extended Euclidean Algorithm to find integers x and y such that $3094x + 2513y = \gcd(3094, 2513)$.

Solution.   The Extended Euclidean Algorithm yields the following tables.

Extended Euclidean Algorithm $3094x + 2513y =$		r
1	0	3094
0	1	2513
1	−1	581
−4	5	189
13	−16	14
−173	213	7
359	−442	0

q_i	Euclidean Algorithm $r_{i-2} = q_i \cdot r_{i-1} + r_i$
1	$3094 = 1 \cdot 2513 + 581$
4	$2513 = 4 \cdot 581 + 189$
3	$581 = 3 \cdot 189 + 14$
13	$189 = 13 \cdot 14 + 7$
2	$14 = 2 \cdot 7 + 0$

The next to the last row shows that $\gcd(3094, 2513) = 7$ and also gives the identity $-173 \cdot 3094 + 213 \cdot 2513 = 7$. Hence one solution to $3094x + 2513y = 7$ is $x = -173$, and $y = 213$.

Check. $-173 \cdot 3094 + 213 \cdot 2513 = -535262 + 535269 = 7$, $3094 = 7 \cdot 442$, and $2513 = 7 \cdot 359$. Hence $7|3094$ and $7|2513$, and the GCD Characterization Theorem 2.24 guarantees that $\gcd(3094, 2513) = 7$. $\square$

If you make an error in your computations, you can easily check your algorithm since every row must satisfy $ax + by = r$.

Proof of the Extended Euclidean Algorithm. **(i)**   The integers $r_3, r_4, \ldots$, in the third column, are precisely the remainders in the Euclidean Algorithm 2.22 (though the numbering of the subscripts differ by two). Hence they form a strictly decreasing sequence of nonnegative integers and the algorithm must terminate with some $r_{n+1} = 0$. By the proof of the Euclidean Algorithm, the last nonzero remainder r_n will be $\gcd(a, b)$.

(ii)   We shall now show that each row satisfies the equation $ax_i + by_i = r_i$. Notice that the first two rows satisfy this equation. If the $(i-2)$nd and $(i-1)$st rows satisfy this equation, then

$$ax_{i-2} + by_{i-2} = r_{i-2}$$
$$ax_{i-1} + by_{i-1} = r_{i-1}.$$

Multiplying the second equation by q_i and subtracting from the first, we have

$$a(x_{i-2} - q_i x_{i-1}) + b(y_{i-2} - q_i y_{i-1}) = r_{i-2} - q_i r_{i-1}$$
$$ax_i + by_i = r_i.$$

Hence the ith row also satisfies the equation. It follows from the Principle of Strong Induction 4.18 that all the rows satisfy the equation.

(iii)   In particular, the nth row satisfies the equation

$$ax_n + by_n = r_n = \gcd(a, b)$$

and so $x = x_n$, $y = y_n$ is one integer solution to $ax + by = \gcd(a, b)$.   □

If you perform the row operations in the Extended Euclidean Algorithm in precisely the manner indicated, then all the integer solutions to the linear equation $ax + by = \gcd(a, b)$ can be read off from the algorithm. The general solution is

$$x = x_n + mx_{n+1}, \quad y = y_n + my_{n+1} \quad \text{for all } m \in \mathbb{Z}.$$

By adding row n to m times row $(n + 1)$, it is clear that these integers satisfy the equation for all $m \in \mathbb{Z}$.

It is a little more tricky to show that every integer solution is of this form. Suppose that x_0 and y_0 are integers satisfying $ax_0 + by_0 = \gcd(a, b)$. We know that row $i = $ row $(i - 2) - q_i \cdot$ row $(i - 1)$, so row $(i - 2) = $ row $i + q_i \cdot$ row $(i - 1)$. Hence each row is an integer linear combination of the two rows below it. We can work our way up the rows to show that every row is an integer linear combination of the last two rows. In particular, $x_0 \cdot$ row $1 + y_0 \cdot$ row 2 is an integer linear combination of the last two rows, say $\ell \cdot$ row $(n) + m \cdot$ row $(n + 1)$ with $\ell, m \in \mathbb{Z}$. This implies that $x_0 = \ell x_n + mx_{n+1}$, $y_0 = \ell y_n + my_{n+1}$ and $ax_0 + by_0 = \ell \gcd(a, b)$. Hence $\ell = 1$ and $x_0 = x_n + mx_{n+1}$, $y_0 = y_n + my_{n+1}$, for some $m \in \mathbb{Z}$.

Hence, from Example 2.26, we see that all the integer solutions to the equation $3094x + 2513y = \gcd(3094, 2513)$ are

$$x = -173 + 359m, \quad y = 213 - 442m \quad \text{for all } m \in \mathbb{Z}.$$

If $\gcd(a, b) = 1$, then a and b are said to be *relatively prime* or *coprime*. In this case they have no common factors (except 1).

Proposition 2.27.   *Let a and b be integers.*

(i) $\gcd(a, b) = 1$ *if and only if there are integers x and y with $ax + by = 1$.*

(ii) *If $d = \gcd(a, b) \neq 0$, then $\gcd\left(\dfrac{a}{d}, \dfrac{b}{d}\right) = 1$.*

Proof (i)   If $\gcd(a, b) = 1$ then, from the Extended Euclidean Algorithm 2.25, there are integers x and y with $ax + by = 1$. Conversely, if there exist integers x and y satisfying $ax + by = 1$, then, as 1 is always a common divisor of any integer, it follows from the GCD Characterization Theorem 2.24 that $\gcd(a, b) = 1$.

(ii)   Since d is a common divisor of a and b, both $\frac{a}{d}$ and $\frac{b}{d}$ are integers. By the Extended Euclidean Algorithm 2.25 there exist integers x and y such that $ax + by = d$. Dividing by d, we have

$$\frac{a}{d}x + \frac{b}{d}y = 1$$

and so $\gcd\left(\frac{a}{d}, \frac{b}{d}\right) = 1$, by part (i). $\qquad\square$

Proposition 2.28.   *If $a, b, c \in \mathbb{Z}$ with $c|ab$ and $\gcd(a, c) = 1$, then $c|b$.*

Proof.   By the Extended Euclidean Algorithm 2.25 there exist integers x and y such that $ax + cy = 1$. Multiplying by b, we have $abx + cby = b$. Since $c|ab$ and $c|c$, it follows from Proposition 2.11 (ii) that $c|(abx + cby)$; that is, $c|b$. $\qquad\square$

Proposition 2.29.   *An integer d is $\gcd(a, b)$ if and only if*

(i) *$d \geq 0$.*

(ii) *d divides both a and b.*

(iii) *any divisor of both a and b also divides d.*

Proof.   Suppose that a and b are not both zero, and d satisfies the three conditions. Conditions (i) and (ii) imply that d is a positive common divisor of a and b, since $d \neq 0$ as it divides a nonzero number. Condition (iii) says that any other common divisor must divide d. In particular, d must be the greatest of all the common divisors.

If $a = b = 0$, then the only number that will satisfy condition (iii) is $d = 0$. Hence d is $\gcd(0, 0)$ in this case.

Conversely suppose that $d = \gcd(a, b)$, so that conditions (i) and (ii) hold. By the Extended Euclidean Algorithm 2.25 there exist integers x and y such that $ax + by = d$. Any common divisor of a and b must also divide $ax + by$; hence the common divisor must divide d and condition (iii) holds. $\qquad\square$

This result provides an alternate definition for the greatest common divisor that generalizes easily to polynomials.

2.3   LINEAR DIOPHANTINE EQUATIONS

The Extended Euclidean Algorithm 2.25 shows how to find integer solutions to the equation $ax + by = d$, when $d = \gcd(a, b)$. We now consider equations with a more general right side of the form

$$ax + by = c$$

for integers x and y, when a, b, c are given integers. Such an equation is called a **linear Diophantine equation** or a *linear integer equation* in two variables. A Diophantine equation, in general, is an equation in one or more unknowns with integer coefficients, for which integer solutions are sought. They are named after the Greek mathematician, Diophantus of Alexandria, who made a study of such equations. He lived from about 200 to 284 A.D., and was one of the first mathematicians to introduce symbolism into algebra. The significance of this can be appreciated if we try to imagine doing algebraic manipulation without symbols. A *linear* equation is one in which the unknowns appear only to the first power.

The first problem, when confronted by any type of equation, is to discover whether any solution exists. If a solution does exist, we are then faced with the problem of determining how many solutions there are and how to find one, or all of them, explicitly.

Consider the simple linear Diophantine equation in one integer variable x

$$ax \; = \; b,$$

where $a, b \in \mathbb{Z}$. We know that this equation has an integer solution for x if and only if $a|b$. Furthermore, if a solution exists, it is unique unless $a = b = 0$ (in which case, every integer is a solution).

Linear Diophantine Equation Theorem 2.31

(i) *The linear Diophantine equation*

$$ax + by \; = \; c$$

has a solution if and only if $\gcd(a, b)|c$.

(ii) *If* $\gcd(a, b) = d \neq 0$, *and* $x = x_0$, $y = y_0$ *is one particular solution, then the complete integer solution is*

$$x \; = \; x_0 + n\frac{b}{d}, \qquad y \; = \; y_0 - n\frac{a}{d} \qquad \text{for all } n \in \mathbb{Z}.$$

Proof. **(i)**   First suppose that the equation does have an integer solution; that is, there are integers x and y for which $ax + by = c$. If $d = \gcd(a, b)$, then $d|a$ and $d|b$ and so, by Proposition 2.11 (ii), $d|(ax + by)$. Hence, if there is a solution, $d|c$.

Conversely, suppose that $d|c$; that is, $c = dc_1$ for some $c_1 \in \mathbb{Z}$. By the Extended Euclidean Algorithm 2.25 there exist integers, say x_1 and y_1, such that

$$ax_1 + by_1 \; = \; d.$$

Multiplying by c_1, we have

$$ac_1x_1 + bc_1y_1 \; = \; dc_1 \; = \; c$$

and we see that $x_0 = c_1x_1$, $y_0 = c_1y_1$ is one particular solution to $ax + by = c$.

(ii)   Let x, y be an arbitrary solution, and $x = x_0$, $y = y_0$ be a particular solution to the Diophantine equation. Then

$$\begin{aligned} ax + by &= c \\ ax_0 + by_0 &= c. \end{aligned}$$

Eliminate c to obtain

$$a(x - x_0) = -b(y - y_0).$$

We know that $d = \gcd(a, b)$ is a common factor of a and b, so $\frac{a}{d}$ and $\frac{b}{d}$ are integers, and we can divide the above equation by d to obtain

$$\frac{a}{d}(x - x_0) = -\frac{b}{d}(y - y_0).$$

By Proposition 2.27 (ii), $\gcd\left(\frac{a}{d}, \frac{b}{d}\right) = 1$. Since $\frac{b}{d}$ divides $\frac{a}{d}(x - x_0)$, it follows from Proposition 2.28 that $\frac{b}{d}$ divides $x - x_0$. If we put $x - x_0 = n\frac{b}{d}$, where $n \in \mathbb{Z}$, then $anb = bd(y_0 - y)$, and so every solution is of the form

$$x = x_0 + n\frac{b}{d}, \quad \text{and} \quad y = y_0 - n\frac{a}{d}.$$

This shows that the solution is of the required form for some integers n. But is it a solution for *every* integer n? If n is any integer, we have

$$\begin{aligned} ax + by &= a\left(x_0 + n\frac{b}{d}\right) + b\left(y_0 - n\frac{a}{d}\right) \\ &= ax_0 + by_0 + n\frac{ab}{d} - n\frac{ab}{d} \\ &= c \end{aligned}$$

since $x = x_0$, $y = y_0$ is a particular solution. Therefore, the complete solution is $x = x_0 + n\frac{b}{d}$ and $y = y_0 - n\frac{a}{d}$, where $n \in \mathbb{Z}$. $\qquad\square$

For example, the complete solution to the Diophantine equation

$$3094x + 2513y = 7$$

of Example 2.26 is $x = -173 + \frac{2513n}{7}, y = 213 - \frac{3094n}{7}$; that is,

$$x = -173 + 359n, \quad y = 213 - 442n \quad \text{for all } n \in \mathbb{Z}.$$

Putting $n = -1, 0, 1$ and 2, we see that some particular solutions are

$$(x, y) = (-532, 655), \ (-173, 213), \ (186, -229) \ \text{and} \ (545, -671).$$

This method of solving an equation, by first finding one particular solution and then finding the complete solution from it, is very common in mathematics. It is used extensively in the solution of differential equations.

Notice that a linear Diophantine equation in two variables either has an infinite number of solutions, or it has no solutions.

Example 2.32.   Find all the solutions to each of the Diophantine equations.

(i) $28x + 35y = 60$

(ii) $21x + 15y = 12$

Solution. (i)   In this case $\gcd(28, 35) = 7$, which does not divide 60. Hence the first equation has no integer solutions.

(ii)   Here $\gcd(21, 15) = 3$, which does divide 12. Hence the second equation does have integer solutions.

$21x + 15y = r$		
1	0	21
0	1	15
1	−1	6
−2	3	3
5	−7	0

We see from the Extended Euclidean Algorithm that $21(-2) + 15(3) = 3$ and so one solution to $21x + 15y = 3$ is $x = -2$, $y = 3$. Multiplying this equation by 4, to get 12 on the right side, we obtain

$$21(-8) + 15(12) = 12.$$

Hence $x = -8$, $y = 12$ is one solution to $21x + 15y = 12$, and the general solution is $x = -8 + \frac{15}{3}n$, $y = 12 - \frac{21}{3}n$, or

$$x = -8 + 5n, \qquad y = 12 - 7n \qquad \text{for all } n \in \mathbb{Z}.$$

Check. $21(-8 + 5n) + 15(12 - 7n) = -168 + 105n + 180 - 105n = 12$ □

It is always a good idea to check your answers to equations. This can usually be done without much effort, by substituting your answer into the equation, and we often find a calculator useful for this purpose. If the check works, then you are guaranteed that you have an answer, though there could be other answers as well.

Note that the Extended Euclidean Algorithm can always be used to find the greatest common divisor and also one particular solution to the Diophantine equation. However, any method can be used to find the particular solution. For very small numbers, it is often quicker to find one solution by inspection.

If $\gcd(a, b) | c$, and $\gcd(a, b) > 1$, it is often easier to divide the whole equation by $\gcd(a, b)$, as this reduces the size of the numbers involved. We could solve Example 2.32 (ii) by dividing the equation by 3 to obtain an equivalent equation $7x + 5y = 4$. You might notice that one solution of this is $x = 2$, $y = -2$. Hence the general integer solution is

$$x = 2 + 5n, \qquad y = -2 - 7n \qquad \text{for all } n \in \mathbb{Z}.$$

This does not look the same as $x = -8 + 5n$, $y = 12 - 7n$, which was given before. However, both answers are correct; they give the same infinite set of solutions.

Example 2.33.   A customer has a very large quantity of dimes and quarters. In how many different ways can the customer pay exactly for an item that costs (i) \$3.49 or (ii) \$2.65?

Solution. *(i)*   Suppose the customer tenders x dimes and y quarters. To pay for the first item exactly, the customer requires

$$10x + 25y \;=\; 349.$$

But $\gcd(10, 25) = 5$, which does not divide 349, and so the equation has no solutions. Therefore, as experience would tell us, the customer cannot pay the exact amount without using some cents.

(ii)   For the second item the customer requires

$$10x + 25y \;=\; 265.$$

In this case $\gcd(10, 25) = 5$ which does divide 265. Hence the equation has integer solutions, though not necessarily positive integer solutions. If faced with this problem in a shop, we would not use the Euclidean Algorithm to find a solution but would obtain one by inspection. For example, $x = 4$, $y = 9$ is one particular solution. The general solution is therefore

$$x \;=\; 4 + 5n, \quad y \;=\; 9 - 2n \quad \text{for } n \in \mathbb{Z}.$$

If we require x and y to be nonnegative, we must have $4 + 5n \geq 0$ and $9 - 2n \geq 0$; that is, $-\frac{4}{5} \leq n \leq \frac{9}{2}$. But as n is an integer, $0 \leq n \leq 4$ will yield all the nonnegative solutions, namely

$$(x, y) \;=\; (4, 9),\ (9, 7),\ (14, 5),\ (19, 3)\ \text{ or } (24, 1).$$

Hence the second item could be paid for in exactly five different ways, as long as the customer has at least 24 dimes and 9 quarters. $\qquad\square$

What do the negative solutions mean in this case? If we put $n = -1$, we obtain the solution $x = -1$, $y = 11$. This corresponds to the situation in which the customer offers 11 quarters and receives one dime in change.

Example 2.34.   A hallway 5 meters long is to be tiled with strips of tile of widths 8 cm and 18 cm. In how many ways can this be done without cutting some of the tiles to different widths?

Solution.   If x strips of tile of width 8 cm, and y strips of width 18 cm exactly fill the length of the hallway, then

$$8x + 18y \;=\; 500.$$

That is, dividing by the $\gcd(8, 18) = 2$,

$$4x + 9y = 250.$$

By inspection, we see that one solution to the equation $4x + 9y = 1$ is $x = -2, y = 1$. Hence one solution to our desired equation is $x = -500, y = 250$. This is clearly not a feasible solution because it is impossible to have a negative number of tiles. However, this particular solution will allow us to obtain the general integer solution. This is $x = -500 + 9n$, $y = 250 - 4n$, where $n \in \mathbb{Z}$. The only workable solutions occur when x and y are nonnegative. Hence we need $-500 + 9n \geq 0$ and $250 - 4n \geq 0$; that is,

$$\frac{500}{9} \leq n \leq \frac{250}{4} \quad \text{or} \quad 55\tfrac{5}{9} \leq n \leq 62\tfrac{1}{2}.$$

Since n must be an integer, it follows that $56 \leq n \leq 62$ and the seven feasible solutions are given in the following table.

n	56	57	58	59	60	61	62
x	4	13	22	31	40	49	58
y	26	22	18	14	10	6	2

For each of these seven ways of choosing the different widths, the tiles, of course, can be permuted amongst themselves when they are laid.

Check. $8 \cdot 58 + 18 \cdot 2 = 464 + 36 = 500$, so $x = 58, y = 2$ is correct. $\square$

Example 2.35. Find all integer solutions to

$$1249x - 379y = 5.$$

Solution.   First apply the Extended Euclidean Algorithm to 1249 and 379 in order to find one solution to

$$1249x + 379y = \gcd(1249, 379).$$

For interest, we also show the ordinary Euclidean Algorithm.

	Extended Euclidean Algorithm $1249x + 379y =$	r		q_i	Euclidean Algorithm $r_{i-2} = q_i \cdot r_{i-1} + r_i$
1	0	1249			
0	1	379			
1	-3	112		3	$1249 = 3 \cdot 379 + 112$
-3	10	43		3	$379 = 3 \cdot 112 + 43$
7	-23	26		2	$112 = 2 \cdot 43 + 26$
-10	33	17		1	$43 = 1 \cdot 26 + 17$
17	-56	9		1	$26 = 1 \cdot 17 + 9$
-27	89	8		1	$17 = 1 \cdot 9 + 8$
44	-145	1		1	$9 = 1 \cdot 8 + 1$
-379	1249	0		8	$8 = 8 \cdot 1 + 0$

From the next to last row, we have $44(1249) + (-145)(379) = 1$. Multiplying through by 5 gives us $(220)(1249) + (-725)(379) = 5$, and so a particular solution to the original equation is $x = 220$ and $y = 725$. All the solutions are

$$x = 220 + 379n, \quad y = 725 + 1249n \quad \text{for all } n \in \mathbb{Z}.$$

Check. $1249(220 + 379n) - 379(725 + 1249n)$
$= 274780 + 473371n - 274775 - 473371n = 5$ □

Example 2.36.   Find all the positive integer solutions to

$$12x + 31y = 10{,}000.$$

Solution.   Apply the Extended Euclidean Algorithm to 12 and 31.

$12x$	$+ 31y$	$= r$
1	0	12
0	1	31
1	0	12
-2	1	7
3	-1	5
-5	2	2
13	-5	1

Hence $12(13) + 31(-5) = 1$, and so one integer solution to the original equation is $x = 130{,}000$, $y = -50{,}000$. By the Linear Diophantine Equation Theorem 2.31 the general solution is

$$x = 130{,}000 + 31n, \quad y = -50{,}000 - 12n \quad \text{for all } n \in \mathbb{Z}.$$

For positive solutions we require $x > 0$ and $y > 0$, so $130{,}000 + 31n > 0$ and $-50{,}000 - 12n > 0$. Also n must be an integer so, using the symbol $\approx$ to mean approximately,

$$-4193.55 \approx -\frac{130{,}000}{31} < n < -\frac{50{,}000}{12} \approx -4166.67$$
$$-4193 \le n \le -4167.$$

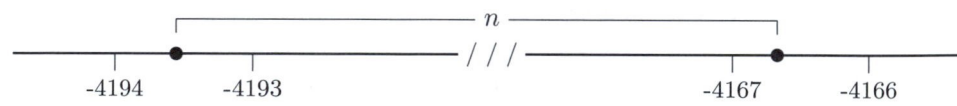

There are 27 positive solutions, namely

$$x = 130{,}000 + 31n, \quad y = -50{,}000 - 12n \text{ for } -4193 \le n \le -4167 \text{ and } n \in \mathbb{Z}.$$

Check.   If $n = -4167$ then $x = 823$ and $y = 4$. This is one positive solution, since $12(823) + 31(4) = 9876 + 124 = 10{,}000$.   □

2.4   INTEGERS IN DIFFERENT BASES

Another application of the Division Algorithm is in the conversion of a number from base 10 to any other base.

In the standard notation for integers, the symbol **52067** stands for the number

$$\mathbf{5} \cdot 10^4 + \mathbf{2} \cdot 10^3 + \mathbf{0} \cdot 10^2 + \mathbf{6} \cdot 10 + \mathbf{7}.$$

In general, any positive integer can be written as the symbol $r_n r_{n-1} \ldots r_1 r_0$, which stands for

$$r_n 10^n + r_{n-1} 10^{n-1} + \cdots + r_1 10 + r_0,$$

where $0 \leq r_i \leq 9$ for $i = 0, 1, 2, \ldots, n$. This system, in which numbers are written in terms of powers of 10, is called the *decimal system* or the *representation of integers in the base 10*.

We use the decimal system because we tend to count on our fingers. However, mathematically, we could use any integer, bigger than one, for the base. The decimal system requires the use of ten symbols as digits, namely 0, 1, 2, 3, 4, 5, 6, 7, 8, and 9. A base b would require the use of b different symbols as digits.

The decimal system is not the only one that has been used in history. The Babylonians wrote their mathematical and astronomical numbers in the base 60, and the remains of this system can still be seen in our division of the hour and minute into 60 parts. In a more modern context, the internal computations in a digital computer are normally performed in the base 2, and this is called the *binary system*.

The binary system only uses the two digits 0 and 1. For example, the symbol 1011011 stands for the number

$$
\begin{aligned}
& 1 \cdot 2^6 + 0 \cdot 2^5 + 1 \cdot 2^4 + 1 \cdot 2^3 + 0 \cdot 2^2 + 1 \cdot 2 + 1 \\
= {} & 64 \;+\; 0 \;+\; 16 \;+\; 8 \;+\; 0 \;+\; 2 \;+ 1 \\
= {} & 91 \qquad \text{in the decimal system.}
\end{aligned}
$$

Notice that a number needs a longer representation in the binary system than in the decimal system. There is a tradeoff in using a small base, such as in the binary system. The arithmetic is much simpler, but the representations are longer.

How can a number in the decimal system be converted to another base? Let us look at the digits $r_n \ldots r_2 r_1 r_0$ of a number x in the decimal system. The rightmost digit, r_0, is the remainder when x is divided by 10, so

$$x \;=\; q_0 10 + r_0.$$

The next digit, r_1, is the remainder when the quotient, q_0, is divided by 10, so

$$
\begin{aligned}
q_0 &= q_1 10 + r_1 \\
q_1 &= q_2 10 + r_2 \qquad \text{etc.}
\end{aligned}
$$

The digits in the representation of an integer to any other base have similar properties, where 10 is replaced by the base in question.

Let us try to convert the decimal 126 to base 4 using this method. Start by dividing 126 by 4 according to the Division Algorithm, and then repeatedly divide the quotients by 4.

$$
\begin{aligned}
126 &= 31 \cdot 4 &+& \mathbf{2} \\
31 &= 7 \cdot 4 &+& \mathbf{3} \\
7 &= 1 \cdot 4 &+& \mathbf{3} \\
1 &= 0 \cdot 4 &+& \mathbf{1}
\end{aligned}
$$

Hence we can write

$$
\begin{aligned}
126 &= 31 \cdot 4 + \mathbf{2} & &= (7 \cdot 4 + \mathbf{3})4 + \mathbf{2} \\
&= 7 \cdot 4^2 + \mathbf{3} \cdot 4 + \mathbf{2} & &= (4 + \mathbf{3})4^2 + \mathbf{3} \cdot 4 + \mathbf{2} \\
&= 1 \cdot 4^3 + \mathbf{3} \cdot 4^2 + \mathbf{3} \cdot 4 + \mathbf{2}.
\end{aligned}
$$

Therefore, 126 can be written as $(1332)_4$ in the base 4. Notice that these digits are just the remainders in the above equations, starting with the bottom equation.

Theorem 2.41. *Let b be a fixed integer greater than 1. Then any positive integer x can be expressed uniquely as $x = r_n b^n + r_{n-1} b^{n-1} + \cdots + r_2 b^2 + r_1 b + r_0$, where $0 \le r_i < b$ for $i = 0, 1, 2, \ldots, n$ and $r_n \ne 0$.*

This expression for x is called the *representation of x in base b* and is written as

$$
x = (r_n r_{n-1} \ldots r_2 r_1 r_0)_b.
$$

Proof. Dividing x by b according to the Division Algorithm and repeatedly dividing the quotients by b, we obtain the following list of equations.

$$
\begin{aligned}
x &= q_0 b + r_0, & &\text{where} \quad 0 \le r_0 < b \\
q_0 &= q_1 b + r_1, & &\text{where} \quad 0 \le r_1 < b \\
q_1 &= q_2 b + r_2, & &\text{where} \quad 0 \le r_2 < b \\
&\ \ \vdots \\
q_{n-2} &= q_{n-1} b + r_{n-1}, & &\text{where} \quad 0 \le r_{n-1} < b \\
q_{n-1} &= 0 \cdot b + r_n, & &\text{where} \quad 0 < r_n < b.
\end{aligned}
$$

Since $b > 1$, it follows that $x > q_0 > q_1 > \ldots$, and the quotients form a strictly decreasing sequence of nonnegative integers, which must eventually reach zero. Using this list of equations, we can write

$$
\begin{aligned}
x &= q_0 b + r_0 \\
&= (q_1 b + r_1)b + r_0 &= q_1 b^2 + r_1 b + r_0 \\
&= (q_2 b + r_2)b^2 + r_1 b + r_0 &= q_2 b^3 + r_2 b^2 + r_1 b + r_0 \\
&\ \ \vdots \\
x &= r_n b^n + r_{n-1} b^{n-1} + \cdots + r_2 b^2 + r_1 b + r_0
\end{aligned}
$$

and this is the required expression.

To show that the expression is unique, suppose that

$$x = r_n b^n + \cdots + r_1 b + r_0 = s_m b^m + \cdots + s_1 b + s_0,$$

where $0 \le r_i < b$, $i = 0, 1, \ldots, n$ and $0 \le s_j < b$, $j = 0, 1, \ldots, m$. Now

$$(r_n b^{n-1} + \cdots + r_1)b + r_0 = (s_m b^{m-1} + \cdots + s_1)b + s_0$$

where each side is an expressions for x in the form of the Division Algorithm under division by b. Hence, by the uniqueness of the Division Algorithm, the remainders and quotients must be equal. That is, $r_0 = s_0$ and

$$r_n b^{n-1} + \cdots + r_1 = s_m b^{m-1} + \cdots + s_1.$$

Repeating this procedure, we can show that $r_1 = s_1$, $r_2 = s_2$, etc. Also, m must equal n and the expansions for x are identical. $\square$

We will use the convention that a number without parentheses and a subscript will be in decimal form.

Example 2.42. Express 545 in base 7 and in base 2.

Solution. Repeatedly dividing by 7, we have

$$
\begin{aligned}
545 &= 77 \cdot 7 + 6 \\
77 &= 11 \cdot 7 + 0 \\
11 &= 1 \cdot 7 + 4 \\
1 &= 0 \cdot 7 + 1.
\end{aligned}
$$

Hence $545 = (1406)_7$

Repeatedly dividing by 2, we have

$$
\begin{aligned}
545 &= 272 \cdot 2 + 1 \\
272 &= 136 \cdot 2 + 0 \\
136 &= 68 \cdot 2 + 0 \\
68 &= 34 \cdot 2 + 0 \\
34 &= 17 \cdot 2 + 0 \\
17 &= 8 \cdot 2 + 1 \\
8 &= 4 \cdot 2 + 0 \\
4 &= 2 \cdot 2 + 0 \\
2 &= 1 \cdot 2 + 0 \\
1 &= 0 \cdot 2 + 1.
\end{aligned}
$$

Hence $545 = (1000100001)_2$.

Check. $(1406)_7 = 1 \cdot 7^3 + 4 \cdot 7^2 + 0 \cdot 7 + 6 = 343 + 196 + 6 = 545$

$$(1000100001)_2 = 2^9 + 2^5 + 2^0 = 512 + 32 + 1 = 545. \qquad \square$$

Each digit in a number in the base b can be any one of the symbols $0, 1, \ldots, b-1$ and therefore numbers in base b require b symbols to represent them. If the base is larger than 10, new symbols are required to represent some of the numbers.

Example 2.43.   Using the symbols A for ten and B for eleven, express $(1AAB9)_{12}$ in the decimal system, and express 1511 in base 12.

Solution.
$$\begin{aligned} (1AAB9)_{12} &= 1(12)^4 + 10(12)^3 + 10(12)^2 + 11 \cdot 12 + 9 \\ &= 20736 + 17280 + 1440 + 132 + 9 \\ &= 39597. \end{aligned}$$

To express 1511 in base 12, repeatedly divide by 12 to obtain
$$\begin{aligned} 1511 &= 125 \cdot 12 + 11 \\ 125 &= 10 \cdot 12 + 5 \\ 10 &= 0 \cdot 12 + 10. \end{aligned}$$

Hence $1511 = (A5B)_{12}$.

Check.
$$\begin{aligned} 39597 &= 3299 \cdot 12 + 9 \\ 3299 &= 274 \cdot 12 + 11 \\ 274 &= 22 \cdot 12 + 10 \\ 22 &= 1 \cdot 12 + 10 \\ 1 &= 0 \cdot 12 + 1 \end{aligned}$$
$$(A5B)_{12} = 10 \cdot 12^2 + 5 \cdot 12 + 11 = 1440 + 60 + 11 = 1511 \qquad \square$$

Arithmetical calculations can be performed in base b without reference to base 10 if the addition and multiplication tables for base b are known. The tables for base 4 are as follows. Remember that only the digits 0, 1, 2, and 3 are used in base 4 arithmetic.

Base 4 Addition Table

$+$	$(1)_4$	$(2)_4$	$(3)_4$
$(1)_4$	$(2)_4$	$(3)_4$	$(10)_4$
$(2)_4$	$(3)_4$	$(10)_4$	$(11)_4$
$(3)_4$	$(10)_4$	$(11)_4$	$(12)_4$

Base 4 Multiplication Table

$\cdot$	$(1)_4$	$(2)_4$	$(3)_4$
$(1)_4$	$(1)_4$	$(2)_4$	$(3)_4$
$(2)_4$	$(2)_4$	$(10)_4$	$(12)_4$
$(3)_4$	$(3)_4$	$(12)_4$	$(21)_4$

Example 2.44.   Calculate the sum and product of $(2031)_4$ and $(332)_4$ in base 4.

Solution.

$$\begin{array}{r} (2031)_4 \\ + \ (332)_4 \\ \hline (3023)_4 \end{array}$$

$$\begin{array}{r} (2031)_4 \\ \times \ (332)_4 \\ \hline (10122)_4 \\ (122130)_4 \\ (1221300)_4 \\ \hline (2020212)_4 \end{array}$$

Check. $(2031)_4 = 2 \cdot 4^3 + 3 \cdot 4 + 1 = 141$ and $(332)_4 = 3 \cdot 4^2 + 3 \cdot 4 + 2 = 62$
$(3023)_4 = 3 \cdot 4^3 + 2 \cdot 4 + 3 = 203 = 141 + 62$
$(2020212)_4 = 2 \cdot 4^6 + 2 \cdot 4^4 + 2 \cdot 4^2 + 4 + 2 = 8742 = 141 \cdot 62$ □

2.5   PRIME NUMBERS

One of the most important class of numbers is that of prime numbers. Most integers can be factored into a product of smaller integers. Numbers that cannot be so factored are called prime numbers. They form the basic building blocks of the number system because any other integer can be written as a product of primes.

Definition.   An integer $p > 1$ is called a **prime** if its only positive divisors are 1 and p; otherwise it is called **composite**.

The first few primes are 2, 3, 5, 7, 11, while $4 = 2 \cdot 2$, $6 = 2 \cdot 3$, $8 = 2 \cdot 2 \cdot 2$, $9 = 3 \cdot 3$, $10 = 2 \cdot 5$ are composite. The integer 1 is neither prime nor composite; a factor 1 is of no interest in any product.

Proposition 2.51.   *Every integer larger than 1 can be expressed as a product of primes.*

We shall prove this by the method of contradiction, though the result could also be proved by induction. The idea behind the proof is that if the result is not true, then there must be a smallest counterexample.

Proof.   Suppose that the result is false, and let N be the smallest integer, greater than 1, that cannot be written as a product of primes. The integer N cannot be prime itself, so we can write $N = r \cdot s$, where $1 < r \le s < N$. By our hypothesis, N is the smallest integer that cannot be written as a product of primes; hence r and s can be written as a product of primes. It follows that $N = r \cdot s$ can also be written as a product of primes. This contradiction shows that our hypothesis is false and the theorem must be true. □

It is natural to wonder whether there are only a finite number of primes, or whether the set of primes is infinite. This question was answered by Euclid in about 300 B.C., and the following proof is essentially the same as his.

Euclid's Theorem 2.52.   *The number of primes is infinite.*

Proof.   Suppose that there is only a finite number of primes, say $p_1, p_2, p_3, \ldots, p_n$. We now show that you cannot generate all the numbers by multiplying just these primes together. We look for a number that obviously does not contain any of these primes as a factor. Consider the integer

$$N = p_1 \cdot p_2 \cdot p_3 \cdots p_n + 1$$

This is not a prime because it is larger than all the primes p_1, p_2, $p_3, \ldots, p_n$. On the other hand, N is not divisible by any of the primes p_i, for $1 \leq i \leq n$; if $p_i | N$, then $p_i | (N - p_1 \cdots p_n)$, and so $p_i | 1$, which is impossible. Therefore, N cannot be written as a product of primes, contrary to the previous theorem.

Hence our original supposition must be false and the theorem is true. $\square$

The following important result shows that if a prime divides a product, then it must divide one of the factors. This is not true for composite numbers; for example, $6 | 15 \cdot 4$ but $6 \nmid 15$ and $6 \nmid 4$.

Theorem 2.53.   *If p is a prime and $p | ab$, then $p | a$ or $p | b$.*

Proof.   Suppose that the prime p divides ab but does not divide a. Since the only positive divisors of the prime p are 1 and p, the only positive common divisor of p and a is 1; hence $\gcd(a, p) = 1$. It now follows from Proposition 2.28 that $p | b$. Therefore, either $p | a$ or $p | b$. $\square$

One of the reasons for introducing primes is to enable us to split numbers into factors that are as small as possible. We shall now show that each number can be written as a product of primes in essentially only one way. This may appear obvious; most probably because you have never seen a number being factored into primes in two different ways. [This can be seen by looking at the set of real numbers $S = \{a + b\sqrt{5} \mid a, b \in \mathbb{Z}\}$ instead of the set of integers $\mathbb{Z}$. In S, 4 can be factored as $(\sqrt{5} + 1)(\sqrt{5} - 1)$, as well as $2 \cdot 2$, and the numbers 2, $\sqrt{5} + 1$ and $\sqrt{5} - 1$ cannot be factored further. Hence they are "primes" in the set S.]

As multiplication is commutative, the prime factors in any factorization can be written in different orders; for example, $45 = 3^2 \cdot 5 = 3 \cdot 5 \cdot 3$. However, up to order, the factorization of integers is unique. This Unique Factorization Theorem is so basic it is often referred to as the *Fundamental Theorem of Arithmetic.*

Unique Factorization Theorem 2.54.   *Every integer, greater than 1, can be expressed as a product of primes and, apart from the order of the factors, this expression is unique.*

Proof.   The existence of the factorization was shown in Proposition 2.51.

Now suppose that an integer x can be factored into primes in two ways as

$$x = p_1 p_2 \cdots p_n = q_1 q_2 \cdots q_m,$$

where all the p's and q's are primes. Since $p_1 | x$, $p_1 | q_1 q_2 \cdots q_m$. By repeated application of Theorem 2.53, it follows that p_1 divides at least one of the q's. If necessary, rearrange the q's so that $p_1 | q_1$. Since q_1 is prime, and p_1 is a nontrivial factor, it follows that $p_1 = q_1$. Cancelling p_1 and q_1, we have

$$p_2 p_3 \cdots p_n = q_2 q_3 \cdots q_m.$$

By continuing in this way, we see that each p must be paired off with one of the q's until there are no factors on either side. Hence $n = m$ and, apart from the order of the factors, the two expressions for x are the same. $\qquad\square$

Note that this theorem does not provide an algorithm for finding the prime factors. The following result shows that we can determine whether an integer x has any factors, other than x or 1, by checking whether it is divisible by all the primes less than or equal to $\sqrt{x}$.

Theorem 2.55.   *An integer $x > 1$ is either prime or contains a prime factor $\leq \sqrt{x}$.*

Proof.   Suppose that p is the smallest prime factor of x. If x is composite, we can write $x = ab$, where a and b are positive integers between 1 and x. Since p is the smallest prime factor, $a \geq p, b \geq p$ and $x = ab \geq p^2$. Hence $p \leq \sqrt{x}$. $\qquad\square$

There is no known efficient procedure for finding large prime numbers. A tedious process known as the *sieve of Eratosthenes* will yield all the primes less than any given number N. This process consists of writing down all the numbers from 2 to $N - 1$. Leave 2 alone and cross out every second number (that is, composite numbers that are multiples of 2). The next remaining number, namely 3, will be prime; keep it and cross out every third number starting from 3 (that is, composite multiples of 3). The next remaining number, namely 5, will be prime; keep it and cross out every fifth number after 5. If we proceed in this way we will eventually cross out all the composite numbers, and all the primes less than N will remain. The above proposition shows that we only have to cross out multiples of prime numbers less than or equal to $\sqrt{N}$. However, this is still a very large task when N is large.

The prime factorization of integers can be used to find all the divisors of an integer and the greatest common divisor of two integers.

Proposition 2.56.   *If $a = p_1^{a_1} p_2^{a_2} \cdots p_n^{a_n}$ is the prime factorization of a into powers of distinct primes $p_1, p_2, \ldots, p_n$, then the positive divisors of a are those integers of the form*

$$d \;=\; p_1^{d_1} p_2^{d_2} \cdots p_n^{d_n}, \qquad \text{where} \quad 0 \leq d_i \leq a_i \text{ for } i = 1, 2, \ldots, n.$$

Proof.   If $d_i \leq a_i$ for each i, then $a \;=\; d(p_1^{a_1 - d_1} p_2^{a_2 - d_2} \cdots p_n^{a_n - d_n})$, and so $d | a$.

On the other hand, if d is a divisor of a, let $a = bd$. Any prime factor of b or d is also a prime factor of a, and so must be one of $p_1, \ldots, p_n$. Hence we can write

$$b \;=\; p_1^{b_1} \cdots p_n^{b_n} \qquad \text{and} \qquad d \;=\; p_1^{d_1} \cdots p_n^{d_n},$$

where some of the exponents of the primes may be zero. By applying the Unique Factorization Theorem 2.54 to the equation $a = bd$, we see that $a_i = b_i + d_i$ for $i = 1, 2, \ldots, n$. In particular, $d_i \leq a_i$ for $i = 1, 2, \ldots, n$. $\qquad\square$

Theorem 2.57.   *If $a = p_1^{a_1} \cdots p_n^{a_n}$ and $b = p_1^{b_1} \cdots p_n^{b_n}$ are prime factorizations of the integers a and b, where some of the exponents may be zero, then*

$$\gcd(a, b) \;=\; p_1^{d_1} \cdots p_n^{d_n},$$

where $d_i = \min(a_i, b_i)$ for $i = 1, \ldots, n$.

Proof.   Let $d = p_1^{d_1} \cdots p_n^{d_n}$, where $d_i = \min(a_i, b_i)$. Since $d_i \leq a_i$ and $d_i \leq b_i$ for each i, it follows from the previous proposition that $d|a$ and $d|b$.

Also, by Proposition 2.56, any other positive divisor of a and b is of the form $c = p_1^{c_1} \cdots p_n^{c_n}$, where $c_i \leq a_i$ and $c_i \leq b_i$ for each i. Hence $c_i \leq \min(a_i, b_i)$ and $c \leq d$. This proves that $d = \gcd(a, b)$. $\square$

For example,

$$
\begin{aligned}
336 &= 2^4 \cdot 3 \cdot 7 &&= 2^4 \cdot 3^1 \cdot 7^1 \cdot 11^0 \\
2156 &= 2^2 \cdot 7^2 \cdot 11 &&= 2^2 \cdot 3^0 \cdot 7^2 \cdot 11^1 \\
\gcd(336, 2156) &= && 2^2 \cdot 3^0 \cdot 7^1 \cdot 11^0 &&= 28.
\end{aligned}
$$

This method does not supersede the Euclidean Algorithm for finding greatest common divisors because it is often too tedious to factor an integer into a product of primes; however, this method is useful when the numbers involved are small.

When we add two fractions together, we normally put them over a common denominator, that is, a common multiple of both the denominators. We now show how to obtain the least common multiple of two integers from their greatest common divisor.

A *common multiple* of two integers is an integer that is a multiple of each. For example, some common multiples of 6 and 9 are 54, 18, 36, and 72. The smallest positive one is 18, and this is called the least common multiple.

Definition.   The **least common multiple** of two positive integers a and b is the smallest positive integer that is divisible by both a and b. It will be denoted by $\operatorname{lcm}(a, b)$.

The number ab is always a common multiple of a and b; hence $\operatorname{lcm}(a, b)$ always exists for positive integers a and b.

Theorem 2.58.   *If $a = p_1^{a_1} \cdots p_n^{a_n}$ and $b = p_1^{b_1} \cdots p_n^{b_n}$ are prime factorizations of the positive integers a and b, where some of the exponents may be zero, then*

$$\operatorname{lcm}(a, b) \;=\; p_1^{e_1} \cdots p_n^{e_n},$$

where $e_i = \max(a_i, b_i)$ for $i = 1, \ldots, n$.

Proof.   Let $e = p_1^{e_1} \cdots p_n^{e_n}$, where $e_i = \max(a_i, b_i)$. By Proposition 2.56, $a|e$ and $b|e$. Also, by Proposition 2.56, any positive integer that is a multiple of a and b

must contain the factor $p_1^{e_1} \cdots p_n^{e_n}$; that is, must contain the factor e and be greater than or equal to e. Hence $e = \text{lcm}(a, b)$. $\qquad\square$

Theorem 2.59.   *For any positive integers a and b*

$$a \cdot b = \gcd(a, b) \cdot \text{lcm}(a, b).$$

Proof.   Let $a = p_1^{a_1} \cdots p_n^{a_n}$ and $b = p_1^{b_1} \cdots p_n^{b_n}$ be prime factorizations of the integers a and b, where some of the exponents may be zero. It is always true that

$$a_1 + b_1 = \min(a_1, b_1) + \max(a_1, b_1)$$

since the minimum and maximum are either a_1 and b_1, or b_1 and a_1, respectively. Hence, if $d_i = \min(a_i, b_i)$ and $e_i = \max(a_i, b_i)$, then

$$p_1^{a_1} \cdots p_n^{a_n} \cdot p_1^{b_1} \cdots p_n^{b_n} = p_1^{d_1} \cdots p_n^{d_n} \cdot p_1^{e_1} \cdots p_n^{e_n}.$$

That is, $a \cdot b = \gcd(a, b) \cdot \text{lcm}(a, b)$. $\qquad\square$

For example, we showed above that $\gcd(336, 2156) = 28$. Hence

$$\text{lcm}(336, 2156) = \frac{336 \cdot 2156}{28} = 25872.$$

Our investigations into prime numbers and the divisibility properties of integers form the beginning of the branch of mathematics known as number theory. Even though the raw material, the set of integers, is apparently elementary, there are many outstanding conjectures about numbers and primes that can be simply stated, but have not yet been solved.

One of these is known as the Goldbach conjecture. Goldbach, in a letter to Euler in 1742, asked if every even number (greater than 2) can be written as the sum of two primes. It is true in every particular case that has been looked at; for example, $4 = 2 + 2$, $6 = 3 + 3$, $8 = 3 + 5, \ldots, 30 = 23 + 7$, $32 = 29 + 3$. However, nobody has proved that it must be true for all even numbers.

Another unsolved problem is that of the number of pairs of primes differing by 2. An examination of a list of prime numbers shows that many primes occur in pairs of the form p and $p + 2$; such pairs are 3 and 5, 11 and 13, 17 and 19, and so on. From circumstantial evidence it appears that the number of such prime pairs is infinite, but no proof has been found.

In 1640, Fermat thought that he had discovered a long-sought-for formula that would yield primes for every value of a variable n. He conjectured that

$$F(n) = 2^{2^n} + 1$$

was a prime for all values of n. Now $F(0) = 3$, $F(1) = 5$, $F(2) = 17$, $F(3) = 257$ and $F(4) = 65537$, which are all primes. However, in 1732, Euler discovered that $F(5)$ contains a factor 641 and hence is not prime. In fact, despite extensive computer searches, no more of these "Fermat numbers" were found to be prime. At present, it is unknown whether $F(n)$ is ever prime if $n > 4$.

Exercise Set 2

1–8. Find the quotient and remainder when b is divided by a in each of the following cases.

1. $a = 3$, $b = 13$
2. $a = 13$, $b = 3$
3. $a = 7$, $b = 7$
4. $a = 7$, $b = 0$
5. $a = 4$, $b = -12$
6. $a = 4$, $b = -10$
7. $a = 11$, $b = -246$
8. $a = 17$, $b = -5$

9. If $3p^2 = q^2$, where $p, q \in \mathbb{Z}$, show that 3 is a common divisor of p and q.
10. If $ac|bc$ and $c \neq 0$, prove that $a|b$.
11. Prove that $\gcd(ad, bd) = |d| \cdot \gcd(a, b)$.

12–18. Find the greatest common divisor of each pair of integers.

12. 5280 and 3600
13. 484 and 451
14. 616 and 427
15. 1137 and -419
16. 19201 and 3587
17. 2^{100} and 100^2
18. $10!$ and 3^{10}

19–26. In each case write $\gcd(a, b)$ in the form $ax + by$, where $x, y \in \mathbb{Z}$.

19. $a = 484$, $b = 451$
20. $a = 5280$, $b = 3600$
21. $a = 17$, $b = 15$
22. $a = 5$, $b = 13$
23. $a = 100$, $b = -35$
24. $a = 3953$, $b = 1829$
25. $a = 51$, $b = 17$
26. $a = 431$, $b = 0$

27. Prove that $\gcd(a, c) = \gcd(b, c) = 1$ if and only if $\gcd(ab, c) = 1$.
28. Prove that any two consecutive integers are relatively prime.
29. Simplify

$$\frac{95}{646} + \frac{40}{391}.$$

30. Gear A turns at 1 rev/min and is meshed into gear B. If A has 32 teeth and B has 120 teeth, how often will both gears be simultaneously back in their starting positions?

31–36. Find one integer solution, if possible, to each Diophantine equation.

31. $21x + 35y = 7$
32. $14x + 18y = 5$
33. $x + 14y = 9$
34. $11x + 15y = 31$
35. $143x + 253y = 156$
36. $91x + 126y = 203$

37–42. Find all the integer solutions to each Diophantine equation.

37. $7x + 9y = 1$
39. $15x - 24y = 9$
41. $243x + 405y = 123$

38. $212x + 37y = 1$
40. $16x + 44y = 20$
42. $169x - 65y = 91$

*43–46. Find all the **nonnegative** integer solutions to each Diophantine equation.*

43. $14x + 9y = 1000$
45. $38x + 34y = 200$

44. $12x + 57y = 423$
46. $11x - 12y = 13$

47. Can 1000 be expressed as the sum of two positive integers, one of which is divisible by 11 and the other by 17?

48. Can 120 be expressed as the sum of two positive integers, one of which is divisible by 11 and the other by 17?

49. Can 120 be expressed as the sum of two positive integers, one of which is divisible by 14 and the other by 18?

50. Find the smallest positive integer x so that $157x$ leaves remainder 10 when divided by 24.

51. The nickel slot of a pay phone will not accept coins. Can a call costing 95 cents be paid for exactly using only dimes and quarters? If so, in how many ways can it be done?

52–54. Convert the following numbers to base 10.

52. $(5613)_7$
53. $(100110111)_2$
54. $(9A411)_{12}$, where A is the symbol for ten.

55. How many seconds are there in 4 hours 27 minutes and 13 seconds?

56–61. Convert the following numbers to the indicated base.

56. 1157 to base 2
58. 433 to base 5
60. 5766 to base 12, writing A for ten and B for eleven
61. 40239 to base 60

57. 1241 to base 9
59. 30 to base 3

62. Add and multiply $(1011)_2$ and $(110110)_2$ together in base 2.

63. Add and multiply $(3130)_4$ and $(103)_4$ together in base 4.

64. Write out the addition and multiplication tables for base 6 arithmetic, and then multiply $(4512)_6$ by $(343)_6$ in base 6.

65. Subtract $(3321)_4$ from $(10020)_4$ in base 4, and check your answer by converting to base 10.

66. If $a = (342)_8$ and $b = (173)_8$, find $a - b$ without converting to base 10. [If you get stuck, listen to the song "The New Math" by Tom Lehrer on the album *That Was the Year That Was.*]

67. How many positive divisors does 12 have?

68. How many positive divisors does 6696 have?

69. If we wish to add the fractions $\frac{1}{132} + \frac{4}{9}$, what is the smallest common denominator we could choose?

70–71. Factor the following numbers into prime factors and calculate the greatest common divisor and least common multiple of each pair.

70. 40 and 144   71. 5280 and 57800

72. Find $\mathrm{lcm}(12827, 20099)$.

Problem Set 2

73. Prove that $\{ax + by \mid x, y \in \mathbb{Z}\} = \{n \cdot \gcd(a, b) \mid n \in \mathbb{Z}\}$.

74. Show that $\gcd(ab, c) = \gcd(b, c)$ if $\gcd(a, c) = 1$. Is it true in general that

$$\gcd(ab, c) = \gcd(a, c) \cdot \gcd(b, c) ?$$

75. Show that the Diophantine equation $ax^2 + by^2 = c$ does not have any integer solutions unless $\gcd(a, b) \mid c$. If $\gcd(a, b) \mid c$, does the equation always have an integer solution?

76. For what values of a and b does the Diophantine equation $ax + by = c$ have an infinite number of positive solutions for x and y?

77. For what values of c does $8x + 5y = c$ have exactly one strictly positive solution?

78. An oil company has a contract to deliver 100000 liters of gasoline. Their tankers can carry 2400 liters and they can attach one trailer carrying 2200 liters to each tanker. All the tankers and trailers must be completely full on this contract, otherwise the gas would slosh around too much when going over some rough roads. Find the least number of tankers required to fulfill the contract. Each trailer, if used, must be pulled by a full tanker.

79. A trucking company has to move 844 refrigerators. It has two types of trucks it can use; one carries 28 refrigerators and the other 34 refrigerators. If it only sends out full trucks and all the trucks return empty, list the possible ways of moving all the refrigerators.

80. Show how to measure exactly 2 liters of water from a river using a 27 liter jug and a 16 liter jug. If you could not lift the larger jug when full but could push it over, could you still measure the 2 liters?

81. Let S be the complete solution set of the Diophantine equation $ax + by = d$. Is

$$cS = \{(cx, cy) \mid (x, y) \in S\}$$

the complete solution set of $ax + by = cd$?

82. Four men and a monkey spend the day gathering coconuts on a tropical island. After they have all gone to sleep at night, one of the men awakens and, not trusting the others, decides to take his share. He divides the coconuts into four equal piles, except for one remaining coconut, which he gives to the monkey. He then hides his share, puts the other piles together, and goes back to sleep. Each of the other men awakens during the night and does likewise, and every time there is one coconut left over for the monkey. In the morning all the men awake, divide what's left of the coconuts into four, and again there is one left over that is given to the monkey. Find the minimum number of coconuts that could have been in the original pile.

83. Let a, b, c be nonzero integers. Their *greatest common divisor* $\gcd(a, b, c)$ is the largest positive integer that divides all of them. Prove that

$$\gcd(a, b, c) = \gcd(a, \gcd(b, c)).$$

84. Prove that the Diophantine equation $ax + by + cz = e$ has a solution if and only if $\gcd(a, b, c) \mid e$.

85. If $\gcd(a, b, c) \mid e$, describe how to find one solution to the Diophantine equation $ax + by + cz = e$.

86. Describe how to find all the solutions to the Diophantine equation

$$ax + by + cz = e.$$

87. Find one integer solution to the Diophantine equation $18x + 14y + 63z = 5$.

88. Find all the ways that \$1.67 worth of stamps can be put on a parcel, using 6 cent, 10 cent, and 15 cent stamps.

89. Given a balance and weights of 1, 2, 3, 5, and 10 grams, show that any integer gram weight up to 21 grams can be weighed. If the weights were 1, 2, 4, 8, and 16 grams, show that any integer weight up to 31 grams could be weighed.

90. If weights could be put on either side of a balance, show that any integer weight up to 121 grams could be weighed using weights of 1, 3, 9, 27, and 81 grams.

91. If numbers (in their decimal form) are written out in words, such as six hundreds, four tens, and three for 643, we require one word for each digit 0, 1, 2, ..., 9, one word for 10, and one word for 10^2, and so on. We can name all the integers below 1000 with twelve words. What base would use the least number of words to name all the numbers below 1000? What base would use the least number of words to name all the numbers below 10^6?

92. Consider the set of all even integers $2\mathbb{Z} = \{2n \mid n \in \mathbb{Z}\}$. We can add, subtract, and multiply elements of $2\mathbb{Z}$, and the result will always be in $2\mathbb{Z}$, but we cannot always divide. We can define divisibility and factorization in $2\mathbb{Z}$ in a similar way to that in $\mathbb{Z}$. (For example, $2|4$ in $2\mathbb{Z}$, but $2 \nmid 6$ even though $6 = 2 \cdot 3$, because $3 \notin 2\mathbb{Z}$.) A prime in $2\mathbb{Z}$ is a positive even integer that cannot be factored into the product of two even integers.

   (a) Find all the primes in $2\mathbb{Z}$.

   (b) Can every positive element of $2\mathbb{Z}$ be expressed as a product of these primes?

   (c) If this factorization into primes can be accomplished, is it unique?

93. Prove that the sum of two consecutive odd primes has at least three prime divisors (not necessarily different).

94. How many zeros are there at the right end of

$$100! = 100 \cdot 99 \cdot 98 \cdot 97 \cdots 2 \cdot 1 \ ?$$

95. Show that

$$1 + \frac{1}{2} + \frac{1}{3} + \cdots + \frac{1}{n}$$

can never be an integer if $n > 1$.

96. If $\lfloor x \rfloor$ is the greatest integer less than or equal to x (that is, the integer part of x), then for which values of n does $\lfloor \sqrt{n} \rfloor$ divide n?

97. Let a and b be integers greater than 1, and let $e = \text{lcm}(a, b)$. Prove that

$$0 < \frac{1}{a} + \frac{1}{b} - \frac{1}{e} < 1.$$

98. If a and b are odd positive integers, and the sum of the integers, less than a and greater than b, is 1000, then find a and b.

99–102.   *Either prove each of the following statements about integers or give a counterexample.*

99. $a^2|b^2$ if and only if $a|b$

100. $\gcd(a, b) = \gcd(a + b, \text{lcm}(a, b))$

101. $\text{lcm}(\gcd(a, b), \gcd(a, c)) = \gcd(a, \text{lcm}(b, c))$

102. If $\gcd(a, b) = 1$ and $ax + by = c$ has a positive integer solution, then so does $ax + by = d$ when $d > c$.

103. Write a computer program to test whether a given number is prime. Use your program to find the smallest positive integer n for which the number $n^2 - n + 41$ fails to be prime.

104. Using a computer, test whether $F(4) = 2^{2^4} + 1$ and $F(5) = 2^{2^5} + 1$ are prime.

105. Show that all the integers, $\mathbb{Z}$, both positive and negative, can be represented in the *negative base* -10 using the digits $0, 1, \ldots, 9$ without using a negative prefix. For example, $-1467 = (2673)_{-10}$ and $10 = (190)_{-10}$.

    (a) What decimal numbers do $(56)_{-10}$ and $(164)_{-10}$ represent?

    (b) Find the negative ten representations of the decimal numbers 1111 and -209.

    (c) Try adding and multiplying some numbers in the base negative ten. Then try adding a number to its negative.

106.   **(a)** Find two consecutive primes that differ by at least 10.

    (b) Prove that there are arbitrarily large gaps between consecutive primes.

107. Let $a < b < c$, where a is a positive integer and b and c are odd primes. Prove that if $a \mid (3b + 2c)$ and $a \mid (2b + 3c)$, then $a = 1$ or 5. Give examples to show that both these values for a are possible.

108. An integer n is *perfect* if the sum of its divisors (including 1 and itself) is $2n$. Show that if $2^p - 1$ is a prime number, then $n = 2^{p-1}(2^p - 1)$ is perfect.

CHAPTER 3

Congruences

3.1 CONGRUENCE

Carl Friedrich Gauss (1777–1855), the German mathematician, physicist, and astronomer, was known as the Prince of Mathematicians to his contemporaries and is one of the greatest mathematicians of all times. His famous work on higher arithmetic and number theory, called *Disquisitiones Arithmeticae*, was completed when he was twenty-one and published in 1801. In this, among many other things, Gauss introduced the notion of congruence, thereby offering a convenient way of dealing with many questions of divisibility.

Definition. Let m be a fixed positive integer. If $a, b, \in \mathbb{Z}$, we say that "a is **congruent** to b **modulo** m" and write

$$a \equiv b \pmod{m}$$

whenever $m|(a-b)$. If $m \nmid (a-b)$, we write $a \not\equiv b \pmod{m}$.

For example, $7 \equiv 3 \pmod 4$, $-6 \equiv 14 \pmod{10}$, $121 \equiv 273 \pmod 2$, but $5 \not\equiv 4 \pmod 3$ and $21 \not\equiv 10 \pmod 2$. In Proposition 3.14, we show that two integers are congruent modulo m if and only if they have the same remainders after division by m.

The condition for a to be congruent to b modulo m is equivalent to the condition that

$$a \;=\; b + km \qquad \text{for some } k \in \mathbb{Z}.$$

Congruences occur in everyday life. The short hand of a clock indicates the hour modulo 12, while the long hand indicates the minute modulo 60. For example, 20 hours after midnight, the clock indicates 8 o'clock because $20 \equiv 8 \pmod{12}$. In determining which day of the week a particular date falls, we apply congruence modulo 7. Two integers are congruent modulo 2 if and only if they have the same parity; that is, if and only if they are both odd or both even.

The idea of congruence is not radically different from divisibility, but its usefulness lies in its notation, and the fact that congruence, with respect to a fixed modulus, has many of the properties of ordinary equality.

Proposition 3.11. *Let a, b, and c be integers. Then*

   (i) $a \equiv a \pmod m$.

  (ii) *If $a \equiv b \pmod m$, then $b \equiv a \pmod m$.*

 (iii) *If $a \equiv b \pmod m$ and $b \equiv c \pmod m$, then $a \equiv c \pmod m$.*

Proof. (i)   Since $a - a = 0$ and $m | 0$, it follows that $a \equiv a \pmod{m}$.

(ii)   If $a \equiv b \pmod{m}$, then $m | (a - b)$ and hence, by Proposition 2.11(ii), $m | (-1)(a - b)$; that is, $m | (b - a)$ and $b \equiv a \pmod{m}$.

(iii)   If $a \equiv b$ and $b \equiv c \pmod{m}$, then $m | (a - b)$ and $m | (b - c)$. Hence, by Proposition 2.11(ii), $m | (a - b) + (b - c)$; that is, $m | (a - c)$ and $a \equiv c \pmod{m}$. $\square$

Proposition 3.12.   If $a \equiv a' \pmod{m}$ and $b \equiv b' \pmod{m}$, then

(i) $a + b \equiv a' + b' \pmod{m}$

(ii) $a - b \equiv a' - b' \pmod{m}$

(iii) $a \cdot b \equiv a' \cdot b' \pmod{m}$.

Proof.   Since $a \equiv a'$ and $b \equiv b' \pmod{m}$ we can write $a = a' + km$ and $b = b' + \ell m$ where $k, \ell \in \mathbb{Z}$. It follows that

$$
\begin{aligned}
a + b &= a' + b' + (k + \ell)m \\
a - b &= a' - b' + (k - \ell)m \\
ab &= a'b' + (kb' + \ell a' + k\ell m)m.
\end{aligned}
$$

The results now follow, since $k + \ell,\ k - \ell,\ kb' + \ell a' + k\ell m \in \mathbb{Z}$. $\square$

Although we can add, subtract, and multiply congruences with respect to the same modulus, we cannot with impunity divide out an integer from either side of a congruence. For example, $6 \equiv 36 \pmod{10}$ but $1 \not\equiv 6 \pmod{10}$. However, the following proposition indicates under what conditions cancellation can occur.

Proposition 3.13.   If $ac \equiv bc \pmod{m}$ and $\gcd(c, m) = 1$, then it follows that $a \equiv b \pmod{m}$.

Proof.   If $ac \equiv bc \pmod{m}$, then $m | c(a - b)$. If $\gcd(c, m) = 1$, it follows from Proposition 2.28 that $m | (a - b)$ and so $a \equiv b \pmod{m}$. $\square$

For example, $35 \equiv 15 \pmod{4}$ and, since $\gcd(5, 4) = 1$, it follows that $7 \equiv 3 \pmod{m}$.

As would be expected from the fact that $a \equiv b \pmod{m}$ is equivalent to $a = b + km$, there is a close relationship between congruences modulo m and remainders under division by m.

Proposition 3.14. $a \equiv b \pmod{m}$ if and only if a and b have the same remainders when divided by m.

Proof.   Divide a and b by m according to the Division Algorithm to obtain

$$
\begin{aligned}
a &= km + r, &&\text{where } 0 \leq r < m \\
b &= \ell m + s, &&\text{where } 0 \leq s < m.
\end{aligned}
$$

Hence $a - b = (k - \ell)m + (r - s)$, where $-m < r - s < m$.

If a and b have the same remainders when divided by m, then $a - b = (k - \ell)m$ and $a \equiv b \pmod{m}$.

Conversely, if $a \equiv b \pmod{m}$, then $m|(a - b)$ and hence $m|(r - s)$. However, $-m < r - s < m$ and so $r - s = 0$. $\qquad\square$

We see from the above proposition that any integer must be congruent to precisely one of $0, 1, 2, \ldots, m - 1$ modulo m.

Example 3.15.   What is the remainder when 2^{37} is divided by 7?

Solution.   It would be very tedious to calculate 2^{37} and then divide by 7. To perform this arithmetic on a calculator, we would need to be able to display all 12 digits. However, we can use the above proposition to find what 2^{37} is congruent to modulo 7.

We know that $2^3 = 8$, and so $2^3 \equiv 1 \pmod{7}$. By repeated application of Proposition 3.12(iii), it follows that $(2^3)^{12} \equiv 1^{12} \equiv 1 \pmod{7}$. Hence

$$2^{37} \equiv 2^{36} \cdot 2 \equiv (2^3)^{12} \cdot 2 \equiv 1 \cdot 2 \equiv 2 \pmod{7}$$

and 2^{37} has remainder 2 when divided by 7. $\qquad\square$

Note that the successive powers of 2 take on a particular form modulo 7. This phenomenon of the remainders cycling will be explained by Fermat's Little Theorem 3.42.

$$2 \equiv 2, \quad 2^2 \equiv 4, \quad 2^3 \equiv 1, \quad 2^4 \equiv 2, \quad 2^5 \equiv 4, \quad 2^6 \equiv 1, \quad 2^7 \equiv 2, \text{ etc. } \pmod{7}.$$

Example 3.16.   What is the remainder when $4^{10} \cdot 7^7$ is divided by 5?

Solution. $4^2 \equiv 16 \equiv 1 \pmod{5}$ and $7^2 \equiv 49 \equiv -1 \pmod{5}$. Hence

$$4^{10} \cdot 7^7 \equiv (4^2)^5 \cdot (7^2)^3 \cdot 7 \equiv 1^5 \cdot (-1)^3 \cdot 7 \equiv -7 \equiv 3 \pmod{5}$$

and $4^{10} \cdot 7^7$ has remainder 3 when divided by 5. $\qquad\square$

3.2   TESTS FOR DIVISIBILITY

Congruences can be used to prove some of the familiar tests for divisibility by certain integers. It is well known that any integer is divisible by 2 if and only if its last digit is even. An integer is divisible by 4 if and only if the number determined by its last two digits is divisible by 4.

This test for divisibility by 4 works because $100 \equiv 0 \pmod{4}$ and so, for example, $56976 \equiv 569 \cdot 100 + 76 \equiv 76 \pmod{4}$. Therefore, the remainder when 56976 is divided by 4 is the same as that of 76 when divided by 4.

Theorem 3.21. *A number is divisible by 9 if and only if the sum of its digits is divisible by 9.*

For example, consider the numbers 5895 and 125942. The sums of their digits are $5 + 8 + 9 + 5 = 27$ and $1 + 2 + 5 + 9 + 4 + 2 = 23$, respectively; since 27 is divisible by 9 but 23 is not, it follows that 5895 is divisible by 9 but 125942 is not.

Proof. Let x be a number with decimal digits $a_r a_{r-1} \ldots a_1 a_0$ so that

$$x = a_r 10^r + a_{r-1} 10^{r-1} + \cdots + a_1 10 + a_0.$$

Now $10 \equiv 1 \pmod 9$, and hence $10^k \equiv 1^k \equiv 1 \pmod 9$ for all $k \geq 0$. Therefore,

$$x \equiv a_r + a_{r-1} + \cdots + a_1 + a_0 \pmod 9.$$

Hence $x \equiv 0 \pmod 9$ if and only if the sum of its digits is congruent to zero modulo 9. □

Note that this not only provides a test for divisibility by 9, it also provides a method for finding the remainder of any number when divided by 9. For example,

$$125942 \equiv 1 + 2 + 5 + 9 + 4 + 2 \equiv 23 \equiv 2 + 3 \equiv 5 \pmod 9$$

and hence 125942 has remainder 5 when divided by 9.

A similar proof also yields the following result for divisibility by 3.

Theorem 3.22. *A number is divisible by 3 if and only if the sum of its digits is divisible by 3.* □

The result on the divisibility by 9 provides the basis for an ancient method of checking arithmetical calculations called *casting out nines*. Suppose we wish to check the calculation

$$43296 \times 1742 - 514376 = 74907256.$$

The check proceeds as follows. For each number involved, add the digits together and throw away any multiples of nine. Then perform the original calculation on these remaining numbers. The calculation checks if this new answer agrees with the original answer, after adding digits and casting out any multiples of nine. If the answers do not agree after casting out nines, an error has occurred in the calculation.

In the above example, we add the digits of 43296 to obtain $4 + 3 + 2 + 9 + 6$ and, after casting out nines, we obtain the number 6. If we do this procedure to the other numbers on the left side of the equation, we get the reduced equation

$$6 \times 5 - 8.$$

Perform this simplified calculation to obtain 22 or, after casting out nines again, 4. The sum of the digits of the original answer, after casting out nines, is also 4; hence this provides a check on the calculation. It does not guarantee that the calculation is correct; it only provides a partial check.

Let us take another example.

$$\begin{array}{lcccccc} \text{Original calculation:} & (442)^3 & + & 5176 & = & 86355064 \\ \text{After casting out nines:} & 1^3 & + & 1 & \equiv & 1 & \pmod 9 \end{array}$$

This reduced congruence is incorrect, so a mistake must have been made in the original calculation.

We see from Theorem 3.21 why this method works. The check just performs the original calculation modulo 9. Therefore, the method works for any calculation involving addition, subtraction, and multiplication. (We can treat exponentiation as repeated multiplication, but we must not reduce the exponent modulo 9.)

Proposition 3.23. *A number is divisible by 11 if and only if the alternating sum of its digits is divisible by 11.*

Proof.   Let $x = a_r 10^r + a_{r-1} 10^{r-1} + \cdots + a_1 10 + a_0$. Now $10 \equiv -1 \pmod{11}$ so that

$$x \equiv (-1)^r a_r + (-1)^{r-1} a_{r-1} + \cdots - a_3 + a_2 - a_1 + a_0 \pmod{11}.$$

Hence any number is congruent modulo 11 to the alternating sum of its digits, and the result follows. □

For example, 2307151 is divisible by 11 because $2-3+0-7+1-5+1 = -11$, which is divisible by 11.

3.3   EQUIVALENCE RELATIONS

Congruence modulo a fixed integer is an example of an important notion in mathematics, namely the concept of an equivalence relation.

Algebra can be considered as the study of operations and relations in sets. Examples of operations are addition, subtraction, multiplication, and exponentiation; these all combine two elements to form a third. Examples of relations are greater than, divisible by, and equals; these all compare two elements. Roughly, R is a *relation* on a set S if, for every ordered pair of elements a and b in S, either a is related to b, in which case we write aRb, or a is not related to b and we write $a \not R b$.

Here are some examples of relations.

Greater Than: For $a, b \in \mathbb{R}$, take aRb to mean $a > b$.

Divisibility: For $a, b \in \mathbb{Z}$, take aRb to mean $a | b$.

Equality: Let S be any set, and take aRb to be $a = b$.

Congruence modulo m: For $a, b \in \mathbb{Z}$, take aRb to be $a \equiv b \pmod{m}$.

Congruence of Triangles: Let S be the set of triangles in the plane and, for two triangles T_1 and T_2, take $T_1 R T_2$ to mean T_1 is congruent to T_2.

Brother: Let S be the set of all people in the country and, if a and b are two people, take aRb to mean a is the brother of b.

Same Surname: Let S again be the set of all people in the country, and take aRb to mean that a has the same surname as b.

Same Day: Let S be the set of all days in a particular year and, for any two days a and b, take aRb to mean that a and b occur on the same day of the week.

Definition.   A relation R on a set S is called an **equivalence relation** if

(i) aRa for all $a \in S$ *(reflexive property)*

(ii) if aRb then bRa *(symmetric property)*

(iii) if aRb and bRc then aRc.   *(transitive property)*

Of the above examples of relations, *equals*, both types of *congruences*, and the *same surname* and *same day* are equivalence relations, while the others are not. Proposition 3.11 shows that congruence modulo m is an equivalence relation.

Definition.   If R is an equivalence relation on a set S, and $a \in S$, write

$$[a] \;=\; \{x \in S \mid xRa\}.$$

This is called the **equivalence class** of a and consists of all elements in S that are equivalent to a. The element a is called a **representative** of the equivalence class $[a]$.

In the equals relation, the equivalence class of an element consists of a alone. In the relation *has the same surname*, the equivalence class containing John Smith consists of all the people with the surname Smith.

In the equivalence relation of congruence modulo m an equivalence class is called a **congruence class**, or sometimes a **residue class**.

In the case of the congruence relation modulo 2

$$
\begin{aligned}
[0] &= \{x \mid x \equiv 0 \pmod{2}\} &= \{\ldots, -4, -2, 0, 2, 4, \ldots\} \\
[1] &= \{x \mid x \equiv 1 \pmod{2}\} &= \{\ldots, -3, -1, 1, 3, 5, \ldots\} \\
[2] &= \{x \mid x \equiv 2 \pmod{2}\} &= \{\ldots, -4, -2, 0, 2, 4, \ldots\} &= [0].
\end{aligned}
$$

In fact, there are only two distinct congruence classes, namely the even integers and the odd integers. We have $[2r] = [0]$ and $[2r + 1] = [1]$, so any even number is a representative of $[0]$, and any odd number is a representative of $[1]$. Furthermore, notice that every integer lies in precisely one congruence class.

Proposition 3.31. *Let R be an equivalence relation on the set S. If $a, b \in S$, then*

(i) $a \in [a]$.

(ii) $[a] = [b]$ if and only if aRb.

(iii) $[a] \cap [b] = \emptyset$ if and only if $a \not\!R b$.

Proof. **(i)** The reflexive property states that aRa, for all $a \in S$, so it follows that $a \in [a]$.

(ii) Suppose $[a] = [b]$. Then, by part (i), $a \in [b]$ and aRb. Conversely, suppose aRb. Let $x \in [a]$ so that xRa. By the transitive property xRb and hence $x \in [b]$. Therefore, $[a] \subseteq [b]$ and, since bRa, it follows similarly that $[b] \subseteq [a]$. Hence $[a] = [b]$.

(iii) Suppose $[a] \cap [b] = \emptyset$. Then $a \notin [b]$ and so $a \not\!R b$. Conversely, suppose $a \not\!R b$ and let $x \in [a] \cap [b]$. That is, xRa and xRb. By the symmetric and transitive properties, aRx and aRb. This is a contradiction, so x cannot exist and $[a] \cap [b] = \emptyset$. $\square$

Therefore, in any equivalence relation, two equivalence classes are either identical or disjoint, and the set of equivalence classes under an equivalence relation R yields a disjoint decomposition of the set S. A decomposition of a set S into such a disjoint union of subsets is called a **partition** of S.

It follows from Proposition 3.14 that the congruence relation modulo m has precisely m distinct congruence classes, namely $[0], [1], [2], \ldots, [m-2], [m-1]$, one corresponding to each remainder under division by m.

The partition of the integers into the m congruence classes modulo m can be visualized as follows. Consider all the integers distance one apart on the number line, and consider a circle whose circumference has length m. If the number line were to be wrapped around this circle, all the integers in one congruence class would fall on the same part of the circle.

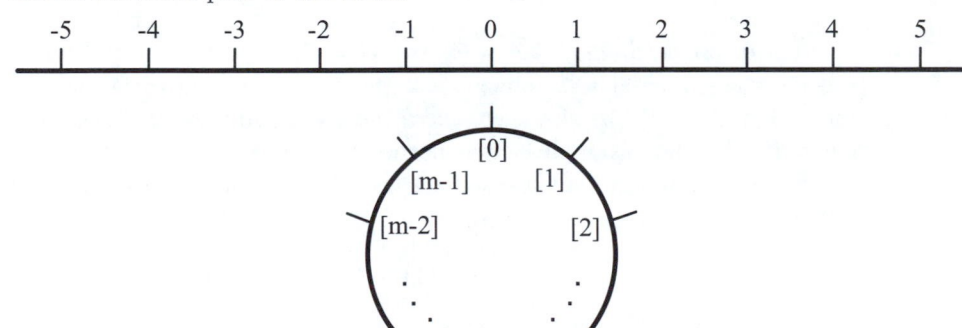

The set of all equivalence classes of a set S under an equivalence relation is called the *quotient set* of S by R. and is often denoted S/R. Therefore,

$$S/R \;=\; \{[a] \mid a \in S\}.$$

3.4 MODULAR ARITHMETIC

Definition.   The **congruence class modulo** m of the integer a is the set of integers

$$[a] \ = \ \{x \in \mathbb{Z} \mid x \equiv a \pmod m\}.$$

The set of congruence classes of integers, under the congruence relation modulo m, is called the set of **integers modulo** m and is denoted by $\mathbb{Z}_m$.

Notice that the modulus m is implicit in the notation $[a]$. Normally this will not cause confusion, since we usually deal with only one modulus at a time. However, if we wish to cope with congruences of different moduli at the same time, we could use the notation $[a]_m$

The set $\mathbb{Z}_m$ is the quotient set of $\mathbb{Z}$ defined by the congruence relation modulo m, and

$$\mathbb{Z}_m \ = \ \{[0], [1], \ldots, [m-1]\}.$$

For example, $\mathbb{Z}_4 = \{[0], [1], [2], [3]\}$, where the four congruence classes are

$$
\begin{aligned}
[0] &= \{\ldots, -8, -4, 0, 4, \ 8, 12, \ldots\} &= \{4k \mid k \in \mathbb{Z}\} \\
[1] &= \{\ldots, -7, -3, 1, 5, \ 9, 13, \ldots\} &= \{4k + 1 \mid k \in \mathbb{Z}\} \\
[2] &= \{\ldots, -6, -2, 2, 6, 10, 14, \ldots\} &= \{4k + 2 \mid k \in \mathbb{Z}\} \\
[3] &= \{\ldots, -5, -1, 3, 7, 11, 15, \ldots\} &= \{4k + 3 \mid k \in \mathbb{Z}\}.
\end{aligned}
$$

It will be useful if we were able to perform the operations of addition, subtraction, multiplication, and, perhaps, division in $\mathbb{Z}_m$. The obvious way of defining addition and multiplication of two congruence classes modulo m would be as follows.

$$
\begin{aligned}
[a] + [b] &= [a + b] \\
[a] \cdot [b] &= [ab]
\end{aligned}
$$

For example, in $\mathbb{Z}_7$, let $A = [4]$ and $B = [5]$ so that we would have $A + B = [4] + [5] = [4 + 5] = [9] = [2]$, since $9 \equiv 2 \pmod 7$, and $A \cdot B = [4] \cdot [5] = [20] = [6]$, since $20 \equiv 6 \pmod 7$. However, these definitions are not as innocuous as they might appear. For example, the congruence class A in $\mathbb{Z}_7$ could equally well be written as $[11]$, and B could be written as $[19]$. Our definitions would then imply that

$$
\begin{aligned}
A + B &= [11] + [19] &= [30] \\
A \cdot B &= [11] \cdot [19] &= [209].
\end{aligned}
$$

Our definitions would lead to trouble, unless $[30] = [9]$ and $[209] = [20]$ in $\mathbb{Z}_7$. In this particular example, it is true that $30 \equiv 9 \pmod 7$ and $209 \equiv 20 \pmod 7$, but how do we know that this will always be the case?

This type of problem arises whenever we define an operation on equivalence classes in terms of representatives. In mathematical language, we say that there is the problem of determining whether the operation is *well defined* on equivalence classes.

In the case of $\mathbb{Z}_m$, a particular congruence class $[a]$ can always be written as $[a']$ if and only if $a' \equiv a \pmod{m}$. Similarly, $[b]$ can be written as $[b']$ if and only if $b' \equiv b \pmod{m}$. Addition and multiplication in $\mathbb{Z}_m$ will be well defined if $[a + b] = [a' + b']$ and $[ab] = [a'b']$. However, Proposition 3.12 guarantees that $a + b \equiv a' + b' \pmod{m}$ and $ab \equiv a'b' \pmod{m}$, and so the corresponding congruence classes are equal. Hence addition and multiplication are well defined in $\mathbb{Z}_m$. This is called **modular arithmetic**.

Example 3.41. Write the addition and multiplication tables for $\mathbb{Z}_4$ and $\mathbb{Z}_5$.

Solution.

Addition in $\mathbb{Z}_4$

+	[0]	[1]	[2]	[3]
[0]	[0]	[1]	[2]	[3]
[1]	[1]	[2]	[3]	[0]
[2]	[2]	[3]	[0]	[1]
[3]	[3]	[0]	[1]	[2]

Multiplication in $\mathbb{Z}_4$

·	[0]	[1]	[2]	[3]
[0]	[0]	[0]	[0]	[0]
[1]	[0]	[1]	[2]	[3]
[2]	[0]	[2]	[0]	[2]
[3]	[0]	[3]	[2]	[1]

Addition in $\mathbb{Z}_5$

+	[0]	[1]	[2]	[3]	[4]
[0]	[0]	[1]	[2]	[3]	[4]
[1]	[1]	[2]	[3]	[4]	[0]
[2]	[2]	[3]	[4]	[0]	[1]
[3]	[3]	[4]	[0]	[1]	[2]
[4]	[4]	[0]	[1]	[2]	[3]

Multiplication in $\mathbb{Z}_5$

·	[0]	[1]	[2]	[3]	[4]
[0]	[0]	[0]	[0]	[0]	[0]
[1]	[0]	[1]	[2]	[3]	[4]
[2]	[0]	[2]	[4]	[1]	[3]
[3]	[0]	[3]	[1]	[4]	[2]
[4]	[0]	[4]	[3]	[2]	[1]

By looking at the above tables, it is seen that addition of $[0]$ leaves an element unchanged, while multiplication by $[0]$ always gives $[0]$. This can be proved true for any modulus, because for all $[a] \in \mathbb{Z}_m$

$$
\begin{aligned}
[0] + [a] &= [0 + a] = [a] \\
[0] \cdot [a] &= [0 \cdot a] = [0].
\end{aligned}
$$

Hence $[0]$ in $\mathbb{Z}_m$ acts just like the zero element of $\mathbb{Z}$. Furthermore, $[1]$ acts like the unit element because

$$[1] \cdot [a] = [a] \qquad \text{for all } [a] \in \mathbb{Z}_m.$$

If $a \in \mathbb{Z}$, $-a$ is the element of $\mathbb{Z}$ for which $a + (-a) = 0$. In a similar way we can define negatives in $\mathbb{Z}_m$ by $-[a] = [-a]$, because $[a] + [-a] = [a - a] = [0]$. For example, in $\mathbb{Z}_5$, $-[1] = [4]$, $-[2] = [3]$, $-[3] = [2]$, $-[4] = [1]$, and $-[0] = [0]$.

Subtraction can be defined by

$$[a] - [b] \ = \ [a - b]$$

and, by Proposition 3.12, this is well defined.

If we compare division in $\mathbb{Z}_m$ with division in $\mathbb{Z}$, interesting differences begin to appear. An element a^{-1} is called the *inverse* of an element a if $a \cdot a^{-1} = 1$. The question of division by an element a is equivalent to the existence of its inverse a^{-1}. In $\mathbb{Z}$, the only elements we can always divide by are 1 and -1, as these are the only two elements with inverses.

However,

$$[2] \cdot [3] \ = \ [1] \qquad \text{in } \mathbb{Z}_5$$

so that [2] has an inverse, and $[2]^{-1} = [3]$. Division by [2] in $\mathbb{Z}_5$ is equivalent to multiplication by [3]. In fact, in $\mathbb{Z}_5$, every nonzero element has an inverse; $[1]^{-1} = [1]$, $[2]^{-1} = [3]$, $[3]^{-1} = [2]$, and $[4]^{-1} = [4]$. We would never expect zero to have an inverse because zero times anything is always zero.

By looking at the multiplication table of $\mathbb{Z}_4$, we see that $[1]^{-1} = [1]$ and $[3]^{-1} = [3]$, but [2] does not have an inverse. Hence division by [2] is not always possible in $\mathbb{Z}_4$.

We shall show in Corollary 3.44 that if p is prime, every nonzero element of $\mathbb{Z}_p$ has an inverse. A set in which we can perform the operations of addition, subtraction, multiplication, and division by nonzero elements, and in which these operations satisfy certain standard properties, is called a *field*. In fact, the set of integers modulo p, $\mathbb{Z}_p$, will form a finite field if and only if p is prime.

We shall now use the concept of congruence to prove the following important theorem that was discovered by Fermat in 1640 and can, incidentally, be used to show the existence of inverses in $\mathbb{Z}_p$. The French mathematician Pierre de Fermat (1601–1665) actually made his living as a lawyer and member of the provincial parliament of Toulouse. However, he devoted most of his spare time to mathematics. Fermat could be considered a founder of modern Number Theory. He stated many results in Number Theory, including the one below. He did not usually provide proofs of his results, because he communicated them in letters to other mathematicians, rather than publishing them in a book or journal. However, nearly all of his results have since been proven. Fermat is best known for his so-called "Last Theorem." This result was written in the margin of his copy of the work of Diophantus, next to the solution of the Diophantine equation $z^2 + y^2 = z^2$. Fermat claimed that $x^n + y^n = z^n$ has no nonzero integer solutions if $n > 2$. He stated that he had a truly marvelous proof that was too long to write in the margin. Over 350 years later, Fermat's Last Theorem was finally proved by Andrew Wiles in 1994, using very advanced mathematics. Most mathematicians are sceptical that Fermat really did have a valid proof. To avoid confusion with his Last Theorem, the following result is usually called Fermat's Little Theorem.

Fermat's Little Theorem 3.42.   *If p is a prime number that does not divide the integer a, then*

$$a^{p-1} \equiv 1 \pmod{p}.$$

Proof.   If $p \nmid a$ we shall first show that no two of the numbers $0a, 1a, 2a, \ldots,$ $(p-1)a$ are congruent modulo p. Suppose that

$$ra \equiv sa \pmod{p}, \qquad \text{where } 0 \leq s \leq r \leq p-1.$$

By the definition of congruence, this implies $p|(r-s)a$ and, by Theorem 2.53, $p|(r-s)$. Hence $r = s$.

Therefore, the congruence classes $[0a], [1a], [2a], \ldots, [(p-1)a]$ are all distinct. But as $\mathbb{Z}_p$ only contains p congruence classes, it follows that

$$\mathbb{Z}_p \;=\; \{[0], [a], [2a], \ldots, [(p-1)a]\}.$$

Hence the nonzero classes $[a], [2a], \ldots, [(p-1)a]$ must be a rearrangement of the classes $[1], [2], \ldots, [p-1]$. In particular, multiplying them together,

$$a \cdot 2a \cdot 3a \cdots (p-1)a \;\equiv\; 1 \cdot 2 \cdot 3 \cdots (p-1) \pmod{p}.$$
$$(p-1)!a^{p-1} \;\equiv\; (p-1)! \pmod{p}.$$

However, $p \nmid (p-1)!$ because the prime p does not divide any of the factors of $(p-1)!$ Hence, by Proposition 3.13, we can cancel $(p-1)!$ and obtain

$$a^{p-1} \;\equiv 1 \qquad \pmod{p}. \qquad\qquad \square$$

Corollary 3.43.   *For any integer a and prime p*

$$a^p \equiv a \pmod{p}.$$

Proof.   If $p \nmid a$ this follows from Fermat's Little Theorem. If $p|a$, both sides are congruent to 0 modulo p. $\square$

Corollary 3.44.   *If $[a]$ is any nonzero element of $\mathbb{Z}_p$, where p is prime, then there exists an element $[b] \in \mathbb{Z}_p$ such that $[a] \cdot [b] = [1]$; that is, every nonzero element of $\mathbb{Z}_p$ has an inverse.*

Proof.   If $[a] \neq [0]$ in $\mathbb{Z}_p$ then $p \nmid a$ and, by Fermat's Little Theorem,

$$[a][a^{p-2}] \;=\; [a^{p-1}] \;=\; [1].$$

Hence

$$[a]^{-1} = [a^{p-2}]. \qquad\qquad \square$$

For example, in $\mathbb{Z}_5$, $[2]^{-1} = [2^3] = [8] = [3]$. If p is large, however, Fermat's Theorem gives a rather cumbersome way of finding inverses, and it will be easier to find the inverses by inspection or to use the methods of the next section. In $\mathbb{Z}_{31}$, Fermat's Theorem gives the inverse of $[2]$ as $[2^{29}]$. However, we can see by inspection that

$$[2] \cdot [16] \quad = \quad [32] \quad = \quad [1] \qquad \text{in } \mathbb{Z}_{31}$$

and so $[2]^{-1} = [16]$ and, incidentally, $2^{29} \equiv 16 \pmod{31}$.

3.5   LINEAR CONGRUENCES

A relation of the form

$$ax \equiv c \pmod{m}$$

is called a *linear congruence* in the variable x. A solution to such a congruence is an integer x_0 for which $ax_0 \equiv c \pmod{m}$. Our problem is to determine whether such a linear congruence has a solution and, if so, how to find all the solutions.

We first notice that if x_0 is any solution and $x_1 \equiv x_0 \pmod{m}$, then x_1 is also a solution; this follows immediately from Proposition 3.12 because $ax_1 \equiv ax_0 \equiv c \pmod{m}$. Therefore, if x_0 is a solution, so is every element of the congruence class x_0 modulo m. Since there are only m distinct congruence classes, the problem reduces to the finite one of determining which of these congruence classes are solutions. Hence any linear congruence $ax \equiv c \pmod{m}$ can be viewed as an equation

$$[a][x] \quad = \quad [c] \quad \text{in } \mathbb{Z}_m$$

and the problem of finding an integer x that satisfies the congruence is equivalent to the problem of finding an equivalence class $[x] \in \mathbb{Z}_m$ that satisfies the equation.

One crude method of solving any congruence modulo m (whether linear or not) is to take one element from each congruence class, say $0, 1, 2, \ldots, m-1$, and test whether they satisfy the congruence. This method is very effective if m is small, but soon becomes tedious for large m.

Example 3.51.   Solve the congruence

$$4x \equiv 2 \pmod{6}.$$

Solution.   We check whether the congruence is satisfied if $x \equiv 0$, 1, 2 , 3, 4, or 5 (mod 6).

Modulo 6						
$x \equiv$	0	1	2	3	4	5
$4x \equiv$	0	4	2	0	4	2

Therefore, $4x \equiv 2 \pmod{6}$ if $x \equiv 2$ or 5 (mod 6). An equivalent way of writing this solution is $x = 6k + 2$ or $6k + 5$, where $k \in \mathbb{Z}$.   □

Example 3.52. Solve the equation

$$[2][x] \ = \ [1] \quad \text{in } \mathbb{Z}_4.$$

Solution. This equation is equivalent to the congruence $2x \equiv 1 \pmod 4$.

Modulo 4				
$x \equiv$	0	1	2	3
$2x \equiv$	0	2	0	2

We see that $2x$ is never congruent to 1 modulo 4 and hence the equation has no solution in $\mathbb{Z}_4$.

The nonexistence of a solution to $[2][x] = [1]$ in $\mathbb{Z}_4$ expresses the fact that $[2]$ has no inverse in $\mathbb{Z}_4$. $\quad\square$

How do we solve a linear congruence if the modulus is large? The following results show that a one-variable linear congruence is equivalent to a Diophantine equation in two variables.

Proposition 3.53. *The linear congruence $ax \equiv c$ (mod m) has a solution $x = x_0$, if and only if the linear Diophantine equation $ax + my = c$ has a solution $x = x_0$, $y = y_0$ for some integer y_0.*

Proof. By the definition of congruence, the relation $ax \equiv c \pmod m$ holds if and only if $m | (c - ax)$ or, equivalently, if and only if $my = c - ax$ for some $y \in \mathbb{Z}$. Hence $ax_0 \equiv c \pmod m$ if and only if there exists $y_0 \in \mathbb{Z}$ such that $ax_0 + my_0 = c$. $\quad\square$

Linear Congruence Theorem 3.54. *The one-variable linear congruence*

$$ax \equiv c \pmod m$$

has a solution if and only if $\gcd(a, m) | c$.

If $x_0 \in \mathbb{Z}$ is one solution, then the complete solution is

$$x \equiv x_0 \quad \left(\text{mod } \frac{m}{d}\right), \qquad \text{where } d = \gcd(a, m)$$

or, equivalently,

$$x \equiv x_0, \ x_0 + \tfrac{m}{d}, \ x_0 + 2\tfrac{m}{d}, \ \dots, \ x_0 + (d-1)\tfrac{m}{d} \pmod m.$$

Hence there are $d = \gcd(a, m)$ noncongruent solutions modulo m.

Proof. By Proposition 3.53 the congruence is equivalent to $ax + my = c$ and, by Theorem 2.31, this has a solution if and only if $\gcd(a, m) | c$.

If x_0 is one particular integer solution, then there exists an integer y_0 such that $ax_0 + my_0 = c$ and, by Theorem 2.31, the complete solution to $ax + my = c$ is

$$x = x_0 + k\frac{m}{d}, \qquad y = y_0 - k\frac{a}{d} \qquad \text{for all } k \in \mathbb{Z}, \qquad \text{where } d = \gcd(a, m).$$

Therefore, the complete solution to $ax \equiv c \pmod{m}$ is $x = x_0 + k\frac{m}{d}$ for every integer k. This is equivalent to $x \equiv x_0 \pmod{\frac{m}{d}}$.

We shall now show that the solution set

$$S = \left\{ x \in \mathbb{Z} \,\Big|\, x \equiv x_0 \left(\text{mod } \frac{m}{d} \right) \right\}$$

to this congruence is the same as the set

$$T = \left\{ x \in \mathbb{Z} \,\Big|\, x \equiv x_0 + k\frac{m}{d} \pmod{m} \quad \text{for } k \in \mathbb{Z} \right\}.$$

If $x \in S$, then there exists an integer k with $x = x_0 + k\frac{m}{d}$; hence $x \in T$. If $x \in T$, then there exist integers k and ℓ with $x = x_0 + k\frac{m}{d} + \ell m = x_0 + (k + \ell d)\frac{m}{d}$; hence $x \in S$. Therefore, we have $S = T$.

We now show that the numbers $x_0 + \frac{m}{d}$, $x_0 + 2\frac{m}{d}, \ldots, x_0 + (d-1)\frac{m}{d}$ are in distinct congruence classes modulo m. We have $x_0 + k_1\frac{m}{d} \equiv x_0 + k_2\frac{m}{d} \pmod{m}$ if and only if $m \mid (k_1 - k_2)\frac{m}{d}$, which happens if and only if $d \mid k_1 - k_2$, or, equivalently, $k_1 \equiv k_2 \pmod{d}$. Hence the solution set consists precisely of the d congruence classes modulo m containing $x_0 + k\frac{m}{d}$, for $k = 0, 1, 2, \ldots, d - 1$. $\qquad \square$

If we cannot find one particular solution to $ax \equiv c \pmod{m}$ by easier means, we can always use the Euclidean Algorithm to solve $ax + my = c$.

Note that if $\gcd(a, m) = 1$, then the congruence $ax \equiv c \pmod{m}$ always has a solution, whatever the value of c. Furthermore, there is exactly one solution modulo m.

In particular, if the modulus is a prime p, then $\gcd(a, p) = 1$ whenever $p \nmid a$. Hence, if p is prime, all congruences of the form $ax \equiv c \pmod{p}$ have solutions, as long as $a \not\equiv 0 \pmod{p}$. Multiplying each side of the congruence by a^{p-2} and applying Fermat's Theorem 3.42, we obtain the theoretical solution

$$x \equiv a^{p-2}c \pmod{p}.$$

Example 3.55. Solve the congruence

$$63x \equiv 20 \pmod{7}.$$

Solution. Since $\gcd(63, 7) = 7$, which does not divide 20, the congruence has no solutions. $\qquad \square$

Example 3.56.   Find the inverse of $[18]$ in $\mathbb{Z}_{31}$, and write it in the form $[r] \in \mathbb{Z}_{31}$, where $0 \leq r < 31$.

Solution.   We have to find the equivalence class $[x]$ for which $[18][x] = [1]$ in $\mathbb{Z}_{31}$ or, equivalently, solve the congruence

$$18x \equiv 1 \pmod{31}.$$

Since $\gcd(18, 31) = 1$, this congruence does have a solution, and there is only one congruence class of solutions modulo 31.

   The congruence is equivalent to the Diophantine equation

$$18x + 31y = 1.$$

Apply the Euclidean Algorithm to 18 and 31.

\multicolumn{3}{c}{$18x + 31y = r$}		
1	0	18
0	1	31
1	0	18
-1	1	13
2	-1	5
-5	3	3
7	-4	2
-12	7	1

From the last row, $(-12)18 + 7(31) = 1$, and so $(-12) \cdot 18 \equiv 1 \pmod{31}$. Hence $x \equiv -12 \equiv 19 \pmod{31}$ is a solution to $18x \equiv 1 \pmod{31}$.

   The inverse of $[18]$ in $\mathbb{Z}_{31}$ is therefore $[19]$.

Check. $18 \cdot 19 = 342 = 11 \cdot 31 + 1$ so $18 \cdot 19 \equiv 1 \pmod{31}$. $\square$

Example 3.57.   Solve $20x \equiv 8 \pmod{44}$.

Solution.   We have $\gcd(20, 44) = 4$, which does divide 8. Therefore, there are exactly 4 noncongruent classes of solutions modulo 44. The congruence is equivalent to the Diophantine equation $20x + 44y = 8$ or $5x + 11y = 2$.

   Now $11 = 2 \cdot 5 + 1$, so $11 - 2 \cdot 5 = 1$, and $2 \cdot 11 - 4 \cdot 5 = 2$. Hence we see by inspection that $x = -4$, $y = 2$ is one solution to the Diophantine equation. By Theorem 3.54, the complete solution to the congruence can either be written as

$$x \equiv -4 \ (\text{mod } \tfrac{44}{4})$$

that is,
$$x \equiv 7 \pmod{11}$$

or written as

$$x \equiv 40, \ 40 + 11, \ 40 + 22, \ 40 + 33 \pmod{44}$$

that is,
$$x \equiv 40, \ 7, \ 18, \ 29 \pmod{44}.$$

Check. $20 \cdot 7 = 140 = 3 \cdot 44 + 8$ so $20 \cdot 7 \equiv 8 \pmod{44}$. $\square$

Although we have a method for completely solving a *linear* congruence, there is no effective method for solving any *polynomial* congruence such as

$$a_n x^n + a_{n-1} x^{n-1} + \cdots + a_1 x + a_0 \equiv 0 \pmod{m}.$$

However, as with linear congruences, if x_0 is one solution, so is any integer congruent to x_0 modulo m. This follows from Proposition 3.12, because if $x_0 \equiv x_1 \pmod{m}$, then $x_0^2 \equiv x_1^2$, $x_0^3 \equiv x_1^3$ and in general $x_0^r \equiv x_1^r \pmod{m}$; hence

$$a_n x_0^n + a_{n-1} x_0^{n-1} + \cdots + a_1 x_0 + a_0 \equiv a_n x_1^n + a_{n-1} x_1^{n-1} + \cdots + a_1 x_1 + a_0 \pmod{m}.$$

Therefore, the solutions to any polynomial congruence occur in congruence classes and, if the modulus is small, we can solve the congruence by exhaustively trying each congruence class.

Example 3.58.   Solve $x^2 \equiv 1 \pmod 8$.

Solution.

Modulo 8							
$x \equiv$	0	1	2	3	4	5	6	7
$x^2 \equiv$	0	1	4	1	0	1	4	1

Hence the solution is $x \equiv 1, 3, 5$ or $7 \pmod 8$. $\square$

Example 3.59.   For which integer x is $x^7 + x^3 + 2x^2 + 4$ divisible by 7?

Solution.   We have to solve the congruence

$$x^7 + x^3 + 2x^2 + 4 \equiv 0 \pmod 7.$$

Since the modulus is prime, it follows from Corollary 3.43 to Fermat's Theorem that $x^7 \equiv x \pmod 7$ for all $x \in \mathbb{Z}$. Therefore, the congruence is equivalent to

$$x^3 + 2x^2 + x + 4 \equiv 0 \pmod 7.$$

Modulo 7							
$x \equiv$	0	1	2	3	4	5	6
$x^2 \equiv$	0	1	4	2	2	4	1
$x^3 \equiv$	0	1	1	6	1	6	6
$x^3 + 2x^2 + x + 4 \equiv$	4	1	1	3	6	2	4

We see from the above table that the congruence has no solution and therefore the original integer polynomial is never divisible by 7. $\square$

3.6 THE CHINESE REMAINDER THEOREM

Around 350 A.D. the Chinese astronomer and mathematician Sun-Tsu posed the problem of finding the two smallest positive integers that have remainders 2, 3, and 2 when divided by 3, 5, and 7 respectively. This is a problem involving three simultaneous congruences. Such questions often arose in ancient calendars that depended on astronomical cycles of different lengths. The solutions to such problems are still useful in the computer age for solving complicated Diophantine equations.

We shall first show how to solve two simultaneous congruences whose moduli are relatively prime and then show how this solution can be extended to any number of simultaneous congruences with relatively prime moduli.

Example 3.61. Solve the simultaneous congruences

$$
\begin{aligned}
x &\equiv 2 \pmod 9 \\
x &\equiv 3 \pmod 7.
\end{aligned}
$$

Solution. The first congruence is equivalent to $x = 2 + 9y$, where $y \in \mathbb{Z}$. Substituting this into the second congruence, we have

$$
2 + 9y \equiv 3 \pmod 7 \quad \text{or} \quad 2y \equiv 1 \pmod 7.
$$

By inspection, we see that this has solution $y \equiv 4 \pmod 7$ or $y = 4 + 7z$ for all $z \in \mathbb{Z}$.

The solution to both congruences is therefore

$$
x = 2 + 9(4 + 7z) = 38 + 63z \qquad \text{for all } z \in \mathbb{Z}
$$

or, equivalently,

$$
x \equiv 38 \pmod{63}.
$$

Check. If $x = 38 + 63z$, then $x \equiv 2 \pmod 9$ and $x \equiv 3 \pmod 7$. □

Chinese Remainder Theorem 3.62. *If $\gcd(m_1, m_2) = 1$, then, for any choice of the integers a_1 and a_2, the simultaneous congruences*

$$
\begin{aligned}
x &\equiv a_1 \pmod{m_1} \\
x &\equiv a_2 \pmod{m_2}
\end{aligned}
$$

have a solution. Moreover, if $x = x_0$ is one integer solution, then the complete solution is

$$
x \equiv x_0 \pmod{m_1 m_2}.
$$

Proof.   The proof follows the previous Example 3.61, replacing the numbers by arbitrary integer constants. It is seen that the only condition on these constants to guarantee a solution is that the moduli are relatively prime.

The integer x satisfies the first congruence if and only if

$$x = a_1 + m_1 y \qquad \text{for some } y \in \mathbb{Z}.$$

This number x also satisfies the second congruence if and only if

$$a_1 + m_1 y \equiv a_2 \pmod{m_2}$$

that is, if and only if

$$m_1 y \equiv a_2 - a_1 \pmod{m_2}.$$

Since $\gcd(m_1, m_2) = 1$, it follows from the Linear Congruence Theorem 3.54 that this congruence always has a solution, say $y = b$, and that the complete solution for y will then be

$$y = b + m_2 z \quad \text{for } z \in \mathbb{Z}.$$

Therefore, $x = a_1 + m_1 b$ is one solution to the simultaneous congruences, and any integer x satisfies the simultaneous congruences if and only if

$$\begin{aligned} x &= a_1 + m_1(b + m_2 z) \\ &= (a_1 + m_1 b) + m_1 m_2 z \qquad \text{for } z \in \mathbb{Z}. \end{aligned}$$

This is exactly one congruence class modulo $m_1 m_2$. Hence, if $x = x_0$ is one solution, then $x \equiv x_0 \pmod{m_1 m_2}$ is the complete solution. $\qquad\square$

Example 3.63.

A small gear with 17 teeth is meshed into a large gear with 60 teeth. The large gear starts rotating at one revolution per minute. How long will it be before the small gear is back to its original position and the large gear is one quarter of a revolution past its initial position?

Solution.   The gears are moving at the rate of one tooth per second. After x seconds the smaller gear will be back to its initial position if $x \equiv 0 \pmod{17}$, and the larger gear will be one quarter of a revolution past its initial position whenever $x \equiv 15 \pmod{60}$.

We can solve these two simultaneous congruences. The second congruence implies that $x = 15 + 60y$, where $y \in \mathbb{Z}$. Substituting this value of x into the first congruence, we have

$$15 + 60y \equiv 0 \pmod{17}.$$

That is, $9y \equiv 2 \pmod{17}$ or $y \equiv 18y \equiv 4 \pmod{17}$. Therefore,

$$y = 4 + 17z \quad \text{for } z \in \mathbb{Z}$$

and

$$x = 15 + 60(4 + 17z) = 15 + 60 \cdot 4 + 60 \cdot 17z.$$

The first positive solution occurs when $z = 0$ and the elapsed time is 4 minutes and 15 seconds. $\quad\square$

Proposition 3.64. If $\gcd(m_1, m_2) = 1$, then

$$x \equiv a \pmod{m_1 m_2} \quad \Longleftrightarrow \quad \begin{cases} x \equiv a \pmod{m_1} \\ x \equiv a \pmod{m_2}. \end{cases}$$

Proof.   Suppose that $x \equiv a \pmod{m_1 m_2}$. This is equivalent to $m_1 m_2 | (x - a)$. Hence $m_1 | (x - a)$ and $m_2 | (x - a)$, which is equivalent to the two simultaneous congruences $x \equiv a \pmod{m_1}$ and $x \equiv a \pmod{m_2}$.

Conversely, given the two simultaneous congruences, it is clear that $x = a$ is one integer solution. If follows from the Chinese Remainder Theorem 3.62 that $x \equiv a \pmod{m_1 m_2}$. $\quad\square$

One of the implications in the above result is not true if $\gcd(m_1, m_2) \neq 1$; for example, $12 \equiv 0 \pmod 4$ and $12 \equiv 0 \pmod 6$, but $12 \not\equiv 0 \pmod{24}$.

Example 3.65.   Find the two smallest positive integers that have remainders 1, 2, and 6, when divided by 3, 5, and 7 respectively.

Solution.   The integers must satisfy the following three congruences.

$$x \equiv 1 \pmod 3$$
$$x \equiv 2 \pmod 5$$
$$x \equiv 6 \pmod 7$$

Let us solve the first two of these congruences. The first implies that $x = 1 + 3y$, where $y \in \mathbb{Z}$ and, substituting this in the second, we have $1 + 3y \equiv 2 \pmod 5$. Hence $3y \equiv 1 \pmod 5$, which has solution $y \equiv 2 \pmod 5$, or $y = 2 + 5z$ for $z \in \mathbb{Z}$. Therefore, the solution to the first two congruences is

$$x = 1 + 3(2 + 5z) = 7 + 15z \quad \text{for } z \in \mathbb{Z}$$

or, equivalently, $x \equiv 7 \pmod{15}$, and we have now reduced the three simultaneous congruences to two simultaneous congruences. We can now solve these two as before.

We have $x = 7 + 15z$, where $z \in \mathbb{Z}$, and when this is substituted into the third congruence $x \equiv 6 \pmod 7$ we obtain

$$7 + 15z \equiv 6 \pmod 7.$$

This reduces to $z \equiv 6 \pmod 7$, or, equivalently, $z = 6 + 7t$, where $t \in \mathbb{Z}$.

The solution to the original three congruences is therefore

$$x \;=\; 7 + 15(6 + 7t) \;=\; 97 + 105t \qquad \text{for } t \in \mathbb{Z}$$

or, equivalently,

$$x \equiv 97 \pmod{105}.$$

The two smallest positive integers satisfying the congruences are 97 and 202.

Check. $97 \equiv 1 \pmod 3$, $97 \equiv 2 \pmod 5$ and $97 \equiv 6 \pmod 7$. $\qquad\square$

We can extend the method of the previous example to solve n simultaneous congruences by repeatedly reducing two congruences modulo m_i and m_j to one modulo $m_i m_j$. We state the result without further proof.

Generalized Chinese Remainder Theorem 3.66.

Let $m_1, m_2, \ldots, m_n$ be positive integers such that $\gcd(m_i, m_j) = 1$ if $i \neq j$. Then for any integers $a_1, a_2, \ldots, a_n$ the simultaneous congruences

$$
\begin{aligned}
x &\equiv a_1 &&\pmod{m_1}\\
x &\equiv a_2 &&\pmod{m_2}\\
&\;\;\vdots\\
x &\equiv a_n &&\pmod{m_n}
\end{aligned}
$$

always have a solution. Moreover, if $x = x_0$ is one solution, then the complete solution is $x \equiv x_0 \pmod{m_1 m_2 \ldots m_n}$. $\qquad\square$

If the modulus of a congruence contains at least two prime factors, then the Chinese Remainder Theorem can be used to break up the congruence into congruences with smaller relatively prime moduli. The congruence does not even have to be linear, as the following example shows.

Example 3.67. Solve the congruence $x^3 \equiv 53 \pmod{120}$.

Solution.   Instead of solving this by trying all 120 congruence classes in turn, we can split the congruence up into congruences with relatively prime moduli. After solving the individual congruences we can fit them together again using the Chinese Remainder Theorem.

The number 120 factors into primes as $2^3 \cdot 3 \cdot 5$ and hence can be written as the product of the numbers 3, 5 and 8, which are relatively prime in pairs. Extending Proposition 3.64 to three relatively prime moduli, we see that the original cubic congruence $x^3 \equiv 53 \pmod{120}$ is equivalent to the three simultaneous congruences

$$\begin{aligned} x^3 &\equiv 53 \equiv 2 &&\pmod 3 \\ x^3 &\equiv 53 \equiv 3 &&\pmod 5 \\ x^3 &\equiv 53 \equiv 5 &&\pmod 8. \end{aligned}$$

We first solve these three individual congruences. By Corollary 3.43 to Fermat's Theorem, $x^3 \equiv x \pmod 3$, so $x \equiv 2 \pmod 3$ is the solution to the first one.

Modulo 5				
$x \equiv$	0	1	2	3	4
$x^3 \equiv$	0	1	3	2	4

Modulo 8							
$x \equiv$	0	1	2	3	4	5	6	7
$x^3 \equiv$	0	1	0	3	0	5	0	7

From the above tables, we see that the only solutions to the second and third congruences are $x \equiv 2 \pmod 5$ and $x \equiv 5 \pmod 8$.

Use the Chinese Remainder Theorem to solve the simultaneous congruences

$$\begin{aligned} x &\equiv 2 \pmod 3 \\ x &\equiv 2 \pmod 5 \\ x &\equiv 5 \pmod 8. \end{aligned}$$

By Proposition 3.64 the solution to the first two is $x \equiv 2 \pmod{15}$, or $x = 2 + 15y$, where $y \in \mathbb{Z}$. Substituting this into the third congruence we have

$$\begin{aligned} 2 + 15y &\equiv 5 \pmod 8 \\ -y &\equiv 3 \pmod 8. \end{aligned}$$

Hence $y \equiv 5 \pmod 8$ or $y = 5 + 8z$ for $z \in \mathbb{Z}$.

The solution to the three simultaneous congruences, and hence to the original problem, is

$$x = 2 + 15(5 + 8z) = 77 + 120z \qquad \text{for } z \in \mathbb{Z}$$

or, equivalently, $x \equiv 77 \pmod{120}$.

Check. $77^2 \equiv 5929 \equiv 49 \pmod{120}$ and $77^3 \equiv 77 \cdot 49 \equiv 3773 \equiv 53 \pmod{120}$. $\square$

The Chinese Remainder Theorem can be used to speed up the solution to a complicated system of Diophantine equations on a computer. The first task is to obtain an estimate of the size of the integer solution required. This allows the moduli that will be used to be chosen judiciously. The system of equations are then solved as congruences with the chosen moduli, and the answer is obtained from the Chinese Remainder Theorem.

For example, if a system of Diophantine equations was known to have positive solutions less than 2000, the system could first be solved modulo 11, then solved modulo 13 and finally modulo 17. By using the Chinese Remainder Theorem the answer can be found modulo $11 \cdot 13 \cdot 17$, that is, modulo 2431. Since the required solution lies between 0 and 2000, it is known exactly. Such a method will often save valuable computing time, especially if the moduli chosen are prime.

3.7   EULER-FERMAT THEOREM

Leonhard Euler (1707–1783) was one the the world's greatest and most productive mathematicians who ever lived. He grew up in Switzerland but worked mainly in St. Petersburg in Russia and also in Berlin. He made fundamental contributions to most of mathematics, as well as astronomy. He introduced or standardized much of today's mathematical notation, including the function notation $f(\ )$, the summation symbol Σ, trigonometric notation, the number e for the base of natural logarithms, and the symbol i for the complex square root of -1. His collected works cover over 70 volumes, much of which was written when he was totally blind.

Euler gave the first published proof of Fermat's Little Theorem 3.42 in 1736. Fermat's Little Theorem is only true when the modulus is prime. One reason it does not work for composite moduli is that, for a nonprime modulus, there are always some nonzero congruence classes without an inverse. In 1760, Euler showed how to generalize Fermat's Little Theorem for composite moduli by looking at only those congruence classes that do have an inverse. It follows from the Linear Congruence Theorem 3.54 that the congruence $ax \equiv 1 \pmod{m}$ has a solution if and only if $\gcd(a, m) = 1$. Hence the congruence class containing a has an inverse if and only if a is relatively prime to the modulus.

Definition.   If m is a positive integer, denote by $\phi(m)$ the number of positive integers less than or equal to m that are relatively prime to m. This is called the **Euler phi function**, as it uses the Greek letter, ϕ, called phi. It is sometimes referred to by the archaic term *Euler totient function*.

For example, $\phi(1) = 1$, $\phi(2) = 1$, $\phi(3) = 2$ and $\phi(4) = 2$. The only numbers between 1 and 12 relatively prime to 12 are 1, 5, 7, and 11, so $\phi(12) = 4$. If p is prime, then all the numbers from 1 to $p-1$ are relatively prime to p, so $\phi(p) = p-1$.

Euler-Fermat Theorem 3.71. *If m is a positive integer and $\gcd(a, m) = 1$, then*

$$a^{\phi(m)} \equiv 1 \pmod{m}.$$

Proof. The proof mimics that of Fermat's Little Theorem 3.42, except that the modulus in not prime now.

Let $b_1, b_2, \ldots, b_{\phi(m)}$ be the positive integers less than or equal to m that are relatively prime to m. If $\gcd(a, m) = 1$ we shall first show that no two of the numbers $ab_1, ab_2, \ldots ab_{\phi(m)}$ are congruent modulo m. Suppose that

$$ab_r \equiv ab_s \pmod{m}.$$

Since $\gcd(a, m) = 1$, it follows from Proposition 3.13 that we can cancel a to obtain $b_r \equiv b_s \pmod{m}$. Therefore, the congruence classes $[ab_1], [ab_2], \ldots, [ab_{\phi(m)}]$ are all distinct modulo m.

By Theorem 2.53, ab_i and m have a common prime factor if and only if a and m have a common prime factor, or b_i and m have a common prime factor. Since $\gcd(a, m) = 1$ and $\gcd(b_i, m) = 1$, all of the numbers $ab_1, ab_2, \ldots, ab_{\phi(m)}$ are relatively prime to m. Hence the congruence classes $[ab_1], [ab_2], \ldots, [ab_{\phi(m)}]$ must be just a rearrangement of the classes $[b_1], [b_2], \ldots, [b_{\phi(m)}]$. In particular, multiplying them together,

$$ab_1 \cdot ab_2 \cdots ab_{\phi(m)} \equiv b_1 \cdot b_2 \cdots b_{\phi(m)} \pmod{m}.$$
$$a^{\phi(m)} b_1 b_2 \cdots b_{\phi(m)} \equiv b_1 b_2 \cdots b_{\phi(m)} \pmod{m}.$$

Also the product $b_1 b_2 \cdots b_{\phi(m)}$ is relatively prime to m, so by Proposition 3.13 again, we can cancel $b_1 b_2 \cdots b_{\phi(m)}$ to obtain

$$a^{\phi(m)} \equiv 1 \pmod{m}. \qquad \square$$

When m is the prime p, then $\phi(p) = p - 1$, and we recover Fermat's Little Theorem 3.42.

We now show how to calculate the Euler phi function for any modulus.

Euler Phi Function Formulas 3.72

(i) $\phi(p^r) = p^{r-1}(p - 1) = p^r \left(1 - \frac{1}{p}\right)$, *if p is prime and $r > 0$.*

(ii) $\phi(mn) = \phi(m)\phi(n)$, *if $\gcd(m, n) = 1$.*

(iii) $\phi(m) = m \left(1 - \frac{1}{p_1}\right) \left(1 - \frac{1}{p_2}\right) \cdots \left(1 - \frac{1}{p_k}\right)$, *where $p_1, p_2, \ldots, p_k$ are the distinct primes that divide m.*

Proof (i) The numbers from 1 to p^r that are relatively prime to p^r are the numbers not divisible by p. Every pth number is divisible by p and there are p^{r-1} of these, so $\phi(p^r) = p^r - p^{r-1} = p^{r-1}(p - 1) = p^r \left(1 - \frac{1}{p}\right)$.

(ii)   Let $U_m = \{x \in \mathbb{Z} \mid 1 \le x \le m$ and $\gcd(x,m) = 1\}$ be the set of positive integers less than or equal to m, that are relatively prime to m. By definition, this set has $\phi(m)$ elements. We shall use the Chinese Remainder Theorem to show that if $\gcd(m,n) = 1$, the set U_{mn} has the same number of elements as the product set $U_m \times U_n = \{(a,b) \in \mathbb{Z} \times \mathbb{Z} \mid a \in U_m, b \in U_n\}$. Since U_{mn} has $\phi(mn)$ elements and $U_m \times U_n$ has $\phi(m)\phi(n)$ elements, this will prove part (ii).

By Theorem 2.53, x and mn have a common prime factor if and only if x and m have a common prime factor, or x and n have a common prime factor. Hence x is relatively prime to mn if and only if x is relatively prime to m, and x is relatively prime to n. Let $\gcd(m,n) = 1$ and let each element $x \in U_{mn}$ correspond to the element $(a,b) \in U_m \times U_n$, where $a \equiv x \pmod{m}$ and $b \equiv x \pmod{n}$. The Chinese Remainder Theorem 3.62 shows that for each pair $(a,b) \in U_m \times U_n$, of integers with $a \in U_m$ and $b \in U_n$, there is exactly one congruence class $[x]$ modulo mn with $x \equiv a \pmod{m}$ and $x \equiv b \pmod{n}$. All the elements of the congruence class $[x]$ are relatively prime to mn, so there is exactly one integer $x \in U_{mn}$ such that $x \equiv a \pmod{m}$ and $x \equiv b \pmod{n}$. This correspondence shows that U_{mn} and $U_m \times U_n$ have the same number of elements. In the terminology of Section 6.5, this defines a one-to-one correspondence between U_{mn} and $U_m \times U_n$.

(iii)   Let $m = p_1^{a_1} p_2^{a_2} \cdots p_k^{a_k}$ be the factorization of m into powers of distinct primes $p_1, p_2, \ldots, p_k$. By parts (i) and (ii),

$$
\begin{aligned}
\phi(m) &= \phi(p_1^{a_1})\phi(p_2^{a_2})\cdots\phi(p_k^{a_k}) \\
&= p_1^{a_1}\left(1 - \tfrac{1}{p_1}\right)p_2^{a_2}\left(1 - \tfrac{1}{p_2}\right)\cdots p_k^{a_k}\left(1 - \tfrac{1}{p_k}\right) \\
&= m\left(1 - \tfrac{1}{p_1}\right)\left(1 - \tfrac{1}{p_2}\right)\cdots\left(1 - \tfrac{1}{p_k}\right). \qquad \square
\end{aligned}
$$

For example,

$$
\begin{aligned}
\phi(594) &= \phi(2 \cdot 3^3 \cdot 11) \\
&= 2 \cdot 3^3 \cdot 11\left(1 - \tfrac{1}{2}\right)\left(1 - \tfrac{1}{3}\right)\left(1 - \tfrac{1}{11}\right) \\
&= 3^2 \cdot 2 \cdot 10 \\
&= 180.
\end{aligned}
$$

The following corollary is the reason why the product $(p-1)(q-1)$ will appear in the RSA cryptographic scheme 7.42.

Corollary 3.73.   *If p and q are distinct primes, and $\gcd(a, pq) = 1$, then*

$$a^{(p-1)(q-1)} \equiv 1 \pmod{pq}.$$

Proof.   As $\phi(pq) = (p-1)(q-1)$, this follows from the previous two results. $\square$

Example 3.74. Find the last two digits of (i) 123^{456} and (ii) 8765^{4321}.

Solution. *(i)* We have to calculate the remainders modulo 100. Now we know $\phi(100) = \phi(2^2 5^2) = 100 \left(1 - \frac{1}{2}\right)\left(1 - \frac{1}{5}\right) = 40$. Hence, if $\gcd(a, 100) = 1$, the Euler-Fermat Theorem 3.71 tells us $a^{40} \equiv 1 \pmod{100}$. Therefore, $a^{40k} \equiv 1 \pmod{100}$, for any positive integer k. Hence we should look at the exponent of 123^{456}, namely 456, modulo 40. Write the exponent as $456 = 11 \cdot 40 + 16$. Now $123 \equiv 23 \pmod{100}$ and

$$
\begin{aligned}
123^{456} &\equiv 23^{456} \pmod{100} \\
&\equiv 23^{11 \cdot 40 + 16} \pmod{100} \\
&\equiv \left(23^{40}\right)^{11} 23^{16} \pmod{100} \\
&\equiv 23^{16} \pmod{100}
\end{aligned}
$$

since $23^{40} \equiv 1 \pmod{100}$ using the Euler-Fermat Theorem. By repeated squaring,

$$
\begin{aligned}
23^2 &\equiv & & & 529 &\equiv 29 \pmod{100}. \\
23^4 &\equiv \left(23^2\right)^2 &\equiv 29^2 &\equiv & 841 &\equiv 41 \pmod{100}. \\
23^8 &\equiv \left(23^4\right)^2 &\equiv 41^2 &\equiv & 1681 &\equiv 81 \pmod{100}. \\
23^{16} &\equiv \left(23^8\right)^2 &\equiv 81^2 &\equiv & 6561 &\equiv 61 \pmod{100}.
\end{aligned}
$$

Hence $123^{456} \equiv 61 \pmod{100}$, and the last two digits are 61.

(ii) In the case of 8765^{4321}, 8765 has a factor 5, so 8765 is not relatively prime to 100. However we can use the Chinese Remainder Theorem after splitting the modulus 100 into the relatively prime moduli 4 and 25. Calculate 8765^{4321} modulo 4 and modulo 25.

Since $8765 \equiv 1 \pmod 4$, we have $8765^{4321} \equiv 1 \pmod 4$.

Now $8765 \equiv 15 \pmod{25}$, and $8765^2 \equiv 15^2 \equiv 225 \equiv 0 \pmod{25}$. Hence $8765^k \equiv 0 \pmod{25}$ for any $k \geq 2$. In particular, $8765^{4321} \equiv 0 \pmod{25}$.

By the Chinese Remainder Theorem 3.62 the congruence

$$
x \equiv 8765^{4321} \pmod{100}
$$

is equivalent to the simultaneous congruences $x \equiv 1 \pmod 4$ and $x \equiv 0 \pmod{25}$. This latter congruence has solution $x \equiv 0, 25, 50, 75 \pmod{100}$, and only 25 is congruent to 1 modulo 4. Hence $x \equiv 25 \pmod{100}$, and the last two digits of 8765^{4321} are 25. $\qquad\square$

Exercise Set 3

1. Which of the following integers are congruent modulo 4?

$$-12, \ -11, \ -9, \ -6, \ -4, \ -1, \ 0, \ 1, \ 2, \ 3, \ 5, \ 7, \ 10$$

2. Which of the following integers are congruent modulo 6?

$$-147, \ -91, \ -22, \ -14, \ -2, \ 2, \ 4, \ 5, \ 21, \ 185$$

3. What is the remainder when 8^{24} is divided by 3?

4. Let $N = 3^{729}$. What is the last digit in the decimal representation of N? What are the last digits in the base 9 and base 8 representations of N?

5. What is the remainder when 10^{45} is divided by 7?

6. Is $6^{17} + 17^6$ divisible by 3 or 7?

7. Show that an integer of the form $5n + 3$, where $n \in \mathbb{P}$, can never be a perfect square.

8–11.   For each of the following congruences, determine whether there exists a positive integer k so that the congruence is satisfied. If so, find the smallest such k.

8. $2^k \equiv 1 \pmod{11}$ 9. $3^k \equiv 1 \pmod{17}$

10. $2^k \equiv 1 \pmod{14}$ 11. $4^k \equiv 1 \pmod{19}$

12–16.   Find tests for determining whether an integer given in the stated base is divisible by the following numbers.

12. Dividing by 8 in base 10   13. Dividing by 12 in base 10

14. Dividing by 7 in base 10   15. Dividing by 7 in base 8

16. Dividing by 13 in base 12

17–20.   Determine whether the following numbers are divisible by 2, 3, 4, 5, 6, 8, 9, 10 or 11.

17. 514000   18. 111111

19. 179652   20. 7654321

21. Check the following calculation by casting out nines.

$$12453 \times 7057 - 84014651 = 3869170$$

22–26.   Determine whether the following relations on $\mathbb{Z}$ are reflexive, symmetric, or transitive. If any are equivalence relations, determine their quotient set.

22. aRb if and only if $a - b \neq 1$

23. aRb if and only if $a \leq b$

24. aRb if and only if $a - b$ is a multiple of 3

25. aRb if and only if $|a - b| < 3$

26. aRb if and only if $a|b$

27–30. Construct addition and multiplication tables for each of the following sets of integers modulo m and find, if possible, multiplicative inverses of each of the elements in the set.

27. $\mathbb{Z}_2$

28. $\mathbb{Z}_3$

29. $\mathbb{Z}_7$

30. $\mathbb{Z}_8$

31. If $d = \gcd(a, m)$ and $d|c$, then show that the congruence $ax \equiv c \pmod{m}$ is equivalent to

$$\frac{a}{d}x \equiv \frac{c}{d} \left(\mathrm{mod}\ \frac{m}{d}\right).$$

32–41. Solve each of the following congruences.

32. $3x \equiv 5 \pmod{13}$

33. $4x \equiv 6 \pmod{14}$

34. $5x \equiv 7 \pmod{15}$

35. $29x \equiv 43 \pmod{128}$

36. $1713x \equiv 871 \pmod{2000}$

37. $1426x \equiv 597 \pmod{2000}$

38. $x^2 \equiv 6x \pmod{8}$

39. $x^2 + 2x \equiv 3 \pmod{8}$

40. $4x^3 + 2x + 1 \equiv 0 \pmod{5}$

41. $x^9 + x^7 + x^6 + 1 \equiv 0 \pmod{2}$

42. Find the inverse of $[4]$ in $\mathbb{Z}_{11}$.

43. Find the inverse of $[2]$ in $\mathbb{Z}_{41}$.

44. Find the inverse of $[23]$ in $\mathbb{Z}_{41}$.

45–47. Solve the following equations in the given set of integers modulo m.

45. $[4][x] + [8] = [1]$ in $\mathbb{Z}_9$

46. $[3][x] = [18]$ in $\mathbb{Z}_{19}$

47. $([x] - [2])([x] - [3]) = [0]$ in $\mathbb{Z}_6$

48. For what integer values of a does $x^2 \equiv a \pmod{7}$ have a solution?

49–54. Solve the following simultaneous congruences.

49. $\quad x \equiv 4 \pmod 5$
$\quad\quad x \equiv 3 \pmod 4$

50. $\quad x \equiv 46 \pmod{51}$
$\quad\quad x \equiv 27 \pmod{52}$

51. $\quad x \equiv 1 \pmod 2$
$\quad\quad x \equiv 2 \pmod 3$
$\quad\quad x \equiv 3 \pmod 7$

52. $\quad 2x \equiv 11 \pmod{13}$
$\quad\quad 3x \equiv 7 \pmod 9$
$\quad\quad 7x \equiv 5 \pmod 8$

53. $\quad 2x \equiv 4 \pmod 7$
$\quad\quad 18x \equiv 43 \pmod{23}$

54. $\quad 161x \equiv 49 \pmod{200}$
$\quad\quad\ 74x \equiv 1 \pmod{53}$

55. Determine the two smallest positive integer solutions of the two simultaneous congruences $x \equiv 5 \pmod 7$ and $x \equiv 24 \pmod{25}$.

Problem Set 3

56. If p is a prime, prove that $x^2 \equiv y^2 \pmod{p}$ if and only if $x \equiv \pm y \pmod{p}$.

57. If p is an odd prime, show that $x^2 \equiv a \pmod{p}$ has a solution for exactly half the values of a between 1 and $p-1$ inclusive. Furthermore, if $1 \le a \le p-1$ and $x^2 \equiv a \pmod{p}$ has a solution, show that it has exactly two congruence classes of solutions modulo p.

58. Does $x^3 \equiv a \pmod{p}$ always have a solution for every value of a, whenever p is prime?

59. Choose any integer larger than 10, subtract the sum of its digits from it, cross out one nonzero digit from the result, and let the sum of the remaining digits be s. From a knowledge of s alone, is it possible to find the digit that was crossed out?

60. Prove that $21 | (3n^7 + 7n^3 + 11n)$ for all integers n.

61. Prove that $n^{91} \equiv n^7 \pmod{91}$ for all integers n. Is $n^{91} \equiv n \pmod{91}$ for all integers n?

62. For which positive values of k is $n^k \equiv n \pmod{6}$ for all integers n?

63. For which positive values of k is $n^k \equiv n \pmod{4}$ for all integers n?

64. For which positive values of k is $n^k \equiv n \pmod{7}$ for all integers n?

65. Prove, without using a calculator or computer, that 641 divides the Fermat number $F(5) = 2^{2^5} + 1$.

66. Show that the product of two numbers of the form $4n + 1$ is still of that form. Hence show that there are infinitely many primes of the form $4n + 3$.

67. Define a relation on the set of real numbers by

$$aRb \text{ if and only if } a - b = 2k\pi \text{ for some } k \in \mathbb{Z}.$$

(a) Prove that this is an equivalence relation.

(b) Which of the following are related?

$$5\pi \text{ and } -10\pi, \quad -\pi \text{ and } \pi, \quad 3 \text{ and } 9, \quad \tfrac{2}{3}\pi \text{ and } -\tfrac{1}{3}\pi, \quad \tfrac{11}{6}\pi \text{ and } \tfrac{23}{6}\pi$$

(c) Two real numbers are equivalent if and only if they represent the same angle in radians. The equivalence classes therefore consist of the different angles. Denote the equivalence class containing a by $[a]$.

Show that addition of angles is well defined by

$$[a] + [b] \;=\; [a + b].$$

(d) Show, by a counterexample, that multiplication of angles is *not* well defined by

$$[a] \cdot [b] \;=\; [ab].$$

68. **(a)** Find a relation R, on a set S, that is symmetric and transitive, but not reflexive.

  (b) If there is an example to part (a), the following "proof," that every symmetric and transitive relation is reflexive, must be fallacious. Find the error. "Let R be a symmetric and transitive relation on the set S. For any $a, b \in S$, aRb implies that bRa, because R is symmetric. But aRb and bRa imply that aRa, because R is transitive. Since aRa, R must also be reflexive."

69. If $m = pq$ is a composite number, where $1 < p \le q < m$, show that $\mathbb{Z}_m$ is not a field by showing that division by nonzero elements is not always possible in $\mathbb{Z}_m$.

70. Solve the following system of simultaneous equations in $\mathbb{Z}_{12}$.

$$\begin{aligned} [8][x] \ + \ [3][y] \ &= \ [9] \\ [6][x] \ + \ [5][y] \ &= \ [2] \end{aligned}$$

71. Solve the following system of simultaneous equations in $\mathbb{Z}_{11}$.

$$\begin{aligned} [3][x] \ + \ [4][y] \ &= \ [5] \\ [7][x] \ + \ [5][y] \ &= \ [4] \end{aligned}$$

72. One common error in copying numbers is the transposition of adjacent digits. For example, 9578 might be copied as 9758. Will the method of casting out nines discover such an error? Discuss other methods of checking for errors.

73–76.   Each new book published is given an **International Standard Book Number** *(ISBN), which consists of 10 digits arranged in four groups, such as 0-123-45678-9. The first group of digits is a code for the language of the book; 0 stands for English, 2 for French, and so on. The second group is a code for the publisher and the third group is the publisher's number for the book. The final digit is a* **check digit**. *This digit provides a check on the other digits to ensure that they are copied correctly. This check digit is chosen so that for any ISBN $a_1 a_2 a_3 \ldots a_9 a_{10}$*

$$1a_1 + 2a_2 + 3a_3 + \cdots + 9a_9 + 10a_{10} \equiv 0 \pmod{11}$$

or, equivalently,

$$a_1 + 2a_2 + 3a_3 + \cdots + 9a_9 \equiv a_{10} \pmod{11}.$$

The check digit can be any one of the digits 0, 1, 2, $\ldots$, 9 or X, where X stands for the number 10.

73. Is 0-467-51402-X a valid ISBN?

74. Is 1-56-004151-5 a valid ISBN?

75. What is the check digit for 14-200-0076-?

76. What is the check digit for 0-4101-1286-?

77. If $\phi(m)$ is the Euler ϕ-function, show that $\phi(m) = \phi(2m)$ if and only if m is odd.

78. Prove that $\phi(m) = m - 1$ if and only if m is prime.

79. (*Wilson's Theorem*) If p is prime, prove that

$$(p-1)! \equiv -1 \pmod{p}.$$

80. If p and q are integers, not divisible by 3 or 5, prove that $p^4 \equiv q^4 \pmod{15}$.

81. Solve the simultaneous congruences

$$\begin{aligned} 9x &\equiv 21 &&\pmod 6 \\ 4x &\equiv 9 &&\pmod{13}. \end{aligned}$$

82. Solve the simultaneous congruences

$$\begin{aligned} 3x &\equiv 7 &&\pmod{11} \\ 8x &\equiv 3 &&\pmod 9. \end{aligned}$$

83. Two watches, one of which gains 2 minutes per day and the other of which loses 3 minutes per day, read the correct time. When will both watches next give the same time? When will they next both give the correct time?

84. Solve $x^3 \equiv 17 \pmod{99}$.

85. Solve $x^2 \equiv 7 \pmod{99}$.

86. If $\gcd(m, n) = d$, when do the simultaneous congruences

$$\begin{aligned} x &\equiv a &&\pmod m \\ x &\equiv b &&\pmod n \end{aligned}$$

have a solution?

87. Let $M = m_1 m_2 \ldots m_n$, where $\gcd(m_i, m_j) = 1$ whenever $i \neq j$, and let $M_i = M/m_i$. Let $y \equiv b_i \pmod{m_i}$ be a solution to $M_i y \equiv 1 \pmod{m_i}$. Prove that the simultaneous congruences

$$\begin{aligned} x &\equiv a_1 &&\pmod{m_1} \\ x &\equiv a_2 &&\pmod{m_2} \\ &\vdots \\ x &\equiv a_n &&\pmod{m_n} \end{aligned}$$

have the solution

$$x \equiv a_1 b_1 M_1 + a_2 b_2 M_2 + a_3 b_3 M_3 + \cdots + a_n b_n M_n \pmod M.$$

88. Solve the simultaneous equations

$$\begin{aligned} 100x &- 9y &= 4264 \\ 11x &+ 109y &= 909 \end{aligned}$$

(a) modulo 9

(b) modulo 11

(c) in integers, using (a) and (b), given the fact that x and y have unique solutions and both are positive integers less than 100.

89. A basket contains a number of eggs and, when the eggs are removed 2, 3, 4, 5, and 6 at a time, there are 1, 2, 3, 4, and 5, respectively, left over. When the eggs are removed 7 at a time, there are none left over. Assuming none of the eggs broke during the preceding operations, determine the minimum number of eggs there were in the basket.

90. Use Problem 87 to solve each of these three simultaneous congruences.

(a) $x \equiv 2 \,(\mathrm{mod}\ 7)$, $x \equiv 5 \,(\mathrm{mod}\ 11)$, $x \equiv 11 \,(\mathrm{mod}\ 17)$

(b) $x \equiv 0 \,(\mathrm{mod}\ 7)$, $x \equiv 8 \,(\mathrm{mod}\ 11)$, $x \equiv 10 \,(\mathrm{mod}\ 17)$

(c) $x \equiv 5 \,(\mathrm{mod}\ 7)$, $x \equiv 6 \,(\mathrm{mod}\ 11)$, $x \equiv 14 \,(\mathrm{mod}\ 17)$

91–92. *Use Problem 87 to find the solution to these simultaneous congruences.*

91. $x \equiv a_1 \,(\mathrm{mod}\ 9)$, $x \equiv a_2 \,(\mathrm{mod}\ 11)$

92. $x \equiv a_1 \,(\mathrm{mod}\ 3)$, $x \equiv a_2 \,(\mathrm{mod}\ 8)$, $x \equiv a_3 \,(\mathrm{mod}\ 25)$

93. Find positive integers a, b, m_1, m_2 such that

$$\begin{aligned} a &\equiv b \quad (\mathrm{mod}\ m_1) \\ a &\equiv b \quad (\mathrm{mod}\ m_2) \\ a &\not\equiv b \quad (\mathrm{mod}\ m_1 m_2). \end{aligned}$$

94. Find all the integer solutions to the Diophantine equation $5x^2 + x + 6 = 7y$.

95. (a) Prove that if p and q are relatively prime and x is an integer such that

$$\begin{aligned} x &\equiv p \quad (\mathrm{mod}\ q) \\ x &\equiv q \quad (\mathrm{mod}\ p), \end{aligned}$$

then $x \equiv p + q \,(\mathrm{mod}\ pq)$.

(b) Show by means of a counterexample that the condition that p and q are relatively prime is necessary.

96. Solve the congruence

$$x^3 - 29x^2 + 35x + 38 \equiv 0 \quad (\mathrm{mod}\ 195).$$

97. If p is prime and k is the smallest positive integer such that $a^k \equiv 1 \pmod{p}$ then prove that k divides $p - 1$.

98. Find the remainder when 17^{40} is divided by 27.

99. Find the remainder when 5^{183} is divided by 99.

100. Find the remainder when $2^{2^{405}}$ is divided by 23.

101–104. Find the last two digits of these numbers.

101. 747^{130}

102. 287^{449}

103. 95^{95}

104. 2554^{3333}

CHAPTER 4

Induction and the Binomial Theorem

4.1 MATHEMATICAL INDUCTION

In this book we are assuming that the reader knows the properties of the positive integers. However, we will study explicitly the property known as the Principle of Mathematical Induction, as this has several important consequences.

In everyday parlance, induction means the inference of a general law from particular cases. Empirical induction is commonly used in the sciences to formulate a general theory. For example, tables for the times and heights of future high tides, at particular points on the coast, are constructed from many previous observations.

In quite a different way, Mathematical Induction allows us to *prove* a statement true for all positive integers. This depends on the following property of the positive integers $\mathbb{P}$. You should convince yourself that it is a plausible property.

Inductive Property of the Positive Integers 4.11. *Let S be a subset of the positive integers $\mathbb{P}$.*

If   *(i)* $1 \in S$
and   *(ii)* $k \in S \Longrightarrow k + 1 \in S$,

then S is the entire set $\mathbb{P}$.

We do not prove this but take it as one of the defining properties of $\mathbb{P}$. This property is equivalent to the *well-ordering principle*, which states that every nonempty set of positive integers has a smallest element.

As an example, we now show how this can be used to prove that the sum of the first n positive integers is $n(n+1)/2$.

Example 4.12. Prove that $1 + 2 + \cdots + n = \dfrac{n(n+1)}{2}$ for all $n \in \mathbb{P}$.

Solution. Let $P(n)$ be the statement "$1 + 2 + \cdots + n = \frac{n(n+1)}{2}$."

We would like to prove that the statement $P(n)$ is true for all $n \in \mathbb{P}$. Let S be the set of positive integers for which $P(n)$ is true; that is,

$$S \;=\; \{n \in \mathbb{P} \mid P(n) \text{ is true}\}.$$

(i) $1 \in S$ because $P(1)$ is "$1 = \frac{1 \cdot 2}{2}$," which is true.

(ii)   Now suppose that $k \in S$, so that $P(k)$ is true and $1 + 2 + \cdots + k = \frac{k(k+1)}{2}$. We then have

$$
\begin{aligned}
1 + 2 + \cdots + k + (k+1) &= \frac{k(k+1)}{2} + (k+1) \\
&= \frac{(k+1)(k+2)}{2},
\end{aligned}
$$

which shows that $P(k+1)$ is true, and so $k+1 \in S$.

It now follows from the Inductive Property of $\mathbb{P}$ that $S = \mathbb{P}$ and so $P(n)$ is true for all $n \in \mathbb{P}$. $\square$

We can formulate the above process into the following principle.

Principle of Mathematical Induction 4.13.   *Let $P(n)$ be a statement that depends on the positive integer n.*

If   *(i) $P(1)$ is true*
and   *(ii) $P(k)$ is true $\implies P(k+1)$ is true,*

then $P(n)$ is true for all $n \in \mathbb{P}$.

Proof.   We shall show that this principle follows from the Inductive Property of $\mathbb{P}$. If $S = \{n \in \mathbb{P} \mid P(n) \text{ is true}\}$, then

         (i) $1 \in S$
and   (ii) $k \in S \implies k+1 \in S$.

Hence, by the Inductive Property 4.11, $S = \mathbb{P}$ and $P(n)$ is true for all $n \in \mathbb{P}$. $\square$

The Principle of Mathematical Induction can be visualized by means of the following "Domino Principle." Suppose a line of dominoes, numbered by the positive integers, are standing on one end on a table.

If   (i)   the first domino is pushed over
and   (ii)   the dominoes are close enough that whenever the kth domino falls
              it knocks over the $(k+1)$st domino,
then all the dominoes will fall down.

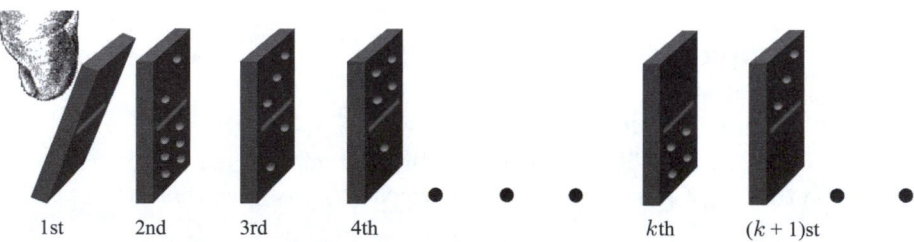

1st   2nd   3rd   4th kth $(k+1)$st

The Domino Principle

A proof by Mathematical Induction always requires two steps. First, the *base case* of the induction has to be proved; that is, we have to prove that

(i) $P(1)$ is true.

This is normally very easy. Second, the *induction step* has to be proved; that is, we have to prove that

(ii) $P(k)$ is true $\implies P(k+1)$ is true.

To do this, we assume that $P(k)$ is true. This is called the *induction hypothesis* or *induction assumption*. We now use this induction hypothesis to prove that $P(k+1)$ is true. The Principle of Mathematical Induction then guarantees that $P(n)$ is true for all $n \in \mathbb{P}$.

This principle of induction has already been implicitly used in the Euclidean Algorithm 2.22, the Extended Euclidean Algorithm 2.25, Theorem 2.41 on base b representations, twice in the Unique Factorization Theorem 2.54, and in the generalized Chinese Remainder Theorem 3.66.

Besides being useful in *proving* results, induction can be used in making *definitions*. For example, for each positive integer n, "n factorial" is defined to be the product of the first n positive integers. It is written as

$$n! \;=\; 1 \cdot 2 \cdot 3 \cdot 4 \cdots (n-1) \cdot n.$$

Alternatively, n factorial can be defined by

(i) $1! = 1$

and (ii) $n! = n(n-1)!$ for all integers $n > 1$.

The factorial of any positive integer is defined in terms of the factorial of the previous integer. To calculate a particular factorial, we have to keep applying this definition until we reach the base case, $1!$.

Example 4.14.   Prove that $n! \geq 2^{n-1}$ for all positive integers n.

Solution. *(i)*   For the base case $n = 1$, we have $1! = 1$ and $2^{1-1} = 1$. Hence the statement $n! \geq 2^{n-1}$ is true when $n = 1$.

(ii)   As induction hypothesis, suppose that $k! \geq 2^{k-1}$. Then

$$(k+1)! \;=\; (k+1)(k!) \geq (k+1)2^{k-1}.$$

But $k + 1 \geq 2$ whenever $k \in \mathbb{P}$ so

$$(k+1)! \geq 2 \cdot 2^{k-1} \;=\; 2^k.$$

That is, the statement $n! \geq 2^{n-1}$ is true for $n = k+1$ whenever it is true for $n = k$.

Therefore, by the Principle of Mathematical Induction, the result if true for all $n \in \mathbb{P}$. $\square$

Example 4.15. Find the sum of the odd positive integers from 1 to $2n - 1$.

Solution. Since n can be any positive integer, there is an infinite number of sums that we have to calculate. Let us look at the first few to see if there is any pattern to the sums.

$$\begin{aligned}
1 &= 1 &= 1^2 \\
1 + 3 &= 4 &= 2^2 \\
1 + 3 + 5 &= 9 &= 3^2 \\
1 + 3 + 5 + 7 &= 16 &= 4^2 \\
1 + 3 + 5 + 7 + 9 &= 25 &= 5^2
\end{aligned}$$

By looking at the above pattern, we can formulate a general result. The above five cases suggest that

$$1 + 3 + 5 + \cdots + (2n - 1) = n^2.$$

However, we have not yet proved that this formula is true for all n; we have only verified it in five cases. We must apply some general argument, such as Mathematical Induction, to verify that the formula always holds.

(i). We have seen that the formula holds if $n = 1$.

(ii). Suppose that the formula holds when $n = k$; that is, suppose

$$1 + 3 + 5 + \cdots + (2k - 1) = k^2.$$

Then

$$\begin{aligned}
1 + 3 + 5 + \cdots + (2k - 1) + (2k + 1) &= k^2 + (2k + 1) \\
&= (k + 1)^2.
\end{aligned}$$

Therefore, the formula also holds for $n = k + 1$.

Hence, by the Principle of Mathematical Induction,

$$1 + 3 + 5 + \cdots + (2n - 1) = n^2 \quad \text{for all } n \in \mathbb{P}. \qquad \square$$

Mathematicians have a notation that often simplifies the writing of the sum of a long series of terms. It is called the *summation notation* or *sigma notation* as it employs the capital Greek letter sigma, $\sum$.

The sum

$$1 + 3 + 5 + \cdots + (2n - 1),$$

whose rth term is $(2r - 1)$, is written as

$$\sum_{r=1}^{n} (2r - 1)$$

and is read as "the sum, from $r = 1$ to $r = n$, of $(2r - 1)$."

In general, if $k \leq \ell$ and $f(r)$ is an expression involving the integer r, then

$$\sum_{r=k}^{\ell} f(r) = f(k) + f(k+1) + \cdots + f(\ell-1) + f(\ell),$$

that is, the sum of all the terms obtained by substituting the integers from k to ℓ for r in the expression $f(r)$.

For example,

$$\sum_{i=5}^{8} (2^i + 3) \;=\; 35 + 67 + 131 + 259.$$

The result proved in Example 4.12 can be written as

$$\sum_{r=1}^{n} r \;=\; \frac{n(n+1)}{2}.$$

Example 4.16.   If x and y are integers, prove that

$$x^{2n-1} + y^{2n-1}$$

contains the factor $x + y$ for all $n \in \mathbb{P}$.

Solution.   We shall prove this result by induction on n.

(i)   When $n = 1$ the result is true, because $(x+y)$ contains the factor $(x+y)$.

(ii)   Suppose that $(x^{2k-1} + y^{2k-1})$ contains the factor $(x+y)$. Then

$$\begin{aligned}
x^{2k+1} + y^{2k+1} \;&=\; x^2(x^{2k-1} + y^{2k-1}) - x^2 y^{2k-1} + y^{2k+1} \\
&=\; x^2(x^{2k-1} + y^{2k-1}) - y^{2k-1}(x^2 - y^2).
\end{aligned}$$

Since $(x^{2k-1} + y^{2k-1})$ contains the factor $(x+y)$ by the induction hypothesis, and $(x^2 - y^2)$ contains a factor $(x+y)$, it follows that $(x^{2k+1} + y^{2k+1})$ contains the factor $(x+y)$. Therefore, by the Principle of Mathematical Induction, $(x^{2n-1} + y^{2n-1})$ contains the factor $(x+y)$ for all $n \in \mathbb{P}$. $\square$

Example 4.17.   The *Tower of Hanoi* puzzle was published by the French mathematician Edouard Lucas in 1883. It consists of three pegs on a stand, and n punctured disks of different sizes that are placed in decreasing order on one of the pegs. The object of the puzzle is to transfer the pile of disks to another peg, by moving one disk at a time, and without placing any disk on top of a smaller disk. Show that it is possible to solve this puzzle in $2^n - 1$ moves.

The Tower of Hanoi Puzzle

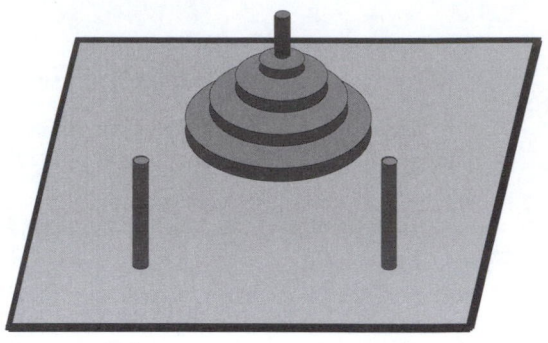

Proof.   We shall prove that it is possible to solve this puzzle in $2^n - 1$ moves, by induction on the number of disks.

(i)   If $n = 1$, then one disk can be transferred to another peg in $2^1 - 1 = 1$ move.

(ii)   Suppose that it is possible to transfer k disks to another peg in $2^k - 1$ moves.

Now suppose that there are $k + 1$ disks on one peg. Leave the largest disk alone and transfer the k other disks to another peg. Since any of the disks can be placed on the largest one, the large disk does not affect the problem and the transfer of the k smaller disks can be accomplished in $2^k - 1$ moves. Now move the largest disk to the vacant peg and finally move the k smaller disks back onto the largest disk in $2^k - 1$ more moves.

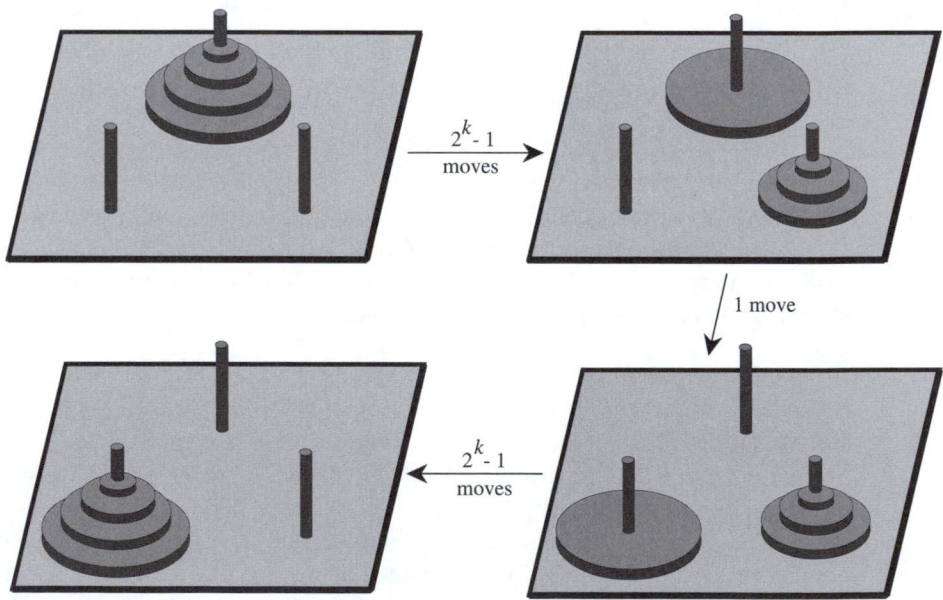

The total number of moves required is $(2^k - 1) + 1 + (2^k - 1) = 2^{k+1} - 1$.

It therefore follows from Mathematical Induction that the problem can always be solved in $2^n - 1$ moves. $\qquad\square$

Slight variations of the Principle of Mathematical Induction given in 4.13 are sometimes useful. For example, the base case could be any integer n_1. If $P(n)$ is a statement depending on the integer n and

(i) $P(n_1)$ is true for some fixed integer n_1 (positive, negative, or zero)

and (ii) $P(k)$ is true $\Longrightarrow P(k+1)$ is true, for $k \geq n_1$,

then it follows that $P(n)$ is true for all integers $n \geq n_1$.

Sometimes it is not possible to prove the statement $P(k+1)$ from the statement $P(k)$, although $P(k+1)$ could be proved from $P(k)$ together with some or all of the statements $P(k-1), P(k-2), \ldots, P(2), P(1)$. In this case the following variation called the *second principle of mathematical induction*, or *strong induction*, is useful. This principle is apparently stronger because the hypotheses in the induction step are stronger, and this sometimes makes it easier to check the induction step. However the Principle of Strong Induction is equivalent to the Principle of Mathematical Induction; one can be deduced from the other.

Principle of Strong Induction 4.18.   *Let $P(n)$ be a statement that depends on the positive integer n.*

If   (i) $P(1)$ *is true*

and   (ii) $P(1), P(2), P(3), \ldots, P(k)$ *are all true* $\Longrightarrow P(k+1)$ *is true*,

then $P(n)$ is true for all $n \in \mathbb{P}$.

Proof.   We show that this principle follows from the Principle of Mathematical Induction. Let $Q(n)$ be a statement "$P(r)$ is true for $1 \leq r \leq n$." Then the hypotheses on $P(n)$ imply that

(i) $Q(1)$ is true

and   (ii) $Q(k)$ is true $\Longrightarrow Q(k+1)$ is true.

Hence, by the Principle of Mathematical Induction 4.13, $Q(n)$ is true for all $n \in \mathbb{P}$ and, in particular, $P(n)$ is true for all $n \in \mathbb{P}$. $\qquad\square$

Let us prove the following result using strong induction without invoking the Unique Factorization Theorem 2.54.

Example 4.19.   Every integer, larger than 1, contains a prime factor.

Solution. *(i)*   The base case in this example is the integer 2. Since this is prime, it contains a prime factor $2 = 2 \cdot 1$.

(ii)   As strong induction hypothesis, suppose that every integer r contains a prime factor, if $2 \leq r < k$; that is, 2, 3, 4, ..., $k - 1$ all contain a prime factor (not the same one, of course). Now consider the integer k. If k is prime, it contains a prime factor. If k is not prime, then it can be factored as $k = s \cdot t$, where $2 \leq s < k$ and $2 \leq t < k$. By the induction hypothesis, s contains a prime factor, and hence k also contains this prime factor.

By the Principle of Strong Induction 4.18, it follows that every integer, larger than 1, contains a prime factor.   □

4.2   RECURSION

One example of mathematical induction that is frequently used in computer science is recursion. Most programming languages allow the use of recursive functions or procedures.

A *recursive routine* is one that calls itself. It obviously must call itself with different arguments, otherwise it would enter an infinite loop. The routine evaluates itself with the new, usually smaller arguments, which would then call the routine with different arguments again. This recursion continues until the routine reaches some *base* or *stopping case*. There may be one or more of these stopping cases.

For example, consider the function $f(n)$, which is the sum of the first n integers. This can be defined in several different ways.

$$
\begin{array}{lll}
\textbf{Iteration:} & f(n) & = & 1 + 2 + 3 + \cdots + n \\
\textbf{Recursion:} & f(n) & = & f(n-1) + n \quad \text{if } n > 1; \text{ and } f(1) = 1 \\
\textbf{Closed form:} & f(n) & = & n(n+1)/2
\end{array}
$$

In this case, the recursive definition of the function $f(n)$ calls the same function but with its argument reduced by one. There is one stopping case, namely $f(1)$.

In a similar way the *factorial function* can be defined in two ways.

$$
\begin{array}{lll}
\textbf{Iteration:} & n! & = & 1 \cdot 2 \cdot 3 \cdots n \quad \text{for } n > 0 \\
\textbf{Recursion:} & n! & = & n(n-1)! \quad \text{if } n > 1; \text{ and } 1! = 1
\end{array}
$$

The *greatest common divisor* of n integers can be defined recursively for $n > 2$ by

$$
\gcd(x_1, x_2, \ldots, x_n) \;=\; \gcd(\gcd(x_1, x_2, \ldots, x_{n-1}), x_n).
$$

In this case the number of arguments change. The stopping case occurs when $n = 2$; we already know the definition of $\gcd(x_1, x_2)$.

To prove that a recursive function produces the correct value, we usually have to use induction. We have to

(i) check the *base* or *stopping cases*; that is, show that the function is correct when there are no recursive calls;

(ii) check the *recursive steps*, assuming inductively that the function gives the correct values for the smaller arguments.

Example 4.21. Prove that the recursive function $f(n)$ defined by

$$f(n) \;=\; f(n-1) + n \quad \text{if } n > 1; \quad \text{and } f(1) = 1$$

is equal to $\frac{n(n+1)}{2}$ for all $n \geq 1$.

Solution. *(i)* When $n = 1$, $f(1) = 1$ and also $\frac{n(n+1)}{2} = 1$. Hence the result is true when $n = 1$.

(ii) Assume, as induction hypothesis, that $f(k) = \frac{k(k+1)}{2}$, for some $k > 0$. Then

$$
\begin{aligned}
f(k+1) \;&=\; f(k) + k + 1 \\
&=\; \frac{k(k+1)}{2} + k + 1 \\
&=\; \frac{(k+1)(k+2)}{2}.
\end{aligned}
$$

Hence the result is true for $n = k+1$. Therefore, by the Principle of Mathematical Induction, $f(n) = \frac{n(n+1)}{2}$ for all $n \geq 1$. $\qquad\square$

Example 4.22. A sequence of integers $x_1, x_2, x_3, x_4 \ldots$ is defined recursively by $x_1 = 2$, $x_2 = 5$, and $x_k = x_{k-1} + 2x_{k-2}$ for all $k \geq 3$. Prove that

$$x_n \;=\; \frac{7 \cdot 2^{n-1} + (-1)^n}{3} \qquad \text{for all } n \in \mathbb{P}.$$

Solution. The next few terms in the sequence are $x_3 = x_2 + 2x_1 = 5 + 2 \cdot 2 = 9$, $x_4 = x_3 + 2x_2 = 9 + 2 \cdot 5 = 19$ and $x_5 = x_4 + 2x_3 = 19 + 2 \cdot 9 = 37$. Strong induction is more suitable than regular induction for proving this example, because the general term x_k depends on the previous *two* terms, x_{k-1} and x_{k-2}, not just the previous term.

(i) When $n = 1$, $x_1 = (7 \cdot 2^0 - 1)/3 = 2$ and the result is true.

We cannot use the recursive formula $x_k = x_{k-1} + 2x_{k-2}$ until $k = 3$. We shall therefore have to verify the result for x_2 separately. When $n = 2$, we have $x_2 = (7 \cdot 2 + 1)/3 = 5$, and the result is true.

(ii) Suppose inductively that $x_r = (7 \cdot 2^{r-1} + (-1)^r)/3$, whenever $1 \leq r \leq k$. Then if $k \geq 2$,

$$
\begin{aligned}
x_{k+1} \;&=\; x_k + 2x_{k-1} \\
&=\; \frac{7 \cdot 2^{k-1} + (-1)^k}{3} + \frac{2[7 \cdot 2^{k-2} + (-1)^{k-1}]}{3} \\
&=\; \frac{7 \cdot 2^{k-1} + (-1)^k + 7 \cdot 2^{k-1} - 2(-1)^k}{3} \\
&=\; \frac{7 \cdot 2^k - (-1)^k}{3} \;=\; \frac{7 \cdot 2^k + (-1)^{k+1}}{3}.
\end{aligned}
$$

Therefore, the result is true for $n = k + 1$.

By the Principle of Strong Induction 4.18, it follows that

$$x_n = \frac{7 \cdot 2^{n-1} + (-1)^n}{3} \qquad \text{for all } n \in \mathbb{P}. \qquad \square$$

In computer science, the integer function MOD denotes the remainder when one integer is divided by another using the Division Algorithm 2.12. If a is a nonnegative integer and b a positive integer, then a MOD $b = r$, where $a = qb + r$, with $0 \le r < b$. Proposition 2.21 yields the following recursive algorithm for computing the greatest common divisor. Since a MOD b is always smaller than b, the algorithm will terminate, as the second variable always decreases.

Recursive Euclidean Algorithm 4.23. *If a and b are nonnegative integers, their greatest common divisor can be calculated recursively using*

$$\gcd(a, b) = \left\{ \begin{array}{ll} a & \text{if } b = 0 \\ \gcd(b, a \text{ MOD } b) & \text{if } b \ne 0. \end{array} \right.$$

The pseudocode for the recursive GCD function is as follows.

Function $\text{GCD}(a, b)$

if $b = 0$

    then return a

else return $\text{GCD}(b, a \text{ MOD } b)$

For example,

$$\begin{array}{llll} \gcd(39, 66) & = & \gcd(66, 39) & = & \gcd(39, 27) & = & \gcd(27, 12) \\ & = & \gcd(12, 3) & = & \gcd(3, 0) & = & 3. \end{array}$$

In the above examples, the recursive function calls *one* copy of itself; this is called *one-way recursion*. It is also possible for the function to call multiple copies of itself. In *two-way recursion* the function calls two copies of itself. One example of this is the *Quicksort Algorithm*, which is used to sort large lists, say to alphabetize a long list of words. This algorithm splits the list into two parts and then applies the Quicksort Algorithm to both parts separately.

4.3 THE BINOMIAL THEOREM

A binomial is a sum of two quantities, such as $a + b$. The square and cube of a binomial have the following familiar expansions.

$$(a + b)^2 = a^2 + 2ab + b^2$$
$$(a + b)^3 = a^3 + 3a^2b + 3ab^2 + b^3$$

The Binomial Theorem yields the expansion of $(a + b)^n$ for each positive integer exponent. The coefficients that occur in the binomial expansions are called binomial coefficients and can be conveniently written in terms of factorials.

Definition.   If $0 \leq r \leq n$, then the **binomial coefficient** $\binom{n}{r}$ is defined by

$$\binom{n}{r} = \frac{n!}{r!(n-r)!}$$

where $0!$ is defined to be 1, so that $\binom{n}{n} = 1$.

The binomial coefficient $\binom{n}{r}$ is pronounced "n choose r" because it is also the number of *combinations of n objects chosen r at a time*; that is, the number of different r element subsets in a set containing n elements. Other common notations for $\binom{n}{r}$ are $_nC_r$ and $C(n,r)$.

The intuitive reason for $n!/r!(n-r)!$ being the number of ways of choosing r objects from n objects is as follows. The number of different ways of ordering n objects is $n!$. Now count this number in a different way, by focusing on the first r terms in the order. Denote the number of ways that the first r objects can be chosen from all the n objects by (n choose r). The first r objects can be ordered among themselves in $r!$ ways. The last $n - r$ objects can be ordered among themselves in $(n - r)!$ ways. Since the choice of the first r, and the orderings of the first r and last $(n - r)$ are independent, the total number of ways of ordering the n objects is (n choose r)$r!(n-r)!$. Hence

$$n! = (n \text{ choose } r)r!(n-r)!$$

and we obtain the desired formula for (n choose r).

Proposition 4.31.   *The following properties of the binomial coefficients follow directly from the definition, when* $0 \leq r \leq n$.

(i) $\dbinom{n}{r} = \dfrac{n(n-1)(n-2)\cdots(n-r+2)(n-r+1)}{1 \cdot 2 \cdot 3 \cdots (r-1) \cdot r}$

(ii) $\dbinom{n}{0} = \dbinom{n}{n} = 1$

(iii) $\dbinom{n}{n-r} = \dbinom{n}{r}$ $\square$

Let us write down the first few cases of the expansion of $(a + b)^n$, and add $(a + b)^0$.

$$
\begin{aligned}
(a+b)^0 &= 1 \\
(a+b)^1 &= a + b \\
(a+b)^2 &= a^2 + 2ab + b^2 \\
(a+b)^3 &= a^3 + 3a^2b + 3ab^2 + b^3 \\
(a+b)^4 &= a^4 + 4a^3b + 6a^2b^2 + 4ab^3 + b^4 \\
(a+b)^5 &= a^5 + 5a^4b + 10a^3b^2 + 10a^2b^3 + 5ab^4 + b^5
\end{aligned}
$$

If the coefficients alone are written out, as shown below, we obtain what is known as *Pascal's Triangle*, named after the famous French mathematician and writer Blaise Pascal (1623–1662).

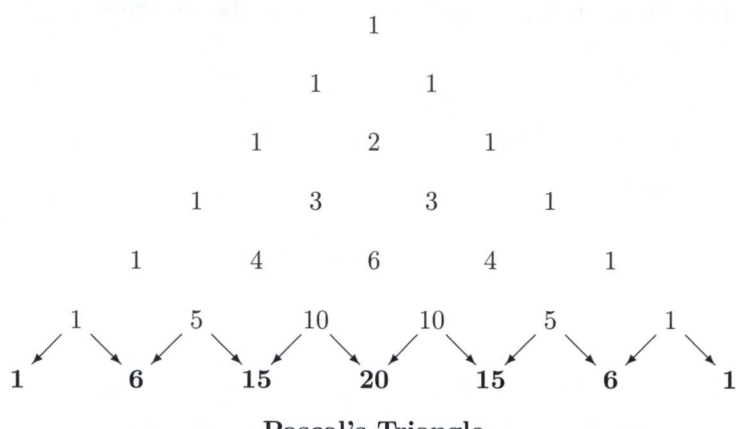

Pascal's Triangle

Notice that that each row in Pascal's Triangle can be obtained from the previous row in the following way. Each element is the sum of the two elements to its immediate left and right in the preceding row. For example, **15** $= 5 + 10$, and **20** $= 10 + 10$. This suggests the following important identity between the binomial coefficients.

Proposition 4.32. *If* $1 \leq r \leq n$, *then*

$$\binom{n}{r-1} + \binom{n}{r} = \binom{n+1}{r}.$$

Proof. The proof is by straightforward calculation.

$$\begin{aligned}
\binom{n}{r-1} + \binom{n}{r} &= \frac{n!}{(r-1)!(n-r+1)!} + \frac{n!}{r!(n-r)!} \\
&= \frac{r(n!) + (n-r+1)(n!)}{r!(n-r+1)!} \\
&= \frac{(n+1)(n!)}{r!(n-r+1)!} = \frac{(n+1)!}{r!(n+1-r)!} \\
&= \binom{n+1}{r} \qquad \qquad \qquad \square
\end{aligned}$$

This identity has a nice combinatorial interpretation. The number $\binom{n+1}{r}$ is the number of different r element subsets that can be chosen from an $n+1$ element set. Label the elements of the set as $L_1, L_2, \ldots, L_n, L_{n+1}$. If L_{n+1} is in the chosen subset, then there are $\binom{n}{r-1}$ ways of choosing the remaining $r - 1$ elements from $L_1, L_2, \ldots, L_n$. If L_{n+1} is not in the chosen subset, then there are $\binom{n}{r}$ ways of choosing an r element subset from $L_1, L_2, \ldots, L_n$. Hence the total number of r element subsets in an $n + 1$ element set is $\binom{n}{r-1} + \binom{n}{r}$.

It is not clear from our definition that the binomial coefficients are always integers.

Proposition 4.33. $\displaystyle\binom{n}{r}$ *is an integer for* $0 \le r \le n$.

Proof.   If $n = 0$, then $\binom{0}{0} = 1$. For $n > 0$, we shall prove the result by induction.

(i).   If $n = 1$, then $\binom{1}{0} = 1$ and $\binom{1}{1} = 1$, which are both integers.

(ii).   Suppose that $\binom{k}{r}$ is an integer if $0 \le r \le k$. Then it follows from Proposition 4.32 that if $1 \le r \le k$,

$$\binom{k+1}{r} = \binom{k}{r-1} + \binom{k}{r},$$

which is the sum of two integers and so is also an integer. Finally, $\binom{k+1}{0} = \binom{k+1}{k+1} = 1$ are integers.

Therefore, by Mathematical Induction, $\displaystyle\binom{n}{r}$ is an integer for all $n \in \mathbb{P}$ and $0 \le r \le n$. $\square$

Binomial Theorem 4.34.   *If a and b are any numbers, and $n \in \mathbb{P}$, then*

$$(a+b)^n = \binom{n}{0}a^n + \binom{n}{1}a^{n-1}b + \cdots + \binom{n}{r}a^{n-r}b^r + \cdots + \binom{n}{n-1}ab^{n-1} + \binom{n}{n}b^n.$$

Proof.   We shall prove the theorem by induction on n.

(i)   If $n = 1$, then

$$\binom{1}{0}a + \binom{1}{1}b = a + b = (a+b)^1$$

and the theorem holds.

(ii)   Suppose the theorem is true for $n = k$; that is,

$$(a+b)^k = \binom{k}{0}a^k + \binom{k}{1}a^{k-1}b + \cdots + \binom{k}{r}a^{k-r}b^r + \cdots + \binom{k}{k}b^k.$$

Now

$$(a+b)^{k+1}$$

$$= \quad (a+b)(a+b)^k$$

$$= \quad a(a+b)^k + b(a+b)^k$$

$$= \quad \binom{k}{0}a^{k+1} + \binom{k}{1}a^k b + \cdots + \binom{k}{r}a^{k-r+1}b^r + \cdots + \binom{k}{k}ab^k$$

$$\quad + \quad \binom{k}{0}a^k b + \cdots + \binom{k}{r-1}a^{k-r+1}b^r + \cdots + \binom{k}{k-1}ab^k + \binom{k}{k}b^{k+1}$$

$$= \quad a^{k+1} + \binom{k+1}{1}a^k b + \cdots + \binom{k+1}{r}a^{k-r+1}b^r + \cdots + \binom{k+1}{k}ab^k + \quad b^{k+1}$$

using Proposition 4.32. Hence the theorem is true for $n = k + 1$. By the Principle of Mathematical Induction, the theorem is true for all $n \in \mathbb{P}$. $\quad\square$

Using the sigma notation, the Binomial Theorem can be written as

$$(a+b)^n \quad = \quad \sum_{r=0}^{n} \binom{n}{r} a^{n-r}b^r.$$

Example 4.35. Expand $(3x - 2y)^4$.

Solution. By the Binomial Theorem,

$$(3x - 2y)^4 \quad = \quad (3x)^4 + 4(3x)^3(-2y) + 6(3x)^2(-2y)^2 + 4(3x)(-2y)^3 + (-2y)^4$$

$$= \quad 81x^4 - 216x^3 y + 216x^2 y^2 - 96xy^3 + 16y^4. \quad\square$$

Example 4.36. Calculate $(2.01)^5$ to 3 decimal places.

Solution. By the Binomial Theorem,

$$(2.01)^5 \quad = \quad (2 + .01)^5$$

$$= \quad 2^5 + 5 \cdot 2^4(.01) + 10 \cdot 2^3(.01)^2 + 10 \cdot 2^2(.01)^3 + 5 \cdot 2(.01)^4 + (.01)^5$$

$$= \quad 32 + .8 + .008 + .00004 + .0000001 + .0000000001$$

$$= \quad 32.808 \quad \text{to 3 decimal places.} \quad\square$$

Example 4.37. If k is an integer and $kx^3 y^4$ is a term in the expansion of $(2x + 5y^2)^n$, then find the values of n and k.

Solution. The rth term in the expansion of $(2x + 5y^2)^n$ is

$$\binom{n}{r}(2x)^{n-r}(5y^2)^r \quad = \quad \binom{n}{r}2^{n-r} \cdot 5^r x^{n-r}y^{2r}.$$

If this equals kx^3y^4, then we must have $n - r = 3$ and $2r = 4$; that is, $r = 2$ and $n = 5$. Hence

$$\begin{aligned} k &= \binom{n}{r}2^{n-r}5^r = \binom{5}{2}2^3 \cdot 5^2 = \frac{5 \cdot 4}{1 \cdot 2} \cdot 2^3 \cdot 5^2 = 2^4 \cdot 5^3 \\ &= 2000. \end{aligned}$$

$\square$

The intuitive reason why the binomial coefficient $\binom{n}{r}$ is equal to the number of combinations of n objects taken r at a time can be seen by writing out $(a + b)^n$ as

$$(a + b)^n = (a + b)(a + b)(a + b) \cdots (a + b),$$

where there are n factors. When the factors on the right side are multiplied out, the term $a^{n-r}b^r$ will occur whenever we choose an a from $n - r$ factors and a b from the remaining r factors. Hence the number of times $a^{n-r}b^r$ will occur is equal to the number of ways of choosing r b's from the n factors.

We have proved the Binomial Theorem for positive integer values of the exponent. It is possible to extend the Binomial Theorem to a negative integer power or, in fact, any real power. In these cases we cannot obtain an expansion involving only a finite number of terms, but instead we obtain an infinite series. For example, if n is a real number, then

$$(1 + x)^n = 1 + nx + \frac{n(n-1)}{2!}x^2 + \frac{n(n-1)(n-2)}{3!}x^3 + \cdots .$$

If n is a positive integer, this series terminates and we obtain the terms given in the Binomial Theorem. If n is not a positive integer, the coefficient of x^r is also

$$\frac{n(n-1)(n-2)\cdots(n-r+1)}{r!}$$

and this never vanishes; hence we obtain an infinite series. It can be shown that the sum of the series on the right side approaches the value $(1 + x)^n$ *provided that* $-1 < x < 1$. If n lies outside this range, then the binomial expansion is meaningless.

Consider

$$\begin{aligned} \frac{1}{(1+x)^2} &= (1+x)^{-2} = 1 - 2x + \frac{(-2)(-3)}{2!}x^2 + \frac{(-2)(-3)(-4)}{3!}x^3 + \cdots \\ &= 1 - 2x + 3x^2 - 4x^3 + \cdots . \end{aligned}$$

This expansion is valid if $-1 < x < 1$. However, if we put $x = -2$ in the above expression, the right side becomes

$$1 + 4 + 12 + 32 + \cdots ,$$

which certainly does not approach $\frac{1}{(1-2)^2} = 1$.

Exercise Set 4

1–4. Calculate the following.

1. $\dbinom{5}{3}$

2. $\dbinom{10}{6}$

3. $\dfrac{8!}{(4!)^2}$

4. $100! - 99!$

5. Show that $\dfrac{1}{n}\dbinom{n}{r} = \dfrac{1}{r}\dbinom{n-1}{r-1}$.

6. Show that $\dbinom{n}{r}\dbinom{r}{s} = \dbinom{n}{s}\dbinom{n-s}{r-s}$.

7. Find n if $\dbinom{n+2}{n} = 36$.

8–10. Write the following in sigma notation.

8. $\dfrac{1}{2} + \dfrac{3}{4} + \dfrac{5}{6} + \cdots + \dfrac{99}{100}$

9. $8 + 15 + 24 + 35 + \cdots + (n^2 - 1)$

10. $a^k + a^{2k} + a^{4k} + a^{8k} + a^{16k} + \cdots + a^{256k}$

11–18. Prove, by induction, the following results for all $n \in \mathbb{P}$.

11. $1^2 + 2^2 + 3^2 + \cdots + n^2 = \dfrac{n(n+1)(2n+1)}{6}$

12. $1^3 + 2^3 + 3^3 + \cdots + n^3 = \left[\dfrac{n(n+1)}{2}\right]^2$

13. $1^4 + 2^4 + 3^4 + \cdots + n^4 = \dfrac{n(n+1)(6n^3 + 9n^2 + n - 1)}{30}$

14. $1^2 + 3^2 + 5^2 + \cdots + (2n-1)^2 = \dfrac{n(2n-1)(2n+1)}{3}$

15. $1 \cdot 2 + 2 \cdot 3 + 3 \cdot 4 + \cdots + n(n+1) = \dfrac{n(n+1)(n+2)}{3}$

16. $\dfrac{1}{2} + \dfrac{2}{2^2} + \dfrac{3}{2^3} + \cdots + \dfrac{n}{2^n} = 2 - \left(\dfrac{n+2}{2^n}\right)$

17. A set with n elements contains 2^n subsets (including the set itself and $\emptyset$).

18. $6 \mid (2n^3 + 3n^2 + n)$

19. Find an expression for $\sum_{r=1}^{n} r(r!)$ and prove that it is correct.

20–23. *Are the following true for all positive integer values of n? If so, prove the result; if not, give a counterexample.*

20. $n! \geq 2n$

21. $3|(2^{2n} - 1)$

22. $7|(5^n + n + 1)$

23. $(a + b)|(a^{2n} - b^{2n})$

24. Prove that the sum of the first n terms of the *arithmetic progression*

$$a + (a + d) + (a + 2d) + \cdots + [a + (n - 1)d]$$

is $\frac{n}{2}[2a + (n - 1)d]$; that is, $\frac{n}{2}$ times the sum of the first and last terms.

25. Prove that the sum of the first n terms of the *geometric progression*

$$a + aq + aq^2 + \cdots + aq^{n-1}$$

is $\dfrac{a(1 - q^n)}{1 - q}$ when $q \neq 1$.

26. *(Fermat's Little Theorem)* If p is a prime, use induction on n and the Binomial Theorem to prove that $n^p \equiv n \pmod{p}$ for all $n \in \mathbb{P}$.

27. Use induction to prove that $a^m \cdot a^n = a^{m+n}$ for all $n \in \mathbb{P}$.

28. A sequence of integers $x_1, x_2, x_3, \ldots$ is defined by $x_1 = 3$, $x_2 = 7$, and

$$x_k = 5x_{k-1} - 6x_{k-2} \quad \text{for } k \geq 3.$$

Prove that $x_n = 2^n + 3^{n-1}$ for all $n \in \mathbb{P}$.

29. If n points lie in a plane and no three are collinear, prove that there are $\frac{1}{2}n(n - 1)$ lines joining these points.

30. Find an expression for

$$1 - 3 + 5 - 7 + 9 - 11 + \cdots + (-1)^{n-1}(2n - 1)$$

and prove that it is correct.

31. The notation $\prod_{r=1}^{n} a_r$, which uses the capital Greek letter pi, stands for the *product* of all the terms obtained by substituting the integers from 1 to n for r in the expression a_r. That is,

$$\prod_{r=1}^{n} a_r = a_1 \cdot a_2 \cdot a_3 \cdots a_n.$$

Prove that $\displaystyle\prod_{r=1}^{n}(1 + x^{2^r}) = \dfrac{1 - x^{2^{n+1}}}{1 - x^2}$, if $x^2 \neq 1$.

32. Find an expression for $\displaystyle\prod_{r=2}^{n}\left(1 - \dfrac{1}{r^2}\right)$ and prove that your expression is correct.

33. Use induction to prove that

$$\frac{1}{1} + \frac{1}{2} + \frac{1}{3} + \frac{1}{4} + \frac{1}{5} + \cdots + \frac{1}{2^n} \geq 1 + \frac{n}{2}.$$

(This can be used to show that the infinite harmonic series $\sum_{r=1}^{\infty} \frac{1}{r}$ diverges.)

34–38. Expand the following by the Binomial Theorem.

34. $(2a + b)^6$

35. $(a - 1)^5$

36. $\left(x + \dfrac{1}{x} \right)^8$

37. $\left(4x^2 - 3y^3 \right)^4$

38. $(a + b + c)^3$

39. Calculate $(2.99)^4$ to 3 decimal places.

40. Calculate $(1.02)^{10}$ to 3 decimal places.

41. Find the fifth term in the expansion of $\left(2x^6 - \dfrac{5}{x^5} \right)^{11}$.

42. Find the term containing x^5 in the expansion of $\left(x^2 + \dfrac{3}{x} \right)^6$.

43. If p is a prime, prove that

$$(a + b)^p \equiv a^p + b^p \pmod{p} \qquad \text{for all } a, b \in \mathbb{Z}.$$

44. A man earns a starting salary of \$10,000 a year and receives an annual wage increase of 10%. Find his salary, to the nearest dollar, after 10 years.

45. Prove that

$$\binom{n}{0} + \binom{n}{1} + \binom{n}{2} + \cdots + \binom{n}{n} = 2^n$$

and that

$$\binom{n}{0} - \binom{n}{1} + \binom{n}{2} - \cdots + (-1)^n \binom{n}{n} = 0.$$

Problem Set 4

46. Prove that

$$\sum_{r=1}^{n} rx^r = \frac{x - (n+1)x^{n+1} + nx^{n+2}}{(1 - x)^2}$$

for all $n \in \mathbb{P}$, where x is a number different from 1.

47. Prove that $\sqrt[n]{2} \leq 1 + \dfrac{1}{n}$ for all $n \in \mathbb{P}$.

48. Find an expression for $1^2 - 3^2 + 5^2 - 7^2 + \cdots + (-1)^{n-1}(2n - 1)^2$ and prove that your expression is correct.

49. Prove that a convex n-gon contains $\frac{1}{2}n(n-3)$ diagonals. (A convex n-gon is a polygon with n sides such that all the line segments, joining two nonadjacent vertices, lie inside the polygon.)

50. Generalize the following sequence of identities and prove your generalization.

$$1 = 1, \quad 3 + 5 = 8, \quad 7 + 9 + 11 = 27, \quad 13 + 15 + 17 + 19 = 64$$

51. Prove that the product of r consecutive positive integers is divisible by $r!$.

52. Find an expression for

$$\sum_{i=1}^{n} i(i+1)(i+2) \cdots (i+r-1)$$

and prove by induction that your expression is correct.

53. If $x \equiv 1 \pmod 2$, prove that $x^{2^n} \equiv 1 \pmod{2^{n+2}}$ for all $n \in \mathbb{P}$.

54. Is $2^n - 1$ the minimum number of moves required to solve the Tower of Hanoi puzzle in Example 4.17? Give reasons for your answer.

55. If the three pegs in the Tower of Hanoi puzzle are labeled A, B, and C, and n disks are initially on peg A, then what is the first move in order to transfer the disks to peg B in $2^n - 1$ moves?

56. *(Pigeonhole Principle)* Prove the following statement by induction on n. "If $n + 1$ objects are placed in n boxes (or pigeonholes), then one box must contain at least two objects."

57. *(Leibniz Rule)* If you have some knowledge of the calculus, prove by induction that

$$D^n(f \cdot g) \;=\; \sum_{r=0}^{n} \binom{n}{r} D^{n-r} f \cdot D^r g$$

where Df is the derivative of the function f. Assume that all the necessary derivatives exist.

58. What is wrong with the following proof that all horses have the same color? We shall prove the result by induction on the number of horses. Clearly, in any set consisting of a single horse, all horses have the same color. As induction hypothesis, assume that in any set of h horses, they all have the same color. In any set of $h + 1$ horses, where $h \geq 1$, remove one horse. By the induction hypothesis, the remaining h horses have the same color. Now put the horse back, and remove a different horse. All the h horses left have the same color. Hence all the $h + 1$ horses have the same color. Therefore, by the Principle of Mathematical Induction, all horses have the same color.

59–67. *In 1202 Leonardo of Pisa, also called Fibonacci, published an influential mathematical book that popularized the Indian-Arabic number system in Europe. He used our current Arabic numerals, instead of Roman numerals, and also a positional decimal system with a sign for zero. One of the problems in the book asked "How many pairs of rabbits can be produced in a year from a single pairs of baby rabbits, if every month each mature pair produces a new pair of babies, that mature after a month?" The answer involves the sequence 1, 1, 2, 3, 5, 8, 13,..., in which each term is the sum of the previous two. This sequence is known as the **Fibonacci sequence** and is formally defined as the sequence of integers $f_1, f_2, f_3,...$ such that $f_1 = 1$, $f_2 = 1$, and $f_n = f_{n-1} + f_{n-2}$, for $n \geq 3$.*

59. Find the first 15 numbers in the Fibonacci sequence.

60. Prove that $\gcd(f_n, f_{n+1}) = 1$ for all $n \in \mathbb{P}$.

61. Look at the Euclidean Algorithm applied to f_{14} and f_{15}. What do you notice about the sequence of remainders?

62. Prove that $f_{n+1} < \left(\dfrac{7}{4}\right)^n$ for all $n \in \mathbb{P}$.

63. If $a = \dfrac{1 + \sqrt{5}}{2}$ and $b = \dfrac{1 - \sqrt{5}}{2}$, prove that $f_n = \dfrac{a^n - b^n}{\sqrt{5}}$ for all $n \in \mathbb{P}$.

64. If $a = \dfrac{1 + \sqrt{5}}{2}$, prove that f_n is the nearest integer to $\dfrac{a^n}{\sqrt{5}}$ for all $n \in \mathbb{P}$. Hence compute f_{28}, f_{29}, and f_{30}, and verify that $f_{30} = f_{28} + f_{29}$.

65. Prove that $\sum_{r=1}^{n} f_r^2 = f_n f_{n+1}$ for all $n \in \mathbb{P}$.

66. Prove that $\sum_{r=1}^{n} f_{2r-1} = f_{2n}$ for all $n \in \mathbb{P}$.

67. Prove that $f_{n+5} \equiv 3f_n \pmod{5}$ for all $n \in \mathbb{P}$.

68. The downtown portion of a city consists of a rectangular area m blocks long and n blocks wide. If all the streets in each direction are through streets, find the number of different shortest routes from one corner of the downtown area to the opposite corner.

69. At a party of logicians, the host attached either a gold star or a blue star to the back of each of the guests. After all the guests had arrived and mingled with each other, the host announced that there was at least one gold star, and anybody who could prove they had one on their back, without peeking, should claim the prize. No one came forward to claim the prize. Every five minutes after this, the host again announced anybody who could prove they had a gold star should claim the prize, and no one came forward. Finally, after the twentieth time the host asked, everyone rushed forward to claim the prize. How many guests were at the party?

70. *(Trinomial Theorem)* Prove that

$$(a+b+c)^n \;=\; \sum_{p+q+r=n} \frac{n!}{p!q!r!} a^p b^q c^r$$

for all $n \in \mathbb{P}$, where the sum is taken over all nonnegative values of p, q, and r for which $p+q+r=n$.

71. What is the maximum number of regions that a plane can be divided into by n straight lines?

72. What is the maximum number of regions that three-dimensional space can be divided into by n planes?

(See the one-hour movie by George Pólya entitled *Let Us Teach Guessing*, distributed by The Mathematical Association of America.)

73. Let $a_1, a_2, \ldots, a_n$ be real numbers such that $0 \le a_i \le 1$ for $1 \le i \le n$. Prove that

$$(1-a_1)(1-a_2)\cdots(1-a_n) \;\ge\; 1-(a_1+a_2+\cdots+a_n).$$

74. A sequence $x_1, x_2, x_3, \ldots$ of real numbers is defined by $x_1 = 1$ and

$$x_{n+1} \;=\; \frac{n}{n+1}x_n + 1, \quad \text{if } n \ge 1.$$

Prove that $x_n = \dfrac{n+1}{2}$ for all positive integers n.

75. A sequence $y_1, y_2, y_3, \ldots$ of integers is defined by $y_1 = 4$, and $y_{n+1} = y_n^2 - 2$, if $n \ge 1$. Prove that

$$y_n \;=\; (2+\sqrt{3})^{2^{n-1}} + (2-\sqrt{3})^{2^{n-1}}$$

for all positive integers n.

76. Prove that

$$\left(1-\frac{1}{4}\right)\left(1-\frac{1}{9}\right)\left(1-\frac{1}{16}\right)\cdots\left(1-\frac{1}{n^2}\right) \;=\; \frac{n+1}{2n}$$

for all integers $n \ge 2$.

77. Show that

$$1 + \frac{1}{2^2} + \frac{1}{3^2} + \cdots + \frac{1}{n^2} \;\le\; 2 - \frac{1}{n}$$

for all integers $n \ge 1$.

78–82. *Find the value of each recursive mystery function* $\mathrm{myst}(x_1, x_2, \ldots, x_n)$ *on any n-tuple* $(x_1, x_2, \ldots, x_n)$ *and prove that your value is correct.*

78. $\mathrm{myst}(x_1, x_2, \ldots, x_n) = \begin{cases} x_1 & \text{if } n = 1 \\ x_n - \mathrm{myst}(x_1, x_2, \ldots, x_{n-1}) & \text{if } n > 1 \end{cases}$

79. $\mathrm{myst}(x_1, x_2, \ldots, x_n) = \begin{cases} x_1 & \text{if } n = 1 \\ x_n \cdot \mathrm{myst}(x_1, x_2, \ldots, x_{n-1}) & \text{if } n > 1 \end{cases}$

80. $\mathrm{myst}(x_1, x_2, \ldots, x_n) = \begin{cases} x_1 & \text{if } n = 1 \\ x_n & \text{if } x_n > \mathrm{myst}(x_1, x_2, \ldots, x_{n-1}) \\ \mathrm{myst}(x_1, x_2, \ldots, x_{n-1}) & \text{otherwise} \end{cases}$

81. $\mathrm{myst}(x_1, x_2, \ldots, x_n) = \begin{cases} x_1 & \text{if } n = 1 \\ x_1 - 2\,\mathrm{myst}(x_2, x_3, \ldots, x_n) & \text{if } n > 1 \end{cases}$

82. $\mathrm{myst}(x_1, x_2, \ldots, x_n) = \begin{cases} x_1 & \text{if } n = 1 \\ \mathrm{myst}(x_1, x_2, \ldots, x_{n-1}) & \\ \quad + \mathrm{myst}(x_2, x_3, \ldots, x_n) & \text{if } n > 1 \end{cases}$

83. Find a recursive definition for the function

$$e(n) = 1 + \frac{1}{1!} + \frac{1}{2!} + \cdots + \frac{1}{n!}$$

that gives an approximation to e, the base of natural logarithms.

84. Discuss the following recursive definition of the greatest common divisor for $a \geq 0$ and $b \geq 0$. Is it correct? Is it efficient?

$$\gcd(a, b) = \begin{cases} a & \text{if } b = 0 \\ \gcd(b, a) & \text{if } a < b \\ \gcd(b, a - b) & \text{if } a \geq b > 0 \end{cases}$$

CHAPTER 5

Rational and Real Numbers

5.1 RATIONAL NUMBERS

In the previous chapters, we have been concerned with the properties of the integers. Whole numbers are excellent for counting, but they are not sufficient for measuring quantities such as length or weight. We need to be able to divide such quantities into any number of equal parts. Therefore, we extend our number system to include fractions of the form a/b, where a and b are whole numbers. The fraction a/b represents the number a divided into b equal parts. The integer a is called the *numerator* and the integer b is called the *denominator* of the fraction.

Mathematically, this is equivalent to saying that the integers $\mathbb{Z}$ are inadequate because equations of the form $bx = a$ cannot always be solved in $\mathbb{Z}$. The fraction, or rational number a/b, that we introduce is a solution to this equation $bx = a$.

As is well known, fractions can be added, subtracted, and multiplied by the following rules.

$$\frac{a}{b} + \frac{c}{d} = \frac{ad + bc}{bd}$$
$$\frac{a}{b} - \frac{c}{d} = \frac{ad - bc}{bd}$$
$$\frac{a}{b} \cdot \frac{c}{d} = \frac{ac}{bd}$$

Furthermore, any fraction can be divided by a nonzero fraction by the rule

$$\frac{a}{b} \div \frac{c}{d} = \frac{ad}{bc}.$$

One snag is that fractions, such as $1/2$ and $2/4$, look different but are both solutions to the same equation $2x = 1$. We should like them both to represent the same number. This suggests that the correct mathematical description of a fraction or rational number should be as an equivalence class. These equivalence classes should be such that a/b and c/d are the same whenever $ad = bc$.

Definition. Define the equivalence relation, $\sim$, on the set $\mathbb{Z} \times (\mathbb{Z} - \{0\}) = \{(a, b) \mid a, b \in \mathbb{Z}, b \neq 0\}$ by $(a, b) \sim (c, d)$ if and only if $ad = bc$. The equivalence classes are called **rational numbers**, and the equivalence class containing (a, b) is denoted by $\frac{a}{b}$ or a/b. The set of all rational numbers is denoted by $\mathbb{Q}$, so that

$$\mathbb{Q} = \left\{ \frac{a}{b} \,\middle|\, a, b \in \mathbb{Z}, b \neq 0 \right\}.$$

The reader should check that the relation $\sim$ on $\mathbb{Z} \times (\mathbb{Z} - \{0\})$ is indeed an equivalence relation, so that this is a valid definition.

Since the operations of addition, subtraction, multiplication, and division are all defined in terms of representatives of equivalence classes, we have to check that these operations are all well defined. We have to show that if we take two different forms for the same rational number, then the result of performing the operations on these is the same. For example, using the definition of addition,

$$\frac{1}{2} + \frac{1}{2} = \frac{2+2}{2 \cdot 2} = \frac{4}{4}$$

$$\frac{2}{4} + \frac{-3}{-6} = \frac{2(-6) + 4(-3)}{4(-6)} = \frac{-24}{-24}.$$

In general, we can check that if $a_1/b_1 = a_2/b_2$ and $c_1/d_1 = c_2/d_2$, then

$$\frac{a_1}{b_1} + \frac{c_1}{d_1} = \frac{a_2}{b_2} + \frac{c_2}{d_2}$$

and obtain similar results for the other operations.

When rational numbers of the form $n/1$ are added, subtracted, and multiplied, we see that the result is also a number with denominator 1;

$$\frac{n}{1} + \frac{m}{1} = \frac{n+m}{1}, \qquad \frac{n}{1} - \frac{m}{1} = \frac{n-m}{1}, \qquad \frac{n}{1} \cdot \frac{m}{1} = \frac{nm}{1}.$$

We can identify the integer n with the rational number $n/1$, and, because of the above results, the operations of addition, subtraction, and multiplication of integers can be performed either on the integers themselves or on the corresponding rational numbers with denominator 1. Therefore, the set $\mathbb{Z}$ can be considered as a subset of $\mathbb{Q}$, or $\mathbb{Q}$ can be thought of as an extension of $\mathbb{Z}$.

The zero rational number is $0/1$ or, equivalently, $0/b$ for any $b \neq 0$. The unit is the number $1/1$ or, equivalently, a/a for any $a \neq 0$. Addition, subtraction, and multiplication of rational numbers have the usual arithmetic properties, and division by nonzero rationals is always possible, so the set of rational numbers, $\mathbb{Q}$, forms a field.

The rational numbers were constructed so that linear equations of the form $bx = a$ could be solved, where $a, b \in \mathbb{Z}, b \neq 0$. However, they do more, because any linear equation with *rational* coefficients,

$$\frac{b}{d} x = \frac{a}{c}, \qquad \text{where} \quad \frac{a}{c}, \frac{b}{d} \in \mathbb{Q}, b \neq 0$$

has a rational solution $x = ad/bc$.

Proposition 5.11. *Any nonzero rational number r can be expressed in a unique way as a fraction $r = a/b$ with $b > 0$ and $\gcd(a, b) = 1$.*

In this case, we say that the fraction has been reduced to its *lowest terms*.

Proof.   By the definition of rational numbers, r can be written in the form p/q, where p and q are nonzero integers. Furthermore, we can assume $q > 0$ for otherwise we could replace p/q by $(-p)/(-q)$.

Let $d = \gcd(p,q)$ and $p = ad, q = bd$, where $a, b \in \mathbb{Z}$ and $b > 0$. Then

$$r \;=\; \frac{p}{q} \;=\; \frac{ad}{bd} \;=\; \frac{a}{b}.$$

This is the required form for r because $\gcd(a,b) = \gcd\left(\frac{p}{d}, \frac{q}{d}\right) = 1$, by Proposition 2.27 (ii).

To show that this form is unique, suppose that

$$r \;=\; \frac{a_1}{b_1} \;=\; \frac{a_2}{b_2}, \quad \text{where } a_1, a_2, b_1, b_2 \in \mathbb{Z}, b_1, b_2 > 0$$

and $\gcd(a_1, b_1) = \gcd(a_2, b_2) = 1$. Then $a_1 b_2 = a_2 b_1$ and it follows from Proposition 2.28 that $b_1 | b_2$ and $b_2 | b_1$. Hence, by Proposition 2.11 (iii), $b_1 = b_2$ and so also $a_1 = a_2$. $\qquad\square$

5.2   REAL NUMBERS

We can give a geometrical picture of the rational numbers as points on a *number line* as follows. Mark off two points on a straight line and label the left point 0 and the right point 1. We shall take the distance between these two points as the unit length. The positive integer n is represented by the point that is n units to the right of 0, and the negative integer $-n$ as the point that is n units to the left of 0. The rational number a/b, where $a, b \in \mathbb{Z}$ and $b > 0$, is represented by the point that is one bth part of the distance from 0 to the integer point a.

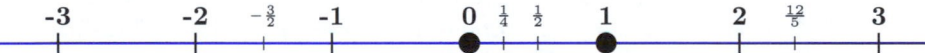

Any interval of the line, however small, contains rational points. We can find such a point by choosing n large enough so that the interval from 0 to $1/n$ is smaller than the interval in question. Then at least one rational point of the form m/n, with $m \in \mathbb{Z}$, must lie in the interval. We express this fact by saying that the rational numbers are dense in the line.

In fact, every interval contains an infinite number of rational points. For example, the interval between $\frac{88}{100}$ and $\frac{89}{100}$ contains the points $\frac{881}{1000}, \frac{882}{1000}, \dots, \frac{889}{1000}$, and $\frac{8801}{10000}, \frac{8802}{10000}, \dots$, and so on.

The surprising fact is that there are points on the line that do not correspond to any rational number. One such point corresponds to the distance of the hypotenuse of a right-angled triangle whose other two sides both have unit length.

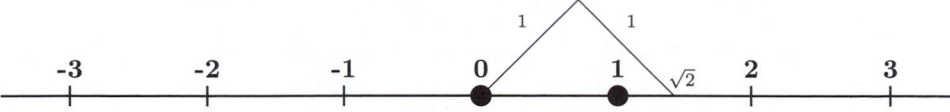

If x is this length, it follows from Pythagoras' Theorem that $x^2 = 2$. We shall now show that this equation has no rational solution.

Theorem 5.21.   *There is no $x \in \mathbb{Q}$ such that $x^2 = 2$.*

Proof.   Suppose that there is a rational number x such that $x^2 = 2$. By Proposition 5.11, write x in its lowest terms as a/b, where $\gcd(a, b) = 1$.

Now $\left(\dfrac{a}{b}\right)^2 = 2$ so that $a^2 = 2b^2$. Therefore $2|a^2$ and, since 2 is prime, $2|a$. Put $a = 2c$, so that $4c^2 = 2b^2$ and $2c^2 = b^2$. It now follows that $2|b$, and so 2 is a common divisor of a and b.

This contradicts the hypothesis on a and b being relatively prime, so it follows that there is no rational number x such that $x^2 = 2$. $\qquad\square$

We can express this result by saying that $\sqrt{2}$ is not a rational number. It is now clear that the rational numbers are not sufficient for many purposes, and we must extend our number system to include all the points on the number line.

The system of numbers that correspond to all the points of the number line is called the set of *real numbers* and is denoted by $\mathbb{R}$. This will include the set of rational numbers $\mathbb{Q}$ as a proper subset.

Real numbers that are not rational numbers are called *irrational numbers*. We have shown that $\sqrt{2}$ is irrational. It is possible, but more difficult, to show that π is an irrational number.

The construction of the system of real numbers, $\mathbb{R}$, in which we can perform the operations of addition, subtraction, multiplication, and division, is beyond the scope of this book. Instead, we shall assume that the real numbers are known, just as we assumed that the integers were known.

Since $\sqrt{2}$ cannot be written in the form a/b for $a, b \in \mathbb{Z}$, it follows that $m\sqrt{2}/n$ is not rational for all nonzero integers m and n. Therefore, by choosing n large enough, we can show that any interval contains irrational numbers of the form $m\sqrt{2}/n$. In fact, every interval contains infinitely many rational numbers and infinitely many irrational numbers, and the rational and irrational numbers are both dense on the number line.

5.3   RATIONAL EXPONENTS

One of the properties of the real numbers is the fact that every positive real number has a unique positive nth root, for each positive integer n.

Definition.   If $a \geq 0$, and n is a positive integer, let $\sqrt[n]{a}$ denote the unique nonnegative nth root of a; that is, $x = \sqrt[n]{a}$ is the positive solution to the equation

$$x^n = a.$$

If $n = 2$, we usually denote $\sqrt[2]{a}$ just by $\sqrt{a}$.

For example, $\sqrt{9} = +3$, $\sqrt[3]{125} = +5$, and $\sqrt[4]{16} = +2$. Note that the square of -3 is also equal to 9, and the fourth power of -2 is equal to 16; however, by definition, $\sqrt{9}$ and $\sqrt[4]{16}$ refer to only the positive roots.

If n is odd and $a > 0$, there is only one real nth root of a, and it is positive. If n is odd and $a < 0$, we can define $\sqrt[n]{a}$ to be the unique real negative number whose nth power is a. For example, $\sqrt[3]{-125} = -5$.

It is often useful to express the root of a number by means of exponents. If n is a positive integer, and $a \in \mathbb{R}$, then we write a^n to stand for $a \cdot a \cdots a$ (n factors) and a^{-n} to stand for $1/a^n$. We also define a^0 to be 1, for all nonzero a. If $a, b \in \mathbb{R}$ and $m, n \in \mathbb{Z}$, then these exponents have the following properties.

$$
\begin{aligned}
a^m \cdot a^n &= a^{m+n} \\
(a^m)^n &= a^{mn} \\
(ab)^m &= a^m b^m
\end{aligned}
$$

These properties can be proved by induction for positive m and n and then proved for negative m and n.

We can extend the definition to fractional exponents. If n is a positive integer, and $a \geq 0$, we would like $x = a^{1/n}$ to satisfy the second exponential law above, so that $x^n = \left(a^{1/n}\right)^n = a$. Hence it is natural to define $a^{1/n}$ to be $\sqrt[n]{a}$.

Definition.   If $a \geq 0$, m is an integer, and n is a positive integer, define

$$
a^{m/n} = \left(\sqrt[n]{a}\right)^m.
$$

For example, $27^{2/3} = (\sqrt[3]{27})^2 = 3^2 = 9$ and $4^{3/4} = (\sqrt[4]{4})^3 = (\sqrt{2})^3 = 2\sqrt{2}$.

Since a rational number is an equivalence class, we have to check that the definition of a^r for rational r does not depend on how we write r. For example, we would want $27^{2/3}$ to equal $27^{6/9}$.

Lemma 5.31.   *If $a \geq 0$, m is an integer, and n is a positive integer, then*

$$
a^{m/n} = \sqrt[n]{a^m}.
$$

Proof.   Let $x = a^{m/n}$ so that

$$
x^n = \left(\left(\sqrt[n]{a}\right)^m\right)^n = \left(\sqrt[n]{a}\right)^{mn} = \left(\left(\sqrt[n]{a}\right)^n\right)^m = a^m.
$$

Since a^m has only one positive nth root, $x = \sqrt[n]{a^m}$. $\qquad\square$

Proposition 5.32.   *If $m/n = r/s$ with n and s positive integers, and $a \geq 0$, then*

$$a^{m/n} \;=\; a^{r/s}.$$

Proof.   Let $x = a^{m/n}$ so that, by Lemma 5.31, $x^n = a^m$. Now $x^{ns} = a^{ms} = a^{nr}$ and $(x^s)^n = (a^r)^n$. Since a positive number has only one positive nth root, $x^s = a^r$. Therefore, $x = \sqrt[s]{a^r}$ and, by Lemma 5.31, $x = a^{r/s}$. $\square$

With the above definition of rational exponents, it can be shown that the exponential laws

$$
\begin{aligned}
a^r \cdot a^s &= a^{r+s} \\
(a^r)^s &= a^{rs} \\
(ab)^r &= a^r b^r
\end{aligned}
$$

still hold when $a, b \in \mathbb{R}$, $a \geq 0$, $b \geq 0$, and $r, s \in \mathbb{Q}$.

Example 5.33.   Simplify $125^{-\frac{2}{3}} + 8^{\frac{7}{6}}$.

Solution.

$$
\begin{aligned}
125^{-\frac{2}{3}} + 8^{\frac{7}{6}} &= (5^3)^{-\frac{2}{3}} + (2^3)^{\frac{7}{6}} = 5^{-2} + 2^{\frac{7}{2}} \\
&= \frac{1}{5^2} + 2^3 \cdot 2^{\frac{1}{2}} = \frac{1}{25} + 8\sqrt{2} \qquad \square
\end{aligned}
$$

Example 5.34.   Expand $(x^{\frac{1}{3}} + \sqrt{2}y^{\frac{1}{6}})^3$.

Solution.   By the Binomial Theorem,

$$
\begin{aligned}
(x^{\frac{1}{3}} + \sqrt{2}y^{\frac{1}{6}})^3 &= (x^{\frac{1}{3}})^3 + 3(x^{\frac{1}{3}})^2(\sqrt{2}y^{\frac{1}{6}}) + 3(x^{\frac{1}{3}})(\sqrt{2}y^{\frac{1}{6}})^2 + (\sqrt{2}y^{\frac{1}{6}})^3 \\
&= x + 3\sqrt{2}x^{\frac{2}{3}}y^{\frac{1}{6}} + 6x^{\frac{1}{3}}y^{\frac{1}{3}} + 2\sqrt{2}y^{\frac{1}{2}}. \qquad \square
\end{aligned}
$$

Example 5.35.   Simplify $\dfrac{2\sqrt{7} - \sqrt{5}}{\sqrt{7} + \sqrt{5}}$.

Solution.   Any expression with a denominator of the form $(\sqrt{a} + \sqrt{b})$ can be simplified by a process called *rationalizing the denominator*. Since

$$(\sqrt{a} + \sqrt{b})(\sqrt{a} - \sqrt{b}) \;=\; (\sqrt{a})^2 - (\sqrt{b})^2 \;=\; a - b$$

we can multiply numerator and denominator by $(\sqrt{a} - \sqrt{b})$ to eliminate the square roots in the denominator.

Hence

$$\frac{2\sqrt{7} - \sqrt{5}}{\sqrt{7} + \sqrt{5}} = \frac{(2\sqrt{7} - \sqrt{5})(\sqrt{7} - \sqrt{5})}{(\sqrt{7} + \sqrt{5})(\sqrt{7} - \sqrt{5})} = \frac{14 - 3\sqrt{35} + 5}{7 - 5} = \frac{19 - 3\sqrt{35}}{2}. \qquad \square$$

5.4   DECIMAL EXPANSIONS

We saw in Section 2.4 that all the positive integers can be written in base 10 using positive powers of 10 and the digits $\{0, 1, 2, 3, \ldots, 9\}$. In a similar way, we can write certain rational numbers as finite decimals, using positive and *negative* powers of 10. For example, the decimal **3905.6402** represents the number

$$\mathbf{3} \times 10^3 + \mathbf{9} \times 10^2 + \mathbf{0} \times 10^1 + \mathbf{5} \times 10^0 + \mathbf{6} \times 10^{-1} + \mathbf{4} \times 10^{-2} + \mathbf{0} \times 10^{-3} + \mathbf{2} \times 10^{-4}.$$

All such finite decimal expansions represent rational numbers whose denominator can be written as some power of 10. For example,

$$2.52 \;=\; 2 + 5 \times 10^{-1} + 2 \times 10^{-2} \;=\; 2 + \frac{5}{10} + \frac{2}{100} \;=\; \frac{252}{100} \;=\; \frac{63}{25}$$

and $3905.6402 = 39056402/10^4$. Therefore, we would only expect to be able to represent in this way fractions that, when written in their lowest terms, have denominators that divide some power of 10.

However, we can still use the idea of decimals to represent all real numbers; but we now have to allow infinite decimal expansions, whose successive finite terms provide better and better approximations to the real number.

Suppose that we wish to find the decimal expansion of a positive real number r lying between 0 and 1.

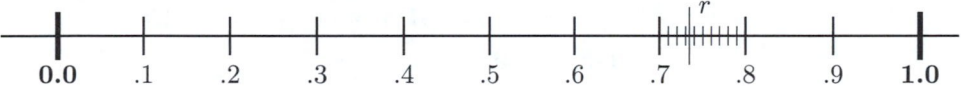

First subdivide the unit interval into 10 equal parts. The number will lie in one of these intervals. Say r lies in the eighth interval so that $0.7 \leq r \leq 0.8$. Subdivide this eighth interval into 10 equal parts, each of length 10^2. Suppose r lies in the fourth such interval so that $0.73 \leq r \leq 0.74$. Proceeding in this way, we obtain an infinite sequence of digits $a_1, a_2, a_3, \ldots$ such that, for each n,

$$0.a_1 a_2 a_3 \ldots a_{n-1} a_n \leq r \leq 0.a_1 a_2 a_3 \ldots a_{n-1} \tilde{a}_n,$$

where $\tilde{a}_n = a_n + 1$, with the appropriate modification when $a_n = 9$.

Definition.   The expression $b.a_1 a_2 \ldots$, where b is a base 10 integer and each a_i is a digit from 0 to 9, is called the **decimal expansion** of the real number r if

$$b.a_1 a_2 \ldots a_n \leq r \leq (b.a_1 a_2 \ldots a_n) + 10^{-n} \quad \text{for all } n \in \mathbb{P}.$$

Decimal expansions of negative numbers can be obtained by placing a minus sign before the expansion.

For example, the decimal expansion of $\sqrt{2}$ begins $1.4142\ldots$ because

$$(1.4142)^2 = 1.99996164 < 2 < 2.00024449 = (1.4143)^2$$

and so
$$1.4142 < \sqrt{2} < 1.4143.$$

A decimal $b.a_1a_2\ldots$ is called *terminating* if there exists an integer n such that $a_i = 0$ for all $i \geq n$. A decimal is called *recurrent* or *periodic* if there exist positive integers p and n such that $a_{i+p} = a_i$ for all $i \geq n$. The smallest such positive integer p is called the *period* of the expansion. Note that a terminating decimal is also a periodic decimal, with period 1.

Certain numbers have two different decimal expansions. For example, the decimal expansion of $1/4$ can be written as $0.2500000\ldots$ and as $0.2499999\ldots$. From our discussion on approximating a real number by a decimal, we see that, when we subdivide an interval into 10 subintervals, it may happen that our real number lies in two subintervals. In that case, there is a choice for the decimal expansion. For example, when the interval from 0.2 to 0.3 is subdivided into 10 equal parts we see that the number $1/4$ lies in the fifth *and* sixth subdivision; hence the first two places of the expansion of $1/4$ are 0.24 *or* 0.25.

It is well known that we can find a decimal expansion of a rational number by a process of long division. For example, $1/11$ has the decimal expansion $0.090909\ldots$, which we shall write as $0.\dot{0}\dot{9}$, where the dots over a string of digits indicate that these digits are to be repeated indefinitely. We also have

$$
\begin{aligned}
1/3 &= 0.333\ldots = 0.\dot{3} \\
1/7 &= 0.1428571428571\ldots = 0.\dot{1}4285\dot{7} \\
1/8 &= 0.125000\ldots = 0.125 \\
32/15 &= 2.1333\ldots = 2.1\dot{3}.
\end{aligned}
$$

We shall show that a rational number always has a periodic (or terminating) decimal expansion. The converse is also true; a periodic expansion always represents a rational number. The following examples show how to find that rational number.

It is possible to justify the addition and subtraction of infinite decimals, and their multiplication by an integer, using infinite series.

Example 5.41.   Find the rational number with the decimal expansion $0.\dot{1}\dot{7}$.

Solution.   Let $x = 0.\dot{1}\dot{7}$. This has period of length 2, so multiply it by 10^2 to obtain $100x = 17.\dot{1}\dot{7}$. Subtract to remove the repeating part.

$$
\begin{aligned}
100x - x &= 17.\dot{1}\dot{7} - 0.\dot{1}\dot{7} \\
99x &= 17 \\
x &= \frac{17}{99}
\end{aligned}
$$

Check.   If we divide 17 by 99, using long division, we see that $17/99 = 0.\dot{1}\dot{7}$. Alternatively, use a calculator to compute $17/99 \approx 0.171717$. $\square$

Example 5.42. Find the rational number whose decimal expansion is $1.1\dot{2}1\dot{6}$.

Solution. Let $x = 1.1\dot{2}1\dot{6}$ so that $1000x = 1121.6\dot{2}1\dot{6}$. Hence

$$
\begin{aligned}
1000x - x &= 1121.6\dot{2}1\dot{6} - 1.1\dot{2}1\dot{6} = 1120.5 \\
999x &= \frac{2241}{2} \\
x &= \frac{2241}{2 \cdot 999} = \frac{249}{2 \cdot 111} = \frac{83}{2 \cdot 37} = \frac{83}{74}.
\end{aligned}
$$

Check. In Example 5.44 we divide 83 by 74 to obtain the expansion $1.1\dot{2}1\dot{6}$. □

Theorem 5.43. *A real number is rational if and only if its decimal expansion is periodic (or terminating). Hence a real number is irrational if and only if it has a nonperiodic infinite decimal expansion.*

Proof. Let us look at the process of long division to find the decimal expansion of the rational number m/n, where $m, n \in \mathbb{P}$.

We first divide m by n and obtain a remainder r_0. We then divide $10r_0$ by n to obtain a remainder r_1 and continue in the following manner.

$$
\begin{aligned}
m &= bn + r_0, & \text{where} \quad 0 \le r_0 < n \\
10r_0 &= a_1 n + r_1, & \text{where} \quad 0 \le r_1 < n \\
10r_1 &= a_2 n + r_2, & \text{where} \quad 0 \le r_2 < n \\
&\;\;\vdots \\
10r_{i-1} &= a_i n + r_i, & \text{where} \quad 0 \le r_i < n \\
&\;\;\vdots
\end{aligned}
$$

The decimal expansion of m/n is $b.a_1 a_2 \ldots$. First notice that $0 \le a_i < 10$ for all $i \in \mathbb{P}$. This follows because $0 \le r_{i-1} < n$, and so $0 \le 10r_{i-1} < 10n$. Now we have

$$
\begin{aligned}
\frac{m}{n} &= b + \frac{r_0}{n} \\
&= b + \frac{a_1}{10} + \frac{r_1}{10n} \\
&= b + \frac{a_1}{10} + \frac{a_2}{10^2} + \frac{r_2}{10^2 n}
\end{aligned}
$$

and, by induction, we can show that

$$
\frac{m}{n} = b + \frac{a_1}{10} + \frac{a_2}{10^2} + \cdots + \frac{a_i}{10^i} + \frac{r_i}{10^i n}
$$

for all $i \in \mathbb{P}$. Since $0 \le r_i < n$, it follows that m/n lies between $b.a_1 a_2 \ldots a_i$ and $(b.a_1 a_2 \ldots a_i) + 10^{-i}$.

Each remainder, r_i, must be one of the n integers $0, 1, 2, \ldots, n-1$ and therefore, after at most n steps, one of the remainders must equal an earlier remainder; say $r_{j+p} = r_j$. Thereafter, the computations just repeat themselves and $a_{i+p} = a_i$

for all $i \geq j$. This shows that the decimal expansion of any rational number is periodic.

Now suppose that we have a periodic decimal expansion $x = b.a_1 a_2 \ldots$, where $a_{i+p} = a_i$ for all $i \geq j$. Then $(10^p - 1)x$ will be a terminating decimal represented by some rational number $k/10^\ell$. Therefore, x will be represented by the rational number

$$\frac{k}{10^\ell (10^p - 1)}. \qquad \square$$

Example 5.44. Illustrate the previous theorem by finding the decimal expansion of 83/74.

Solution. Write out the division of 83 by 74 as in the previous theorem.

$$
\begin{array}{r}
1.1\,2\,1\,6\,2\,1\,6\ldots \\
74\overline{\smash{)}\,8\,3.0\,0\,0\,0\,0\,0\,0\ldots} \\
7\,4 \\
\hline
9\,0 \\
7\,4 \\
\hline
1\,6\,0 \\
1\,4\,8 \\
\hline
1\,2\,0 \\
7\,4 \\
\hline
4\,6\,0 \\
4\,4\,4 \\
\hline
1\,6\,0 \\
1\,4\,8 \\
\hline
1\,2\,0
\end{array}
$$

$83 = 1 \cdot 74 + \mathbf{9}$

$90 = 1 \cdot 74 + \mathbf{16}$ ⟷

$160 = 2 \cdot 74 + \mathbf{12}$ *repeating*

$120 = 1 \cdot 74 + \mathbf{46}$ *remainders*

$460 = 6 \cdot 74 + \mathbf{16}$ ⟷

$160 = 2 \cdot 74 + \mathbf{12}$

The decimal expansion of 83/74 is therefore $1.1\dot{2}1\dot{6}$.

Check. This agrees with Example 5.42. $\qquad \square$

If you are familiar with infinite series, the decimal expansion $b.a_1 a_2 \ldots$ can be interpreted as the infinite series

$$b + \frac{a_1}{10} + \frac{a_2}{10^2} + \cdots + \frac{a_r}{10^r} + \cdots .$$

This series always converges if $0 \leq a_r \leq 9$. (Why?)

For example, $0.\dot{3}$ is the infinite geometric series

$$\frac{3}{10} + \frac{3}{10^2} + \frac{3}{10^3} + \cdots = \frac{3}{10}\left(1 + \frac{1}{10} + \frac{1}{10^2} + \cdots\right)$$

whose sum is

$$\frac{3}{10}\left(\frac{1}{1 - \frac{1}{10}}\right) = \frac{3}{10} \cdot \frac{10}{9} = \frac{1}{3}.$$

All infinite periodic decimals can be written as an infinite geometric series, after a certain point.

Exercise Set 5

1. Verify that the relation $\sim$ on $\mathbb{Z} \times (\mathbb{Z} - \{0\})$, defined by $(a, b) \sim (c, d)$ if and only if $ad = bc$, is indeed an equivalence relation.

2. Verify that addition, subtraction, multiplication, and division are all well defined in $\mathbb{Q}$. For example, for addition, it has to be verified that if $(a_1, b_1) \sim (a_2, b_2)$ and $(c_1, d_1) \sim (c_2, d_2)$, then $(a_1 d_1 + b_1 c_1, b_1 d_1) \sim (a_2 d_2 + b_2 c_2, b_2 d_2)$.

3. Prove that $\sqrt{3}$ is irrational.

4. Prove that $\sqrt[3]{4}$ is irrational.

5. Prove that $\sqrt{6}$ is irrational.

6. Is $\sqrt{2} + \sqrt{3}$ rational or irrational? Give reasons.

7. If a is rational and b is irrational, prove that $a + b$ is irrational.

8. If a is rational and b is irrational, prove that ab is irrational, except for one case. What is the exceptional case?

9. Define a relation $\perp$ on the real numbers $\mathbb{R}$ by $a \perp b$ if and only if $a - b \in \mathbb{Q}$. Prove that this is an equivalence relation. Which of the following are related?

    (a) 3 and 4/5
    (b) $1 + \sqrt{5}$ and $1 - \sqrt{5}$
    (c) $\sqrt{2}$ and $\sqrt{2} + 1$
    (d) $\sqrt{2}$ and $\sqrt{3}$
    (e) $\dfrac{1}{\sqrt{2}}$ and $\dfrac{\sqrt{2} - 1}{2}$

10. If a is a positive rational number, let $\tilde{a} = \frac{2+a}{1+a}$. Prove that $\tilde{a}^2$ is closer to 2 then a^2 is. By starting with $a = 1$, use this to find a sequence of rational numbers approximating $\sqrt{2}$.

11. The amount of light that enters a camera is determined by the f-number of the aperture. An aperture of $f/8$ means that the effective diameter of the lens is $1/8$ of its focal length, and larger f-numbers mean smaller lens openings. The exposure time increases with the square of the f-number. Manual cameras are usually marked with a sequence of f-numbers, so that adjacent f-numbers let in amounts of light that differ by a factor of 2. Hence when the f-number is changed to the next higher number, the exposure time has to be doubled. If the largest aperture of a camera is $f/2$ and the smallest $f/16$, find the sequence of intermediate f-numbers that is needed.

12. The international paper size A0 has an area of 1 square meter and is such that when cut in half, the ratio of the long side to the short side remains unchanged. What is the ratio of the long side to the short side, and what are the dimensions of the A0 paper (to the nearest millimeter.). The standard A4 size typing paper, which is used in Europe, is obtained by cutting A0 into half four times. What are its dimensions?

13–19. Simplify the following.

13. $\sqrt{2^8}$

14. $(49)^{1/4}$

15. $\left(\dfrac{1}{81}\right)^{-3/4}$

16. $\dfrac{7^{-2/5} \cdot 7^{1/2}}{7^{1/12}}$

17. $\dfrac{2\sqrt{3} + 3\sqrt{2}}{\sqrt{3} + \sqrt{2}}$

18. $\dfrac{\sqrt{5}}{3 - \sqrt{5}}$

19. $\sqrt{\sqrt{2}} \cdot \sqrt{8\sqrt{2}}$

20. Rationalize the denominator of $\dfrac{1}{\sqrt{2} + \sqrt{3} + \sqrt{5}}$.

21–22. Simplify the following, assuming $a > 0$.

21. $a^{5/4}(3a^2 - a^{1/4})$

22. $(a^{2/3} + a^{1/2})^3$

23–26. Express the following rational numbers as periodic (or terminating) decimals. Which of these numbers have two different decimal expansions?

23. $1/12$

24. $3/16$

25. $7/40$

26. $5/19$

27–32. Express the following decimals as rational numbers in their lowest terms.

27. 2.105

28. $0.\dot{4}\dot{2}$

29. $0.4\dot{2}$

30. $0.7\dot{6}92\dot{3}0$

31. $0.13\dot{1}6\dot{2}$

32. $1.50\dot{5}\dot{1}$

Problem Set 5

33. We have shown how to construct the rational numbers from the integers by means of an equivalence relation on $\mathbb{Z} \times (\mathbb{Z} - \{0\})$. Show how the integers could be constructed from the positive integers $\mathbb{P}$ by means of an equivalence relation on $\mathbb{P} \times \mathbb{P}$.

34. Could we have introduced positive rational numbers by means of an equivalence relation on $\mathbb{P} \times \mathbb{P}$ before introducing negative numbers?

35. **(a)** Find irrational numbers a and b with $a + b$ rational.

   (b) Find irrational numbers a and b with ab rational.

   (c) The next chapter will indicate how irrational exponents could be defined. Could there be irrational numbers a and b, with a^b rational?

36. Let $a, b, c, d \in \mathbb{Q}$, where $\sqrt{b}$ and $\sqrt{d}$ exist and are irrational. If $a+\sqrt{b} = c+\sqrt{d}$, prove that $a = c$ and $b = d$.

37.   (a) If $m, n \in \mathbb{Z}$ and $\gcd(m, n) = 1$, prove that $\gcd(m^2, n^2) = 1$.

   (b) If $r \in \mathbb{Q}$ and $r^2 \in \mathbb{Z}$, prove that $r \in \mathbb{Z}$.

   (c) Prove that $\sqrt{p}$ is irrational whenever p is prime.

38. Resolve the following contradiction.

$$-1 \; = \; \sqrt[3]{-1} \; = \; (-1)^{1/3} = (-1)^{2/6} \; = \; \sqrt[6]{(-1)^2} \; = \; \sqrt[6]{1} \; = \; 1$$

39. Rationalize the denominator of $\dfrac{\sqrt[3]{12} + 1}{\sqrt[3]{3} - \sqrt[3]{2}}$.

40. When does a/b have a terminating decimal expansion?

41.   (a) Use Fermat's Little Theorem 3.42 to show that, for every prime p other than 2 or 5, there is some positive integer r for which $p|(10^r - 1)$.

   (b) Is it true that, for all integers n, other than multiples of 2 and 5, there is some positive integer r for which $n|(10^r - 1)$?

   (c) What is the relationship between these questions and decimal expansions?

42. Let $\gcd(10, n) = 1$, and let r be the smallest positive integer for which $10^r \equiv 1 \pmod{n}$.

   (a) Prove that $1/n$ has a recurring decimal expansion with period r.

   (b) If n is prime, prove that $r|(n - 1)$.

   (c) Find the periods of $1/13$, $1/17$, $2/31$ and $1/47$.

43. Find a method for converting a rational number m/n into an expansion in base 6. Use your method to expand $1/4$ in base 6.

44. Find the expansions of $1/7$ in (a) base 10, (b) base 9, (c) base 8, and (d) base 7.

45. Convert $(0.125)_6$ from base 6 to a decimal in base 10.

46. Calculate π in base 2 correct to 6 binary places, after the point.

47. Convert the decimal 0.15 to base 3.

48. Calculate the first 5 places of the decimal number 0.241 in base 8. Will the expansion terminate, repeat indefinitely, or be nonperiodic?

49. Convert $(2.4\dot{6}\dot{7})_8$ from base 8 to a rational number in base 10.

50. Convert $(0.11\dot{0}1\dot{0})_2$ from base 2 to a rational number in base 10.

51. Let $x_1 = \sqrt{44}$ and $x_{n+1} = \sqrt{3x_n + 1}$, for $n \geq 1$. Prove that x_n is irrational for every $n \geq 1$.

52. Find three positive fractions with denominators 5, 7, and 9 whose sum is $481/315$.

CHAPTER 6

Functions and Bijections

6.1 FUNCTIONS

One of the basic notions of modern mathematics is that of a function. To a mathematician in 1800, a function referred to an algebraic or trigonometrical formula involving one or more variables. Numbers could be substituted for each of the variables and the resulting numerical value of the function could be calculated from the given formula. Although such functions are still very important today, it became necessary to broaden the scope of a function to include relationships that could not be expressed by a simple formula and to allow variables that were not necessarily numbers.

Definition. Let X and Y be sets. A **function**, f, from X to Y is denoted by $f : X \rightarrow Y$ and is a rule that assigns to each element $x \in X$ a unique element $f(x) \in Y$.

The set X is called the **domain** of f, while the set Y is called the **codomain** of f. The element $f(x) \in Y$ is called the *value* of the function f at x.

Whenever a specific element is assigned to the variable x, the corresponding value of $y = f(x)$ is determined by the function; hence x is called the *independent variable* and y is called the *dependent variable*.

For example, if f is the function from the real numbers, $\mathbb{R}$, to the real numbers in which f assigns to each real number its cube, we would say that $f : \mathbb{R} \rightarrow \mathbb{R}$ was defined by $f(x) = x^3$. The value of f at 3 would be 27, while the value of f at $-\sqrt{2}$ would be $-2\sqrt{2}$.

It often happens that the set of values of a function is not the whole codomain, but a proper subset of it. The set of values of a function $f : X \rightarrow Y$ is called the **image** of f and is denoted by $f(X)$. It is a subset of the codomain Y and

$$f(X) = \{f(x) \mid x \in X\} \subseteq Y.$$

The term *range* is also used to mean the image. However, since many authors refer to the codomain as the range, we shall avoid its use.

A telephone directory assigns to each name a telephone number and, assuming that no person has two telephone numbers, it can be considered a function whose domain is the set of subscribers and whose codomain is the set of possible telephone numbers. The image consists of those telephone numbers in actual use.

Let X be the set of all human beings who have ever lived. If x is any person, let $m(x)$ be the mother of x. Then $m : X \to X$ is a function because, assuming that applied genetics has not advanced too far, each person has a unique mother. The image of this function, $m(X)$, consists of the set of all mothers.

Notice however, that the rule $c(x) =$ "child of x" would *not* define a function from X to X, for two reasons. Some parents have many children, while some people have no children. Therefore, for child-free people, c is not defined, while c does not assign a unique element to parents who have more than one child.

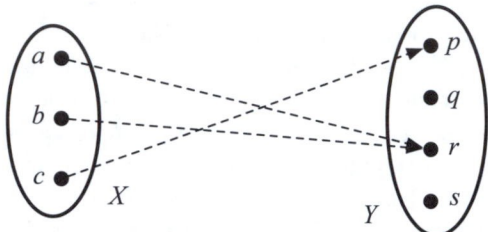

The diagram defines a function

$$f : X \longrightarrow Y$$

from the finite set $X = \{a, b, c\}$ to the finite set $Y = \{p, q, r, s\}$. Here $f(a) = r$, $f(b) = r$ and $f(c) = p$. The codomain is $Y = \{p, q, r, s\}$ but the image is $f(X) = \{p, r\}$.

A table of squares and cubes of integers from 1 to 100 defines a function f whose domain is the set of $\{1, 2, \ldots, 100\}$ and whose codomain is $\mathbb{P} \times \mathbb{P}$. The value of the function at the integer n is the pair of positive integers (n^2, n^3). For example, $f(1) = (1, 1)$, $f(2) = (4, 8)$, $f(3) = (9, 27)$.

The functions in elementary calculus usually have domains and codomains that are subsets of the real numbers. Such functions are called real-valued functions of a real variable. However, in calculus, it is customary not to refer explicitly to the domain or codomain of such functions. The domain of such a function is usually taken to be the largest subset of $\mathbb{R}$ for which the given formula makes sense (that is, gives a real value). For example, if

$$f(x) \;=\; \frac{\sqrt{x - 2}}{x - 4},$$

then the domain of f cannot be the whole of $\mathbb{R}$, because $\sqrt{x - 2}$ is not real if $x < 2$, and $f(x)$ is not defined when $x = 4$. We would take the domain of f to be $X = \{x \in \mathbb{R} \mid x \geq 2, x \neq 4\}$, which is the union of the two intervals $\{x \mid 2 \leq x < 4\}$ and $\{x \mid x > 4\}$. The codomain of a real valued function of a real variable is not very crucial (unless one is dealing with the inverse of the function) and it can usually be taken to be the whole real line $\mathbb{R}$.

Definition.   Two functions f and g are said to be *equal* if they have the same domains, the same codomains and, for every element x in the domain, $f(x) = g(x)$.

For example, $f : \mathbb{R} \to \mathbb{R}$ defined by $f(x) = x^2$, and $g : \mathbb{Z} \to \mathbb{Z}$ defined by $g(x) = x^2$, are not considered equal functions, even though they have the same formula.

However, the functions $f : \mathbb{P} \to \mathbb{P}$, defined by $f(n) = \operatorname{lcm}(2, n)$, and $g : \mathbb{P} \to \mathbb{P}$, defined by $g(n) = (3 - (-1)^n)n/2$, are equal functions because they have the same domains and codomains, and $f(n) = g(n)$ for all $n \in \mathbb{P}$.

There are several synonyms for the word function, the most common being *mapping*, *transformation* and *correspondence*.

6.2   THE GRAPH OF A FUNCTION

The reader will undoubtedly have drawn the graph of a real valued function such as $f : \mathbb{R} \to \mathbb{R}$, when $f(x) = x^2$. This is constructed by plotting the points (x, y) in the plane $\mathbb{R} \times \mathbb{R}$, where $y = f(x) = x^2$. We can, in a similar way, define the graph of a general function. However, we will not always be able to represent such a graph on a two dimensional piece of paper.

Definition.   The **graph** of the function $f : X \to Y$ is the subset of the Cartesian product $X \times Y$ consisting of pairs $(x, f(x))$, for all $x \in X$. That is, the graph of f is the set $\{(x, f(x)) \in X \times Y \mid x \in X\}$.

Whenever f is a real valued function of a real variable, then X and Y are subsets of $\mathbb{R}$, and the preceding definition of the graph of f agrees with the usual notion of the graph as a subset of the plane $\mathbb{R} \times \mathbb{R}$.

For example, if $f : X \to \mathbb{R}$ is the function defined by

$$f(x) \;=\; \sqrt{x + 4}$$

where $X = \{x \in \mathbb{R} \mid x \geq -4\}$, then the graph of f is the subset of $X \times \mathbb{R}$ consisting of

$$\{(x, \sqrt{x + 4}) \mid x \geq -4\}.$$

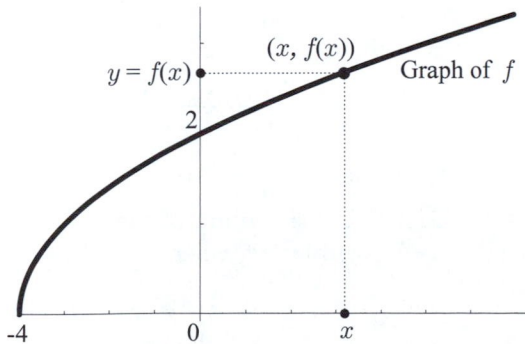

It is customary for the horizontal axis in a graph to represent the domain, and the vertical axis to represent the codomain. A point y, in the codomain, lies in the image of the function if the horizontal line through y intersects the graph. The image is therefore the projection of the graph on the vertical axis.

In the case of the preceding function $f : X \to \mathbb{R}$, where $f(x) = \sqrt{x + 4}$, the projection of the graph onto the vertical axis is the set of nonnegative real numbers; that is, $f(X) = \{y \in \mathbb{R} \mid y \geq 0\}$.

The following illustration tries to depict the graph of the function $f : \mathbb{R} \to \mathbb{R}$ defined by

$$f(x) \;=\; \begin{cases} 1 & \text{if } x \text{ is rational} \\ x & \text{if } x \text{ is irrational.} \end{cases}$$

The image of f consists of the irrational numbers together with the point 1.

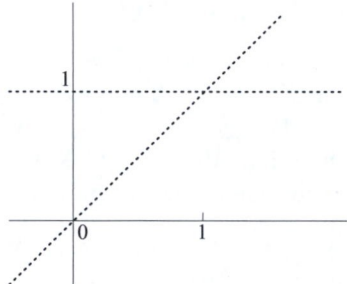

Example 6.21.   Illustrate the graph of the function $f : \mathbb{Z}_4 \to \mathbb{Z}_4$, defined by $f([x]) = [x^2]$, and find the image of f.

Solution.   There are four points in the domain and codomain, and $f([0]) = [0]$, $f([1]) = [1], f([2]) = [0]$ and $f([3]) = [1]$

The graph of f is a four element subset of the sixteen element set $\mathbb{Z}_4 \times \mathbb{Z}_4$. We can represent $\mathbb{Z}_4 \times \mathbb{Z}_4$ by the 4×4 array as shown. The graph of f consists of the solid dots.

The image of f is the set $\{[0], [1]\}$.

$$
\begin{array}{c|cccc}
[3] & \circ & \circ & \circ & \circ \\
[2] & \circ & \circ & \circ & \circ \\
[1] & \circ & \bullet & \circ & \bullet \\
[0] & \bullet & \circ & \bullet & \circ \\
\hline
 & [0] & [1] & [2] & [3]
\end{array}
$$

$\square$

Example 6.22.   Sketch the graph of the *absolute value function* from $\mathbb{R}$ to $\mathbb{R}$ whose value at x is denoted by $|x|$, where

$$|x| \;=\; \begin{cases} x & \text{if } x \geq 0 \\ -x & \text{if } x < 0. \end{cases}$$

Solution.

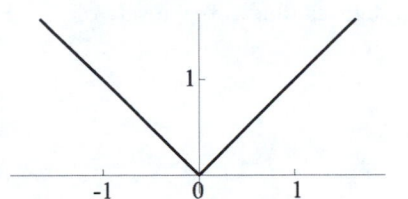

The graph of a function contains all the information about the function. In fact, the concept of a function may be defined by means of its graph. The following is an alternative definition of function, and it is more precise in the sense that it avoids the use of the undefined terms *rule* and *assigns*. However, it is not usually as conceptually useful as the former definition.

Definition.   A *function f* with domain X and codomain Y is a subset of $X \times Y$ in which each element of X occurs precisely once as the first element of an ordered pair in f.

Hence, using this definition, we can write

$$f \;=\; \{(x, f(x)) \in X \times Y \mid x \in X\}$$

though we shall usually still refer to this subset as the graph of f.

For example, the function $f : \mathbb{Z}_4 \to \mathbb{Z}_4$, in Example 6.21, can be written as

$$f \;=\; \{([0], [0]), ([1], [1]), ([2], [0]), ([3], [1])\}.$$

It is usually fairly easy to recognize which subsets of $X \times Y$ do define the graph of some function; each vertical line through a point of the domain must meet the graph precisely once. If there is a vertical line through a point of the domain that does not intercept the subset, or intercepts the subset in more than one point, then the subset cannot be a graph.

The figures below illustrate some graphs of functions from X to Y.

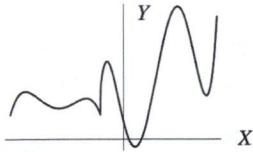 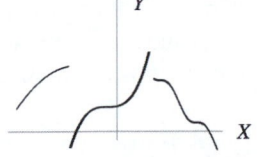 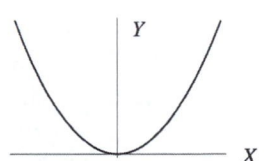

Subsets that Are Graphs

The following figures illustrate subsets of $X \times Y$ that are not the graph of any function from X to Y.

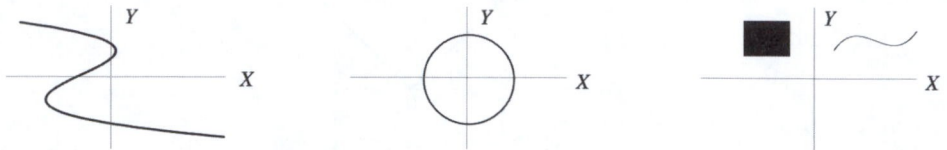

Subsets that Are Not Graphs

We now have various ways in which to describe a function. For example, the function $f : \mathbb{R} \to \mathbb{R}$ defined by $f(x) = x^3$ could also be written as

- $f : \mathbb{R} \longrightarrow \mathbb{R}$,   where $f(x) = y$ with $y = x^3$

- $f = \{(x, x^3) \mid x \in \mathbb{R}\} \subset \mathbb{R} \times \mathbb{R}$

- $f = \{(x, y) \mid y = x^3, x \in \mathbb{R}\} \subset \mathbb{R} \times \mathbb{R}$

- $f : \mathbb{R} \longrightarrow \mathbb{R}$,   where $x \mapsto x^3$, and "$\mapsto$" is a symbol pronounced "maps to."

6.3 COMPOSITION OF FUNCTIONS

If $f : X \to Y$ and $g : Y \to Z$ are two functions such that the codomain of f is equal to the domain of g, then we can form a new function

$$g \circ f : X \longrightarrow Z$$

called the *composite* of f and g, which is defined by

$$(g \circ f)(x) = g(f(x)).$$

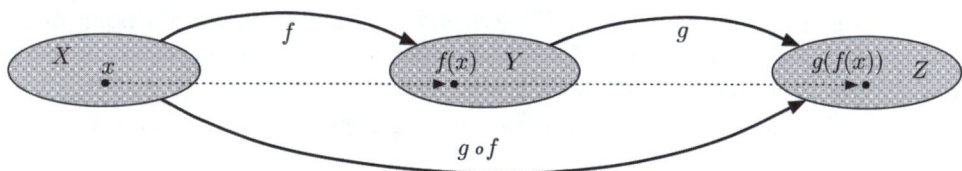

Observe the order of the composite $g \circ f$; the function f is applied first, and then g is applied to the result. It may seem rather odd for the composite to be written in this order, but it appears more natural when one looks at the definition $(g \circ f)(x) = g(f(x))$.

Two functions cannot always be composed; this can only happen in the special case in which the codomain of the first is equal to the domain of the second. Even if $g \circ f$ is defined, where $f : X \to Y$ and $g : Y \to Z$, then the composite in the reverse order, $f \circ g$, will not be defined unless $X = Z$.

Example 6.31. Let $f : X \to \mathbb{R}$ be defined by $f(x) = \sqrt{x}$, where the domain, X, is the set of nonnegative real numbers. Let $g : \mathbb{R} \to \mathbb{R}$ be defined by $g(x) = 3x - 1$. Determine whether the composites $g \circ f$ or $f \circ g$ are defined. If so, find them.

Solution. The codomain of f is equal to $\mathbb{R}$, the domain of g; hence

$$g \circ f : X \longrightarrow \mathbb{R}$$

is defined and $(g \circ f)(x) = g(f(x)) = g(\sqrt{x}) = 3\sqrt{x} - 1$. However, the codomain of g is not equal to the domain of f, so $f \circ g$ is not defined. $\square$

Example 6.32. Let $f : \mathbb{R} \to \mathbb{R}$ and $g : \mathbb{R} \to \mathbb{R}$ be functions defined by $f(x) = x^2$ and $g(x) = 4x - 5$. Are $g \circ f$ and $f \circ g$ defined? If so, find them.

Solution. Since the codomain of f and the domain of g are both equal to $\mathbb{R}$, and the codomain of g and the domain of f are also equal to $\mathbb{R}$, both composites are defined and are functions from $\mathbb{R}$ to $\mathbb{R}$. Now

$$(g \circ f)(x) \;=\; g(f(x)) \;=\; g(x^2) \;=\; 4x^2 - 5$$

while

$$(f \circ g)(x) \;=\; f(g(x)) \;=\; f(4x - 5) \;=\; (4x - 5)^2 \;=\; 16x^2 - 40x + 25. \qquad \square$$

Note that even if $g \circ f$ and $f \circ g$ are defined, they do not, in general, define equal functions. Thus the operation of composition is *not commutative*. However, whenever the composition of three functions is defined, the composition is *associative*.

Theorem 6.33. *Let $f : X \to Y$, $g : Y \to Z$ and $h : Z \to T$ be three functions. Then*

$$h \circ (g \circ f) \;=\; (h \circ g) \circ f : X \longrightarrow T.$$

In other words, the placing of the brackets is unimportant.

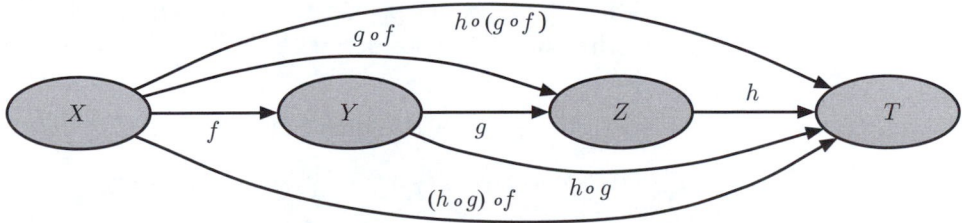

Proof. Notice that the functions $g \circ f$, $h \circ (g \circ f)$, $h \circ g$, and $(h \circ g) \circ f$ are all defined, and that $h \circ (g \circ f)$ and $(h \circ g) \circ f$ have the same domain and codomain.

If $x \in X$, then

$$
\begin{aligned}
[h \circ (g \circ f)](x) &= h(g \circ f(x)) &= h(g(f(x))) \\
[(h \circ g) \circ f](x) &= (h \circ g)(f(x)) &= h(g(f(x))).
\end{aligned}
$$

Hence $h \circ (g \circ f)$ and $(h \circ g) \circ f$ define equal functions. $\square$

For example, let X denote the set of married men and Y denote the set of all people. Let $w : X \to Y$ be the wife function, $m : Y \to Y$ the mother function and $f : Y \to Y$ the father function. Then $m \circ w$ is the mother-in-law' function and $f \circ m$ is the maternal grandfather function. By associativity, $f \circ (m \circ w) = (f \circ m) \circ w$; in other words, one's mother-in law's father is the same as one's wife's maternal grandfather.

Consider the functions

$$
\mathbb{R} \xrightarrow{f} \mathbb{R} \xrightarrow{g} \mathbb{R} \xrightarrow{h} \mathbb{R}
$$

where $f(x) = x + 1$, $g(x) = x^2$ and $h(x) = x - 1$. Then

$$
g \circ f(x) \;=\; g(f(x)) \;=\; g(x+1) \;=\; x^2 + 2x + 1
$$

so

$$
[h \circ (g \circ f)](x) \;=\; h(x^2 + 2x + 1) \;=\; (x^2 + 2x + 1) - 1 \;=\; x^2 + 2x.
$$

Now

$$
h \circ g(x) \;=\; h(g(x)) \;=\; h(x^2) \;=\; x^2 - 1
$$

so

$$
[(h \circ g) \circ f](x) \;=\; (h \circ g)(x+1) \;=\; (x+1)^2 - 1 \;=\; x^2 + 2x.
$$

If X is any set, there is always a function from X to itself that leaves each element unchanged. This is called the *identity function* on X and is denoted by $1_X : X \to X$. It is defined by $1_X(x) = x$ for all $x \in X$.

Proposition 6.34. *If $f : X \to Y$ is any function, and $1_X : X \to X$ and $1_Y : Y \to Y$ are the identity functions on X and Y, respectively, then $f \circ 1_X = f$ and $1_Y \circ f = f$. In other words, composition with the identity leaves any function unchanged.*

Proof. The function $f \circ 1_X : X \to Y$ is defined by

$$
f \circ 1_X(x) \;=\; f(1_X(x)) \;=\; f(x) \quad \text{for all } x \in X.
$$

Therefore, $f \circ 1_X = f$ and, similarly, $1_Y \circ f = f : X \to Y$. $\square$

6.4   INVERSE FUNCTIONS

A function $f : X \to Y$ is a rule that assigns to each element in X, an element in Y. If $g : Y \to X$ is a function that undoes what f does, then g is called the inverse function to f. For example, the function $f : \mathbb{R} \to \mathbb{R}$ defined by $f(x) = x^3$, cubes every number. The function $g : \mathbb{R} \to \mathbb{R}$ defined by $g(x) = \sqrt[3]{x}$ takes the cube root of every number, or uncubes the number, and hence g is the inverse function of f.

What is the precise meaning of undoes? A function $g : Y \to X$ undoes the function $f : X \to Y$ if the composite $g \circ f : X \to X$ restores each element to its original position; that is, if $g \circ f$ is the identity function on X. If g is the inverse of f, then f is the inverse of g. This suggests the following definition.

Definition.   If $f : X \to Y$ and $g : Y \to X$ are functions such that

$$g \circ f = 1_X \quad \text{and} \quad f \circ g = 1_Y,$$

then g is called the **inverse** to f. We write this as $g = f^{-1}$.

The domain of the original function f becomes the codomain of its inverse f^{-1}, while the codomain of f becomes the domain of f^{-1}.

Note that the notation f^{-1} refers to the inverse of the function f, not to its reciprocal. If $f : \mathbb{R} \to \mathbb{R}$ is defined by $f(x) = x^3$, then $f^{-1} : \mathbb{R} \to \mathbb{R}$ is defined by $f^{-1}(x) = \sqrt[3]{x}$ and *not* by $f^{-1}(x) = \frac{1}{x^3}$.

If x is in the domain of the function f, and $y = f(x)$, then

$$x \;=\; 1_X(x) \;=\; f^{-1} \circ f(x) \;=\; f^{-1}(f(x)) \;=\; f^{-1}(y).$$

Similarly, if $x = f^{-1}(y)$, then $y = f(x)$. Hence $y = f(x)$ if and only if $x = f^{-1}(y)$.

Therefore, if the equation $y = f(x)$ defines a function that has an inverse, the formula for the inverse function, $y = f^{-1}(x)$, can be found by

(i)   interchanging x and y in the given equation to obtain $x = f(y)$, and

(ii)   solving the resulting equation for y in terms of x.

There is no real necessity to interchange x and y. We could just as well solve $y = f(x)$ for x in terms of y to obtain the equation $x = f^{-1}(y)$. The reason many people like to interchange x and y is to keep x as the independent variable in the inverse function.

By no means do all functions have inverses. For example, if $f : \mathbb{R} \to \mathbb{R}$ is defined by $f(x) = x^2$, then $f(2) = 4$ and $f(-2) = 4$. This can have no inverse because there is no function that will undo f by sending 4 to 2, and to -2. If we tried to solve the equation $y = x^2$ for x in terms of y, we would obtain $x = \pm\sqrt{y}$. This does not describe a function, because the right side yields two values whenever y is positive, and no real value if y is negative.

Example 6.41. Find, if possible, the inverse of the function $f : \mathbb{Z} \to \mathbb{Z}$ defined by $f(x) = x + 2$.

Solution. Let $y = x + 2$. Interchanging x and y, we have $x = y + 2$ and, solving for y in terms of x, we have $y = x - 2$. The inverse function is therefore $f^{-1} : \mathbb{Z} \to \mathbb{Z}$ defined by $f^{-1}(x) = x - 2$.

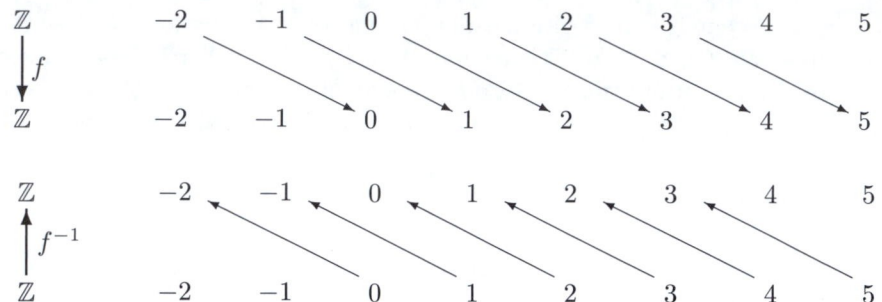

We check that $f(f^{-1}(x)) = f(x - 2) = x - 2 + 2 = x$ so that $f \circ f^{-1} = 1_{\mathbb{Z}}$. Similarly $f^{-1} \circ f = 1_{\mathbb{Z}}$. $\qquad\square$

Example 6.42. Does the function $f : X \to Y$ have an inverse, where $X = \{x_1, x_2, x_3\}$, $Y = \{y_1, y_2, y_3, y_4\}$, and $f(x_i) = y_i$, for $i = 1, 2, 3$?

Solution.

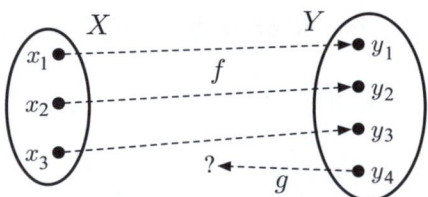

It is clear that if f did have an inverse, $g : Y \to X$, then g must send y_i to x_i for $i = 1, 2, 3$; but where should y_4 be sent to? Whatever the value of $g(y_4)$, $f \circ g(y_4)$ can never equal y_4, because y_4 is not in the image of f. Hence f cannot have an inverse. $\qquad\square$

There is a very close relationship between the graphs of a function $f : X \to Y$ and its inverse $f^{-1} : Y \to X$.

$$\begin{aligned}
\text{Graph of } f &= \{(x, y) \in X \times Y \mid y = f(x)\} \\
\text{Graph of } f^{-1} &= \{(y, x) \in Y \times X \mid x = f^{-1}(y)\} \\
&= \{(y, x) \in Y \times X \mid y = f(x)\}
\end{aligned}$$

The graph of f^{-1} is obtained from that of f by replacing (x, y) by (y, x); that is, by reflecting the graph in the line $y = x$.

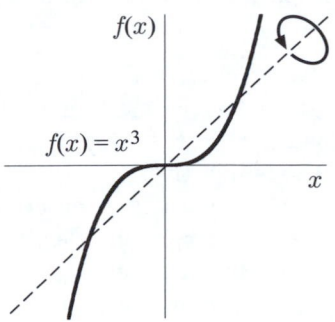

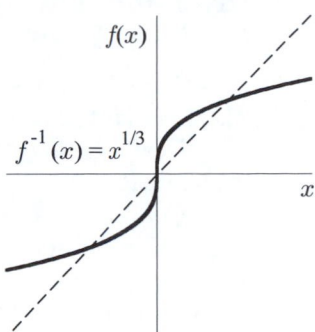

6.5   THE INVERSION THEOREM

We now lead up to the Inversion Theorem, which gives conditions for a function to have an inverse. We shall need to use the following important classes of functions.

Definition.   A function $f : X \to Y$ is called **injective** or **one-to-one** if $f(x_1) = f(x_2)$ implies that $x_1 = x_2$; that is, distinct elements in the domain must have distinct images. An injective function is called an **injection**.

A function $f : X \to Y$ is called **surjective** or **onto** whenever $f(X) = Y$; that is, whenever the image is the whole of the codomain. A surjective function is called a **surjection**.

A function $f : X \to Y$ is called **bijective** or a **one-to-one correspondence** if f is both injective and surjective. A bijective function is called a *bijection*.

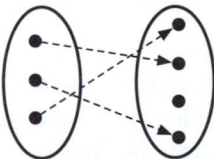   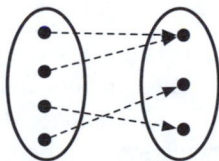   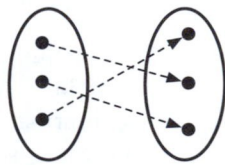

Injective Function   Surjective Function   Bijective Function

A function $f : X \to Y$ is injective if, for each element y in the codomain, there is *at most* one element x in the domain such that $y = f(x)$.

A function $f : X \to Y$ is surjective if, for each element y in the codomain, there is *at least* one element x in the domain such that $y = f(x)$.

A function $f : X \to Y$ is bijective if, for each element y in the codomain, there is *precisely* one element x in the domain, such that $y = f(x)$.

For any function, each element in the domain corresponds to precisely one element in the codomain. In a bijection, the reverse is also true and so there is a one-to-one correspondence between the elements in the domain and elements in the codomain. This correspondence allows us to define an inverse function.

Inversion Theorem 6.51.   *A function has an inverse if and only if the function is bijective.*

Proof.   Let $f : X \to Y$ be a function that has an inverse $f^{-1} : Y \to X$. Suppose $f(x_1) = f(x_2)$. Then $f^{-1}(f(x_1)) = f^{-1}(f(x_2))$ and $x_1 = x_2$. Hence f is injective.

Now let $y \in Y$ and let $x = f^{-1}(y)$. Then $f(x) = f(f^{-1}(y)) = y$, so that y is in the image of f. Hence f is surjective, and so is bijective.

Conversely, suppose $f : X \to Y$ is bijective. Since f is surjective, for any $y \in Y$, there exists $x \in X$ with $f(x) = y$. Moreover, if x_1 is any element of X with $f(x_1) = y$, then, because f is injective, $x_1 = x$. Hence, for each $y \in Y$ there is a unique element $x \in X$ with $f(x) = y$. If we denote x by $g(y)$, then this defines a function $g : Y \to X$.

We shall now show that g is the inverse to f. We have $(f \circ g)(y) = f(g(y)) = f(x) = y$ for all $y \in Y$, so that $f \circ g = 1_Y$.

If $x \in X$, then put $y = f(x)$ so that $(g \circ f)(x) = g(f(x)) = g(y)$. By the definition of g, $g(y)$ is the unique element of X whose image, under f, is y. Since the image of x is y, it follows that $x = g(y)$ and so

$$(g \circ f)(x) \;=\; g(y) \;=\; x.$$

Therefore, $g \circ f = 1_X$, and g is the inverse to f. $\qquad\square$

Example 6.52.   Let $f : \{1, 2, 3, 4, 5\} \to \{A, B, C, D, E\}$ be the function such that $f(i)$ is the ith letter in $DECAB$. Does this function have an inverse?

Solution.   The function f is surjective because all the letters A, B, C, D and E occur in $DECAB$. It is injective because all the letters of $DECAB$ are different. Hence f is bijective and does have an inverse. $\qquad\square$

Suppose that each person in a telephone directory only has one telephone number. This directory then defines a bijective function from the set of listed subscribers to the set of listed numbers. Hence this function does have an inverse, but it is sometimes difficult for members of the general public to obtain a list of subscribers corresponding to each phone number. The telephone company often keeps this list secret, even though all the information is contained in the regular telephone directory.

A glance at the graph of a function is often sufficient to tell whether a function has an inverse or not.

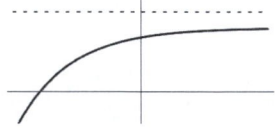

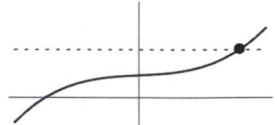

Injective but Not Surjective   Surjective but Not Injective   Bijective

- A function is injective if the horizontal line through each point of the codomain intersects the graph *at most* once.

- A function is surjective if the horizontal line through each point of the co-domain intersects the graph *at least* once.

- A function is bijective if the horizontal line through each point of the codomain intersects the graph *precisely* once.

Example 6.53. Does the function $f : \mathbb{R} \longrightarrow \mathbb{R}$ defined by the formula

$$f(x) = x(x-1)(x-2)$$

have an inverse?

Solution.

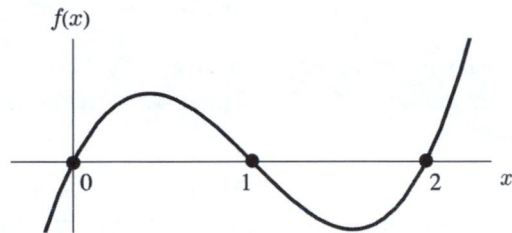

It is clear that this function is not injective because

$$f(0) = f(1) = f(2).$$

Hence it cannot have an inverse. □

Proposition 6.54.   *Let $f : X \to Y$ and $g : Y \to Z$ be functions.*

(i) *If f and g are injections, then $g \circ f$ is an injection.*

(ii) *If f and g are surjections, then $g \circ f$ is a surjection.*

(iii) *If f and g are bijections, then $g \circ f$ is a bijection.*

Proof. **(i)** Suppose that $g \circ f(x_1) = g \circ f(x_2)$. Then $g(y_1) = g(y_2)$, where $y_1 = f(x_1)$ and $y_2 = f(x_2)$. If g is an injection, then $y_1 = y_2$; that is, $f(x_1) = f(x_2)$. If f is also an injection, $x_1 = x_2$ and it follows that $g \circ f$ is an injection.

(ii) Let f and g be surjections and let $z \in Z$. Then there exists $y \in Y$ with $g(y) = z$, and there exists $x \in X$ with $f(x) = y$. Hence, $g \circ f(x) = g(f(x)) = g(y) = z$ and so $g \circ f$ is surjective.

(iii) This follows from parts (i) and (ii). □

6.6   CARDINALITY

You may have noticed that if $f : X \to Y$ is a bijection whose domain, X, contains three elements say, then the codomain Y must also contain three elements. We will now use bijections to define the notion of sets having equal number of elements.

Definition.   If there exists a bijection from the set X to the set Y, we write

$$\#X \;=\; \#Y$$

and we say that the sets X and Y have the same **cardinality**. Another common notation is $|X| = |Y|$.

Let $\mathbb{P}_n$ denote the subset of $\mathbb{P}$ consisting of all positive integers less than or equal to n. Hence $\mathbb{P}_0 = \emptyset$, $\mathbb{P}_1 = \{1\}$, $\mathbb{P}_2 = \{1, 2\}$, $\mathbb{P}_3 = \{1, 2, 3\}$ and $\mathbb{P}_n = \{1, 2, 3, \dots, n\}$. If there exists a bijection between a set X and $\mathbb{P}_n$, we write

$$\#X \;=\; n$$

and say that the **number of elements** in X is n. If such a bijection exists for some integer n, we say that X is a **finite set**; if no such bijection exists, X is called an **infinite set**.

For example, $\#\{a, b\} = \#\{8, 5\}$, and these both have two elements. Furthermore, these two sets are finite, while $\mathbb{Z}$ is an infinite set. If $\#X = 0$, then there is a bijection between X and $\emptyset$, and it follows that X must be the empty set $\emptyset$; there is only one empty set.

This notion of the number of elements of a set agrees with our everyday idea of counting elements. A child, wishing to count five oranges, usually sets up a one-to-one correspondence or bijection between the oranges and a known set, such as the five fingers on a hand, or the words one, two, three, four, five. Our definition does precisely the same, by setting up a one-to-one correspondence between a set and one of the known sets $\mathbb{P}_n$.

This definition extends this idea of sets with the same number of elements to include infinite sets, but, as we shall see later, our intuition often fails us when dealing with the cardinality of infinite sets.

Proposition 6.61.   *If A and B are disjoint finite sets, then*

$$\#(A \cup B) \;=\; \#A + \#B.$$

Proof.   Suppose that $\#A = m$ and $\#B = n$, where $m \geq 0$ and $n \geq 0$. Then there exist bijections $f : A \to \mathbb{P}_m$ and $g : B \to \mathbb{P}_n$. Construct the function $h : A \cup B \to \mathbb{P}_{m+n}$ by

$$h(x) \;=\; \begin{cases} f(x) & \text{if } x \in A \\ g(x) + m & \text{if } x \in B. \end{cases}$$

This has the inverse $h^{-1} : \mathbb{P}_{m+n} \to A \cup B$, where

$$h^{-1}(k) = \begin{cases} f^{-1}(k) & \text{if } 1 \le k \le m \\ g^{-1}(k-m) & \text{if } m+1 \le k \le m+n. \end{cases}$$

Hence h is a bijection, and $\#(A \cup B) = m + n = \#A + \#B$.   □

We now extend this result to sets that are not necessarily disjoint. We obtain the result we would expect if we look at the Venn diagram of two sets.

Theorem 6.62.   *If A and B are any finite sets, then*

$$\#(A \cup B) = \#A + \#B - \#(A \cap B).$$

Proof.   Since the previous proposition describes how to count the number of elements in the union of sets when they are disjoint, we split $A \cup B$ into disjoint subsets.

We can write
$$A \cup B = A \cup (B - A),$$

where $B - A$ denotes the set of elements in B that are not in A. The union is disjoint; hence by Proposition 6.61

$$\#(A \cup B) = \#A + \#(B - A).$$

We can also write B as the disjoint union

$$B = (A \cap B) \cup (B - A)$$

so that $\#B = \#(A \cap B) + \#(B - A)$. Eliminating $\#(B - A)$, we obtain

$$\#(A \cup B) = \#A + \#B - \#(A \cap B).$$ □

Example 6.63.   In a survey of 100 people on their source of news, 72 said they got their news from watching TV, 42 said they obtained it from reading a newspaper, while 8 said they got it from both TV and a newspaper. Are these figures consistent?

Solution. Let T be the set of people who got their news from TV and N be the set of people who got their news from a newspaper. Then, by Theorem 6.62,

$$
\begin{aligned}
\#(T \cup N) &= \#T + \#N - \#(T \cap N) \\
&= 72 + 42 - 8 \\
&= 106.
\end{aligned}
$$

Since this number is greater than the number of people in the survey, the figures must be inconsistent. $\qquad\square$

We shall now look at the cardinality of some infinite sets.

Theorem 6.64. $\#\mathbb{P} = \#\mathbb{Z}$

Proof. We have to show that there is a bijection between the positive integers, $\mathbb{P}$, and all the integers, $\mathbb{Z}$. We can define such a bijection as follows.

$$
\begin{array}{ccccccccccc}
\mathbb{P} & & 1 & 2 & 3 & 4 & 5 & 6 & 7 & 8 & \cdots \\
f\downarrow & & \downarrow & \downarrow & \downarrow & \downarrow & \downarrow & \downarrow & \downarrow & \downarrow & \\
\mathbb{Z} & & 0 & 1 & -1 & 2 & -2 & 3 & -3 & 4 & \cdots
\end{array}
$$

The function $f : \mathbb{P} \to \mathbb{Z}$ can be defined by

$$
f(n) = \begin{cases} n/2 & \text{if } n \text{ is even} \\ -(n-1)/2 & \text{if } n \text{ is odd.} \end{cases}
$$

This is a bijection because it has an inverse $f^{-1} : \mathbb{Z} \to \mathbb{P}$ defined by

$$
f^{-1}(n) = \begin{cases} 2n & \text{if } n > 0 \\ 1 - 2n & \text{if } n \leq 0. \end{cases}
$$

Therefore, $\#\mathbb{P} = \#\mathbb{Z}$. $\qquad\square$

This result may seem surprising, because $\mathbb{P}$ is a proper subset of $\mathbb{Z}$ but still has the same cardinality (that is, the same number of elements) as $\mathbb{Z}$. Such a situation can never occur with finite sets, but it is a characteristic of infinite sets that there is always a proper subset of any infinite set that has the same cardinality as the whole set.

The following two results may be even more surprising.

Proposition 6.65. $\#(\mathbb{P} \times \mathbb{P}) = \#\mathbb{P}$

Proof. $\mathbb{P} \times \mathbb{P}$ is the set of all pairs of positive integers. We can write out all the elements of $\mathbb{P} \times \mathbb{P}$ in rows as follows.

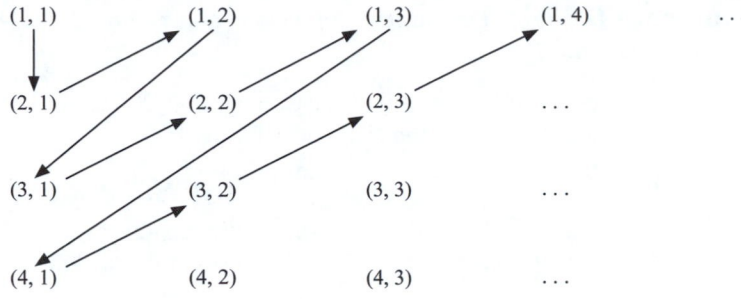

The sequence of arrows in the previous diagram will eventually pass through every pair of positive integers precisely once, and we can use this fact to define a bijection

$$f : \mathbb{P} \times \mathbb{P} \longrightarrow \mathbb{P}$$

where $f(1,1) = 1$, $f(2,1) = 2$, $f(1,2) = 3$, $f(3,1) = 4$, and so on. With a little work we can show that a formula for this function is

$$f(i,j) \;=\; \frac{(i+j-1)(i+j-2)}{2} + j.$$

This bijection shows that $\#(\mathbb{P} \times \mathbb{P}) = \#\mathbb{P}$. □

Theorem 6.66. $\#\mathbb{Q} \;=\; \#\mathbb{P}$

Proof.   A bijection from the rational numbers, $\mathbb{Q}$, to the positive integers, $\mathbb{P}$, can be constructed in a similar way to that of the previous proposition by considering the following diagram.

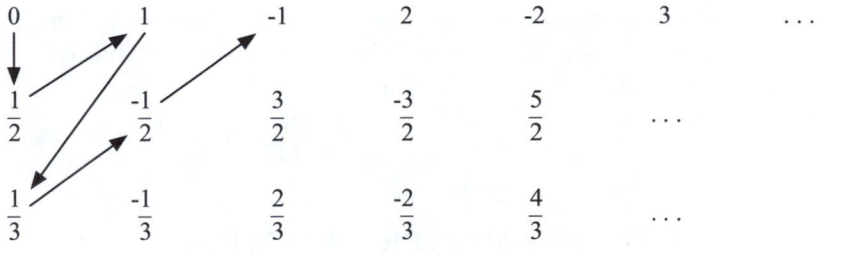

The last three examples might suggest that all infinite sets have the same cardinality. However, this is not the case. The following theorem, proved by Cantor in 1874, shows that there is no bijection between $\mathbb{P}$ and the set of all real numbers $\mathbb{R}$. The proof is called *Cantor's diagonal argument* after the German mathematician Georg Cantor (1845–1918), who was the founder of modern set theory. He was the first mathematician to seriously study infinite sets.

Theorem 6.67.   *The positive integers* $\mathbb{P}$ *and the real numbers* $\mathbb{R}$ *do not have the same cardinality.*

Proof.   Suppose that there is a surjection $f : \mathbb{P} \to \mathbb{R}$. Write each element of $\mathbb{R}$ as an infinite decimal, and list the elements of $\mathbb{R}$ as follows.

$$
\begin{aligned}
f(1) &= & b_{1r} \ldots \ldots b_{12}b_{11}.a_{11}a_{12}a_{13}\ldots \\
f(2) &= b_{2s} \ldots \ldots \ldots b_{22}b_{21}.a_{21}a_{22}a_{23}\ldots \\
f(3) &= & b_{3t}\ldots b_{32}b_{31}.a_{31}a_{32}a_{33}\ldots \\
&\vdots &
\end{aligned}
$$

Since f is supposed to be a surjection, all the elements of $\mathbb{R}$ must appear in the above list.

Construct the decimal $c = 0.c_1c_2c_3\ldots$ in the following way. Each digit is chosen from the digits 1 through 8 and $c_1 \neq a_{11}$, $c_2 \neq a_{22}$, $c_3 \neq a_{33}$ and, in general $c_r \neq a_{rr}$. Then c is a real number that does not contain a repeated sequence of zeros or nines and so has only one decimal expansion. Now $c \neq f(1)$ because they differ in the first decimal place; $c \neq f(2)$ because they differ in the second decimal place and, in general, $c \neq f(r)$ because they differ in the rth decimal place. Hence $c \notin f(\mathbb{P})$, which contradicts the assumption that f is a surjection.

Therefore, there is no surjection from $\mathbb{P}$ to $\mathbb{R}$, and certainly no bijection from $\mathbb{P}$ to $\mathbb{R}$. $\qquad\square$

Since there is no surjection from $\mathbb{P}$ to $\mathbb{R}$, in some sense, $\mathbb{R}$ contains more elements than $\mathbb{P}$. In fact, this notion of sets of larger infinite cardinality can be made precise, and it can be shown that, for each set X, it is possible to construct a set of cardinality larger than X. (See Problem 101.)

Sets that have the same cardinality as $\mathbb{P}$ have the smallest infinite cardinality. Such sets are called *countable* because a one-to-one correspondence between $\mathbb{P}$ and the set has the effect of counting the elements of the set. Infinite sets that are not countable are called *uncountable*. The above theorem shows that the real numbers are uncountable.

6.7   INVERSE TRIGONOMETRIC FUNCTIONS

As an example of the use of the Inversion Theorem 6.51, we shall show that, by suitably restricting the domain and codomain of the trigonometric functions, it is possible to find inverse trigonometric functions.

Consider the graph of the sine function

$$\sin : \mathbb{R} \longrightarrow \mathbb{R},$$

where $\sin x$ is the sine of x radians.

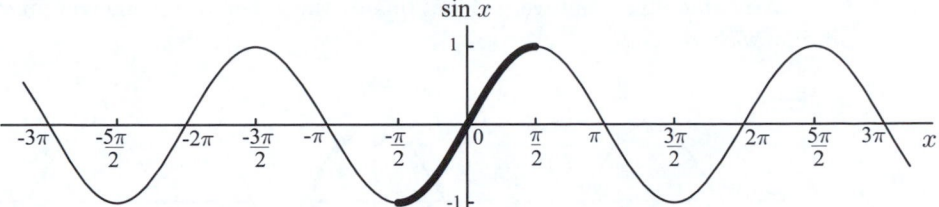

If we are given the sine of an angle and need to know the angle, then we require an inverse function to the sine function. For example, if $\sin\theta = 1/2$, then θ could be $\pi/6$ or $5\pi/6$ or $13\pi/6$ or, in fact, one of an infinite number of values. It is clear from the above graph that the sine function is not bijective and hence cannot have an inverse.

Note that the sine function is increasing on the interval $[-\pi/2, \pi/2]$. Hence if we restrict the domain to the interval $[-\pi/2, \pi/2]$ and restrict the codomain to $[-1, 1]$, then the function will be a bijection. (See Problem 110.)

Definition.   Denote the restricted sine function by

$$\text{Sin} : [-\pi/2, \pi/2] \longrightarrow [-1, 1]$$

and the **inverse sine function** by

$$\text{Sin}^{-1} : [-1, 1] \longrightarrow [-\pi/2, \pi/2].$$

An alternative notation for $\text{Sin}^{-1}x$ is Arcsin x.

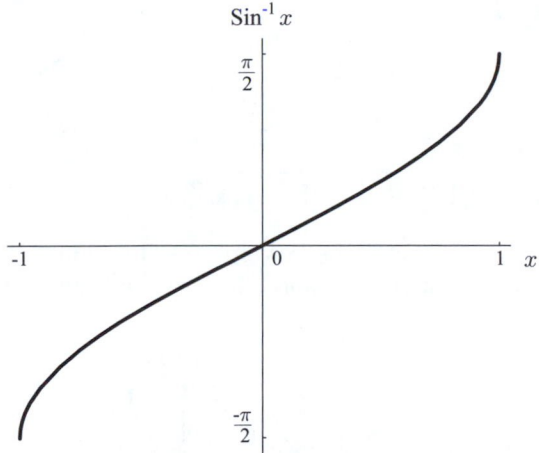

Hence $y = \text{Sin}^{-1}x$ if and only if $x = \sin y$ and $-\pi/2 \leq y \leq \pi/2$.

For example, $\text{Sin}^{-1}(1/2) = \pi/6$, $\text{Sin}^{-1}(-1/2) = -\pi/6$ and $\text{Sin}^{-1}(1) = \pi/2$.

Observe that the notation $\text{Sin}^{-1}x$ is the inverse function notation, and it does *not* refer to the reciprocal $\frac{1}{\text{Sin }x}$.

We can define the inverse functions to the other trigonometric functions in a similar way.

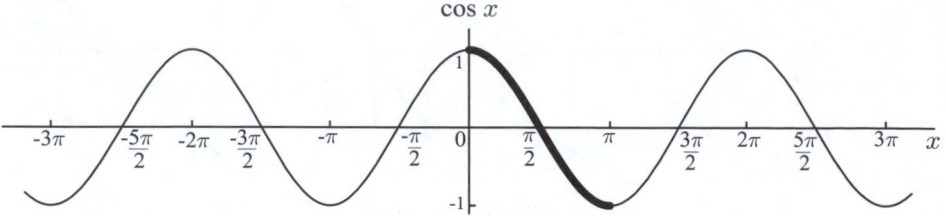

Definition. Denote the restricted cosine function by

$$\text{Cos} : [0, \pi] \longrightarrow [-1, 1]$$

and the **inverse cosine function** by

$$\text{Cos}^{-1} : [-1, 1] \longrightarrow [0, \pi]$$

or by Arccos.

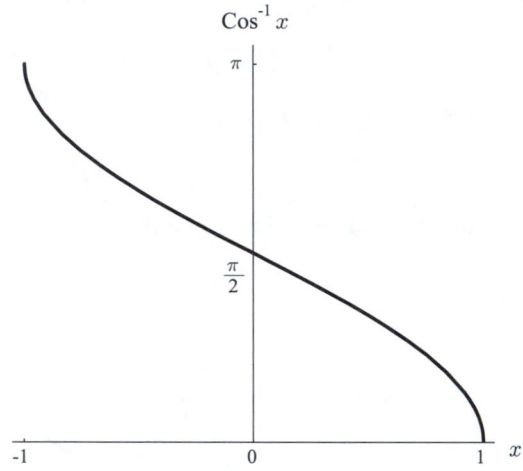

The tangent function has asymptotes at odd multiples of $\pi/2$, and each branch has an inverse function. We normally use the branch through the origin to define the inverse.

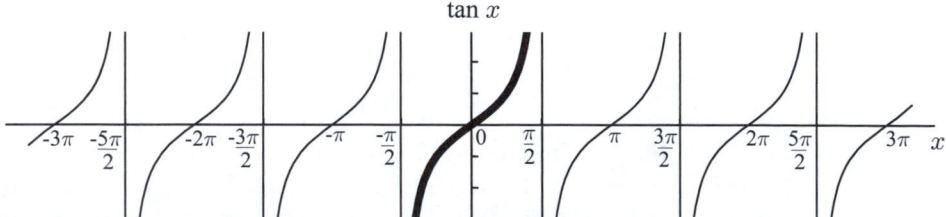

Definition.   Denote the restricted tangent function by

$$\mathrm{Tan} : (-\pi/2, \pi/2) \longrightarrow \mathbb{R}$$

and the **inverse tangent function** by

$$\mathrm{Tan}^{-1} : \mathbb{R} \longrightarrow (-\pi/2, \pi/2)$$

or by Arctan.

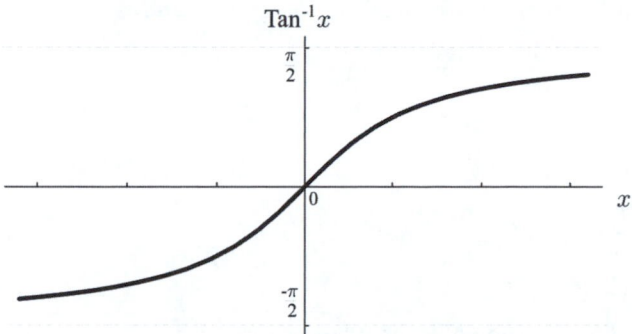

6.8   EXPONENTIAL AND LOGARITHMIC FUNCTIONS

A very important class of functions in the physical and biological sciences is that of the exponential function and its inverse, the logarithmic function.

A function such as $f(x) = x^2$ is called a power function, whereas a function such as $g(x) = 2^x$, in which the independent variable appears as an exponent, is called an exponential function.

These functions occur in science whenever a rate of growth (or decay) of some quantity is proportional to its own size. For example, a colony of bacteria will start growing at an exponential rate, because each bacteria will divide after a certain time, and hence the rate of growth will depend on the size of the colony. Of course, after a while, the colony will run out of space and food and will stop growing so rapidly; if this were not so, we would soon be overrun by bacteria.

An **exponential function** is a function of the form

$$f : \mathbb{R} \longrightarrow \mathbb{R}, \quad \text{where } f(x) = b^x$$

for some fixed positive real number $b \neq 1$.

We have only defined what we mean by b^x, if x is a rational number. What does b^x mean if x is not rational? What is $2^{\sqrt{2}}$ for instance?

An irrational number can be approximated by the first n terms of its decimal expansion. Hence we would expect an irrational power to be approximated by rational powers obtained from successive terms of the decimal expansion. For example, we would hope that

$$2^{1.4} < 2^{\sqrt{2}} < 2^{1.5} \quad \text{since} \quad 1.4 < \sqrt{2} < 1.5$$
$$2^{1.41} < 2^{\sqrt{2}} < 2^{1.42} \quad \text{since} \quad 1.41 < \sqrt{2} < 1.42.$$

We shall not give a precise definition of irrational powers here, as it is too complicated, and it depends on the definition of the real numbers. But it can be shown, by using the properties of the real numbers, that for each $b > 0$ there is an exponential function $f : \mathbb{R} \to \mathbb{R}$, defined by $f(x) = b^x$, which has the following properties.

Properties of Exponents 6.81.

(i) $b^{m/n} = \sqrt[n]{b^m}$ for positive integers m and n

(ii) $b^{-x} = \dfrac{1}{b^x}$

(iii) $b^0 = 1$

(iv) $b^x \cdot b^y = b^{x+y}$

(v) $\dfrac{b^x}{b^y} = b^{x-y}$

(vi) $(b^x)^y = b^{xy}$

(viia) If $b > 1$, then $b^x < b^y$ whenever $x < y$.

(viib) If $0 < b < 1$, then $b^x > b^y$ whenever $x < y$.

(viic) If $b = 1$, then $b^x = 1$ for all $x \in \mathbb{R}$.

The graphs of these exponential functions take the following forms when $b > 1$, and when $0 < b < 1$.

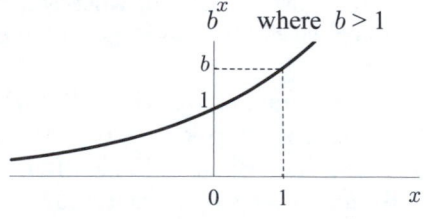
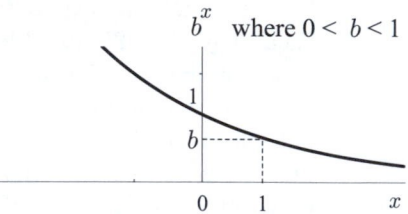

A real-valued function, f, of a real variable is said to be *monotone increasing* if $f(x) > f(y)$ whenever $x > y$. The function, f, is called *monotone decreasing* if $f(x) < f(y)$ whenever $x > y$.

The exponential function $f(x) = b^x$ is monotone increasing if $b > 1$. This means that the function increases as the independent variable increases and that the graph always slopes upward. It is this property that shows that $2^{\sqrt{2}}$ is sandwiched between $2^{1.41}$ and $2^{1.42}$.

If $0 < b < 1$, then the exponential function $f(x) = b^x$ is monotone decreasing, and its graph always slopes downward.

A radioactive substance is a material that emits subatomic particles and changes to another substance. If it is assumed that the probability of an individual atom disintegrating is unaffected by the number of other atoms present, then the amount of radioactive substance remaining decreases exponentially with time. This yields an example of a function of the form b^t, where $0 < b < 1$.

The rate of decay of a substance is usually measured by its *half-life*. This is the length of time required for the radioactive substance to decay to half its original amount. For example, the isotope strontium 90 has a half-life of 28 years. If we started with one gram of strontium 90, there would be one half of a gram remaining after 28 years, and one quarter of a gram after 56 years. In general, there would be

$$\left(\frac{1}{2}\right)^{t/28} \quad \text{grams}$$

remaining after t years.

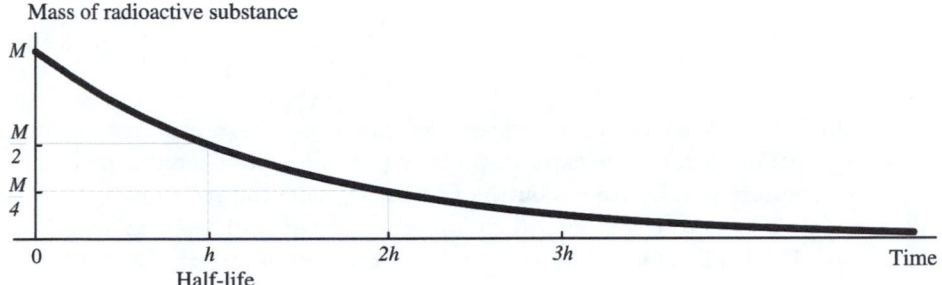

Example 6.82. Archaeologists use the isotope carbon 14 to date ancient remains. Carbon 14 occurs naturally in all living tissue and, while the plant or animal is alive, its carbon 14 content is maintained at a constant level. However, when the plant or animal dies, the carbon 14 is not replenished and decays with a half-life of 5700 years.

A piece of charcoal from an ancient campfire was discovered at an archaeological dig and was found to contain 25% of its original carbon 14. Estimate the age of the campfire.

Solution. If there was c_0 milligrams of carbon 14 in the charcoal when it was burnt, after t years the amount remaining would be

$$c = c_0 \left(\frac{1}{2}\right)^{t/5700}.$$

If the amount remaining is $c_0/4$, then

$$\frac{c_0}{4} = c_0 \left(\frac{1}{2}\right)^{t/5700}$$

$$\left(\frac{1}{2}\right)^2 = \left(\frac{1}{2}\right)^{t/5700}.$$

Hence $2 = t/5700$ and $t = 11400$.

The campsite would therefore be approximately 11 thousand years old and would date from 9000 B.C. □

There is one exponential function that plays a very important role in mathematics. It is the function whose rate of growth is not only proportional to its size but is actually equal to its size. That is, it is a solution to the differential equation

$$\frac{dy}{dx} = y.$$

It can be shown that all the solutions to this equation are of the form $y = ke^x$, where k is a constant that depends on the initial conditions, and e is a certain fixed irrational number, whose value to five decimals places is 2.71828. The function

$$f(x) = e^x$$

is often referred to as *the* exponential function. This exponential function e^x is sometimes denoted by $\exp x$, especially if x is a complicated expression. Scientific calculators usually have a button for this exponential function.

A monotone function (increasing or decreasing) is an injection. (See Problem 110.) Therefore, if we restrict the codomain of an exponential function to the positive real numbers, X, then $f : \mathbb{R} \to X$ defined by $f(x) = b^x$ is a bijection, as long as $b \neq 1$.

Definition.   The inverse of the general exponential function b^x is the **logarithmic function**

$$\log_b : X \to \mathbb{R}$$

with the positive real numbers as domain. This is called the *logarithm of x to the base b*, or just $\log x$ to the base b.

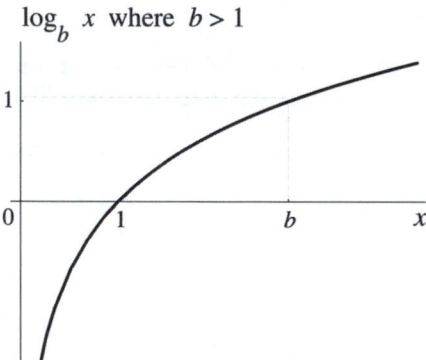

Hence $y = \log_b x$ if and only if $x = b^y$; that is, $\log_b x$ is the power that b has to be raised to in order to obtain the number x. In particular,

$$b^{\log_b x} = x \quad \text{and} \quad \log_b b^y = y.$$

For example, $\log_{10} 1000 = 3$ because $1000 = 10^3$, $\log_3 81 = 4$ because $81 = 3^4$, $\log_2(1/32) = -5$ because $1/32 = 2^{-5}$ and $\log_{10} \sqrt{10} = 1/2$.

For any base b, $\log_b 1 = 0$, since $1 = b^0$, $\log_b b = 1$, since $b = b^1$ and $\log_b b^r = r$.

The logarithmic function has the following important properties that follow from the Properties of Exponents 6.81.

Properties of Logarithms 6.83.   *If x and y are positive real numbers and r is real, then*

(i) $\log_b(xy) = \log_b x + \log_b y$

(ii) $\log_b\left(\dfrac{x}{y}\right) = \log_b x - \log_b y$

(iii) $\log_b(x^r) = r\log_b x$

Proof. (i)   Let $k = \log_b x$, $\ell = \log_b y$ and $m = \log_b(xy)$, so that $x = b^k$, $y = b^\ell$ and $xy = b^m$. Now $b^m = b^k b^\ell$ and it follows from 6.81 (iv) that $b^m = b^{k+\ell}$. Hence $m = k + \ell$, which proves (i).

(iii)   Let $k = \log_b x$ again, so that $x = b^k$. Then, by 6.81 (vi), $x^r = (b^k)^r = b^{kr}$ and $\log_b(x^r) = kr = r\log_b x$.

(ii)   Since $x/y = x \cdot y^{-1}$, part (ii) follows from parts (i) and (iii). $\Box$

Logarithms to the base 10 are called *common logarithms*. These are very useful when dealing with numbers in scientific notation. The sun is about 149,000,000 kilometers from the earth and, in scientific notation, this would be written as 1.49×10^8 kilometers. A number in *scientific notation* consists of a real number between 1 and 10 followed by a multiplication sign and the appropriate integer power of 10. For example, the mass of the hydrogen atom is about 1.66×10^{-24} grams; it would be very inconvenient to write this number as a decimal.

The common logarithms for the numbers in the following table can be calculated once it is known that $\log_{10} 2.63$ is approximately 0.42.

Number x	In Scientific Notation	As a power of 10	$\log_{10} x$
263000	2.63×10^5	$10^{0.42} \times 10^5 = 10^{5.42}$	5.42
26.3	2.63×10	$10^{0.42} \times 10 = 10^{1.42}$	1.42
2.63	2.63	$10^{0.42}$	0.42
0.263	2.63×10^{-1}	$10^{0.42} \times 10^{-1} = 10^{0.42-1}$	$0.42 - 1$
0.000263	2.63×10^{-4}	$10^{0.42} \times 10^{-4} = 10^{0.42-4}$	$0.42 - 4$

Logarithms to the base e are called *natural logarithms* and these are often used in mathematics, because this base appears naturally in the calculus. In higher mathematics books, when no base is explicitly mentioned, it can be assumed that the logarithms are to the base e.

Logarithms to the base 2 are often used in computer science, as machine numbers are usually represented in the binary system.

The Properties of Logarithms form the basis for the method of calculation used by log tables and slide rules, since they convert multiplication and division into addition and subtraction. A *log table* is a list of values of logarithms of numbers to a given base. Any positive integer, other than 1, can be used as the base of a system of logarithms. Before electronic calculators, from the 1600s to the 1970s, log tables and slide rules were the main tools used by scientists and engineers to perform complicated calculations. For example, to calculate an expression such as

$$s \;=\; \frac{327\sqrt{140.2}}{(57.6)^3}$$

a scientist would take logs (to any fixed base) and obtain

$$\log s \;=\; \log 327 + \tfrac{1}{2} \log 140.2 - 3 \log 57.6.$$

The scientist would compute the right side using log tables and paper and pencil and then use the inverse of the log table to find s.

It is fairly easy to convert from one base to another using the following result.

Theorem 6.84. $\log_a x \;=\; \log_a b \cdot \log_b x$

Proof.   Let $k = \log_a b$ and $\ell = \log_b x$ so that $b = a^k$ and $x = b^\ell$. Now

$$x \;=\; b^\ell \;=\; (a^k)^\ell \;=\; a^{k\ell}$$

and so $\log_a x = k\ell$ and the theorem is true.   □

Corollary 6.85. $\log_a b \;=\; \dfrac{1}{\log_b a}$

Proof.   This corollary follows from the above theorem by putting $x = a$ and using the fact that $\log_a a = 1$.   □

For example, to convert natural logarithms to common logarithms you just have to multiply by $\log_{10} e \approx 0.4343$ and, conversely, to convert common logarithms to natural logarithms you multiply by $\frac{1}{0.4343} \approx 2.3026$.

In many calculus books, the natural logarithm is defined before the exponential function, by

$$\log_e x \;=\; \int_1^x \frac{dt}{t}.$$

It is then shown that this function is monotone increasing and satisfies the Properties of Logarithms 6.83. The inverse of this function is then defined to be the function e^x, which is shown to have the usual properties of an exponential.

6.9 PERMUTATIONS

Bijective functions appear in another disguise as permutations of a set.

Definition.   Let S be a finite nonempty set. A **permutation** σ on the set S is a bijection

$$\sigma : S \to S.$$

If $S = \{a, b, c, d\}$, then one permutation on S is defined by $\sigma(a) = b$, $\sigma(b) = d$, $\sigma(c) = a$, and $\sigma(d) = c$. We can think of the permutation σ as a rearrangement of the elements a, b, c, d, to form b, d, a, c. A convenient way of writing this permutation is

$$\sigma = \begin{pmatrix} a\ b\ c\ d \\ b\ d\ a\ c \end{pmatrix},$$

where the elements of S are written in the top row, and their corresponding images under σ are written below. The set $\mathbb{P}_3 = \{1, 2, 3\}$ has six different permutations.

$$\sigma_1 = \begin{pmatrix} 1\ 2\ 3 \\ 1\ 2\ 3 \end{pmatrix} \qquad \sigma_2 = \begin{pmatrix} 1\ 2\ 3 \\ 2\ 3\ 1 \end{pmatrix}$$

$$\sigma_3 = \begin{pmatrix} 1\ 2\ 3 \\ 3\ 1\ 2 \end{pmatrix} \qquad \sigma_4 = \begin{pmatrix} 1\ 2\ 3 \\ 1\ 3\ 2 \end{pmatrix}$$

$$\sigma_5 = \begin{pmatrix} 1\ 2\ 3 \\ 3\ 2\ 1 \end{pmatrix} \qquad \sigma_6 = \begin{pmatrix} 1\ 2\ 3 \\ 2\ 1\ 3 \end{pmatrix}$$

The permutation σ_1 fixes all the elements and is called the *identity permutation*.

The set $\mathbb{P}_n = \{1, 2, 3, \ldots, n\}$ contains $n!$ permutations. There are n choices for the image of the element 1 and, once this has been chosen, this image cannot be the image of any other element; hence there are $(n - 1)$ choices for the image of 2, $(n - 2)$ choices for the image of 3, and so on. The total number of choices is therefore $n(n - 1)(n - 2) \cdots 2 \cdot 1 = n!$. Denote the set of permutations on $\mathbb{P}_n$ by $\mathcal{S}_n$.

If σ and τ are two permutations in $\mathcal{S}_n$, then $\tau : \mathbb{P}_n \to \mathbb{P}_n$ and $\sigma : \mathbb{P}_n \to \mathbb{P}_n$ are bijections, and their composite $\sigma \circ \tau : \mathbb{P}_n \to \mathbb{P}_n$ can be defined.

Example 6.91.   Compute the composite $\sigma_5 \circ \sigma_2$ of the permutations σ_5 and σ_2 of $\mathcal{S}_3$ given before.

Solution.   To compute $\sigma_5 \circ \sigma_2 : \mathbb{P}_3 \to \mathbb{P}_3$ we must first apply σ_2 and then apply σ_5 to the result. Since $\sigma_2(1) = 2$ and $\sigma_5(2) = 2$ it follows that $\sigma_5 \circ \sigma_2(1) = 2$. Similarly, $\sigma_5 \circ \sigma_2(2) = \sigma_5(\sigma_2(2)) = \sigma_5(3) = 1$ and $\sigma_5 \circ \sigma_2(3) = \sigma_5(1) = 3$. Hence

$$\sigma_5 \circ \sigma_2 \;=\; \begin{pmatrix} 1 \ 2 \ 3 \\ 3 \ 2 \ 1 \end{pmatrix} \circ \begin{pmatrix} 1 \ 2 \ 3 \\ 2 \ 3 \ 1 \end{pmatrix} \;=\; \begin{pmatrix} 1 \ 2 \ 3 \\ 2 \ 1 \ 3 \end{pmatrix} \;=\; \sigma_6$$

and the composite $\sigma_5 \circ \sigma_2$ is the permutation σ_6. $\qquad\qquad\square$

This composition has the following properties, which make the set $\mathcal{S}_n$, under the composition $\circ$, into what is called a *group*.

Proposition 6.92. *If $\sigma, \tau \in \mathcal{S}_n$, then $\sigma \circ \tau \in \mathcal{S}_n$ and*

(i) $\sigma \circ (\tau \circ \rho) = (\sigma \circ \tau) \circ \rho$ *for all $\sigma, \tau, \rho \in \mathcal{S}_n$.*

(ii) *There is an identity permutation $\iota \in \mathcal{S}_n$ such that $\iota \circ \sigma = \sigma \circ \iota = \sigma$ for all $\sigma \in \mathcal{S}_n$.*

(iii) *For each permutation $\sigma \in \mathcal{S}_n$, there is an inverse permutation $\sigma^{-1} \in \mathcal{S}_n$ such that $\sigma \circ \sigma^{-1} = \sigma^{-1} \circ \sigma = \iota$.*

Proof. It follows from Proposition 6.54 that the composite $\sigma \circ \tau$ of two permutations σ and τ is also a permutation.

(i) Associativity follows from the associativity of functions, Theorem 6.33.

(ii) The identity permutation ι is the identity function on $\mathbb{P}_n$ and its properties follow from Proposition 6.34.

(iii) Any permutation $\sigma : \mathbb{P}_n \to \mathbb{P}_n$ is a bijection, and so has an inverse function $\sigma^{-1} : \mathbb{P}_n \to \mathbb{P}_n$ that is also a bijection. Hence σ^{-1} is also a permutation. $\qquad\square$

The composition of the elements $\sigma_1 = \iota$, σ_2, σ_3, σ_4, σ_5, and σ_6 in $\mathcal{S}_3$ is given by the following table, in which $\sigma_i \circ \sigma_j$ is the element in the ith row and the jth column.

$\circ$	σ_1	σ_2	σ_3	σ_4	σ_5	σ_6
σ_1	σ_1	σ_2	σ_3	σ_4	σ_5	σ_6
σ_2	σ_2	σ_3	σ_1	σ_6	σ_4	σ_5
σ_3	σ_3	σ_1	σ_2	σ_5	σ_6	σ_4
σ_4	σ_4	σ_5	σ_6	σ_1	σ_2	σ_3
σ_5	σ_5	σ_6	σ_4	σ_3	σ_1	σ_2
σ_6	σ_6	σ_4	σ_5	σ_2	σ_3	σ_1

For example, $\sigma_3^2 = \sigma_2$, $\sigma_3^{-1} = \sigma_2$, $\sigma_2 \circ \sigma_5 = \sigma_4$ and $\sigma_5 \circ \sigma_2 = \sigma_6$. These last two relations show that this composition is not commutative. This should not be surprising as composition of functions is not commutative in general.

Exercise Set 6

1–4. *Which of the following are functions from* $\{1,2,3\}$ *to* $\{a,b,c\}$?

1.

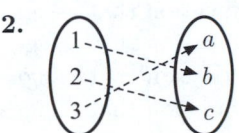

2.

3.

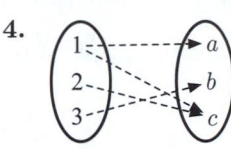

4.

5–14. *For each of the following formulas, find the largest set* $X \subseteq \mathbb{R}$ *for which* $f : X \to \mathbb{R}$ *defines a real-valued function of a real variable. Sketch the graph of each function and find the image* $Y = f(X)$. *State whether or not the function* $f : X \to Y$ *has an inverse* $f^{-1} : Y \to X$, *and give a formula for* $f^{-1}(x)$ *when the inverse exists.*

5. $f(x) = x^2 - 1$

6. $f(x) = \sqrt{x-2}$

7. $f(x) = \dfrac{x}{x^2 + x}$

8. $f(x) = \dfrac{x}{x+1}$

9. $f(x) = \sin 5x$

10. $f(x) = 5^x$

11. $f(x) = \log_{10}(3 - x)$

12. $f(x) = \sqrt[4]{2 - \log_3 x}$

13. $f(x) = \sqrt{\sin x}$

14. $f(x) = \mathrm{Sin}^{-1} 3x$

15–17. *Below are three diagrams defining functions* f, g *and* h *from* $\mathbb{P}_5$ *to itself.*

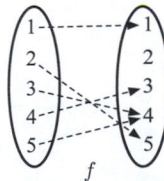 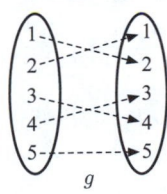 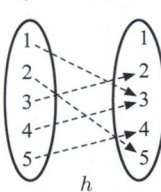

Draw similar diagrams for the following functions.

15. $f \circ g$ and $g \circ f$

16. $f \circ h$ and $h \circ g$

17. $g \circ h$, $f \circ (g \circ h)$ and $(f \circ g) \circ h$

18–23. *For each of the following formulas, specify subsets* X *and* Y *of* $\mathbb{R}$ *so that* $f : X \to Y$ *is a bijective function. Find a formula for each inverse function.*

18. $f(x) = x^4$

19. $f(x) = \sqrt{3 - x}$

20. $f(x) = (\tan x)^2$

21. $f(x) = \dfrac{x+2}{2x-1}$

22. $f(x) = \sqrt{\log_e x}$

23. $f(x) = 3^{\sqrt{x}}$

24–28. Which of the following functions are injections?

24. The function that assigns to everybody their height, to the nearest millimetre.
25. The function that assigns to everybody their maternal grandfather.
26. The function that assigns to each ship at sea, its latitude and longitude.
27. The function that assigns to everybody their name.
28. The function that assigns to every worker their social security number (choose the lowest number if there are several).

29–31. Draw the graph of each of the following functions.

29. $f : \mathbb{P}_4 \longrightarrow \mathbb{P}_6$ defined by $f(n) = n$
30. $g : \mathbb{P}_6 \longrightarrow \mathbb{P}_6$ defined by $g(n) = \gcd(2, n)$
31. $h : \mathbb{P}_3 \longrightarrow \mathbb{P}_{10}$ defined by $h(n) = n^2 + 1$

32. Sketch the graph of the *greatest integer function* or *floor function* $f : \mathbb{R} \to \mathbb{R}$, where $f(x) = \lfloor x \rfloor$, the greatest integer less than or equal to x. What is the image of this function?

33. Let $f : \mathbb{R} \to \mathbb{R}$ and $g : \mathbb{R} \to \mathbb{R}$ be defined by $f(x) = x^2 - 1$ and $g(x) = x - 1$. Find formulas for $f \circ g$, $g \circ f$, $f \circ f$, and $g \circ f \circ g$.

34. Let $X = \mathbb{R} - \{0, 1\}$, the set of all real numbers, except for 0 and 1. Find all the possible functions obtained by taking composites of the two functions $f : X \to X$ and $g : X \to X$, where $f(x) = 1 - x$ and $g(x) = \frac{1}{x}$.

35–36. Find inverses of each of the following bijective functions, $f : \mathbb{Z} \to \mathbb{Z}$.

35. $f(n) = \begin{cases} n + 5 & \text{if } n \text{ is even} \\ n - 5 & \text{if } n \text{ is odd} \end{cases}$

36. $f(n) = \begin{cases} n + 4 & \text{if } n \equiv 0 \pmod{3} \\ -n - 3 & \text{if } n \equiv 1 \pmod{3} \\ n + 1 & \text{if } n \equiv 2 \pmod{3} \end{cases}$

37. If $f : X \to Y$ is a bijective function, prove that its inverse is unique.
38. Let $f : X \to Y$ and $g : Y \to Z$ be bijective functions. Prove that

$$(g \circ f)^{-1} = f^{-1} \circ g^{-1}.$$

39–42. Determine which of the following functions are injective, surjective, and bijective. Find the inverses of the bijective functions.

39. $f : \mathbb{R} \longrightarrow \mathbb{R}$ defined by $f(x) = (x - 2)^3$

40. $g : \mathbb{R} - \{0\} \longrightarrow \mathbb{R}$ defined by $g(x) = \log_2 |x|$

41. $h : \mathbb{Z} \longrightarrow \mathbb{Z}$ defined by $h(n) = n^3$

42. $k : X \longrightarrow X$, where $X = \mathbb{P} - \{1\}$ is the set of integers greater than 1, and $k(n)$ is the greatest prime factor of n

43. Let $f : X \to Y$ and $g : Y \to X$ be functions so that $g \circ f = 1_X$. Prove that f is injective and g is surjective. Need either be bijective?

44. (a) Prove that $\#A = \#A$ for every set A.
  (b) If $\#A = \#B$, prove that $\#B = \#A$.
  (c) If $\#A = \#B$ and $\#B = \#C$, prove that $\#A = \#C$.

45. If A, B and C are finite sets, show that

$$\#(A \cup B \cup C) =$$
$$\#A + \#B + \#C - \#(A \cap B) - \#(A \cap C) - \#(B \cap C) + \#(A \cap B \cap C).$$

46. In a regiment returning from war, 70% of the men had lost an eye, 80% an arm, and 85% a leg. What percentage, at least, must have lost all three? (Adapted from Lewis Carroll, "A Tangled Tale".)

47. Show that the set of even positive integers has the same cardinality as the set of all positive integers. Show that the set of odd positive integers also has the same cardinality.

48. Let $\mathcal{P}(X)$ denote the set of all subsets (including $\emptyset$ and X) of a set X. Write out the elements of $\mathcal{P}(\emptyset)$, $\mathcal{P}(\{a\})$, and $\mathcal{P}(\{r, s, t\})$. If $\#X = n$, a finite number, what is $\#\mathcal{P}(X)$? Prove your assertion.

49. Define the function $\mathrm{Sec}^{-1}x$, giving its domain and image.

50. Define the function $\mathrm{Cot}^{-1}x$, giving its domain and image.

51. Solve the equation $\mathrm{Sin}^{-1}x + \mathrm{Sin}^{-1}1 = \pi$.

52. Solve the equation $\mathrm{Sin}^{-1}\frac{1}{2} + \mathrm{Sin}^{-1}\frac{\sqrt{3}}{2} = x$.

53. What is the relationship between $\mathrm{Cos}^{-1}x$ and $\mathrm{Sin}^{-1}x$ in the domain in which they are both defined?

54. Prove that the shaded segment of the circle of radius r has area
$$r^2\mathrm{Cos}^{-1}\left(\frac{p}{r}\right) - p\sqrt{r^2 - p^2},$$
where p is the perpendicular distance from the center to the segment.

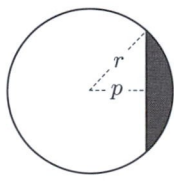

55. Sketch the graph of $y = \dfrac{1}{10^x}$.

56. Sketch the graphs of $y = (1/2)^x$, $y = 2^x$ and $y = 3^x$ in the same diagram.

57. Sketch the graph of $y = \dfrac{2^x + 2^{-x}}{2}$.

58. Sketch the graph of $y = e^{-x^2}$.
(This is a simple form of the normal probability curve in statistics.)

59–64. Solve the following equations.

59. $8 = \log_2 x$

60. $2 = \log_x 10$

61. $\sqrt[x]{2} = e^2$

62. $\log_e(\log_e x) = 5$

63. $10^{-3\log_{10} x} = 5$

64. $5 = 2^{\log_e x}$

65. If $x > 0$, write x^x as a power of e.

66. If a piece of paper 0.2 mm thick could be folded in half 20 times, approximately how thick would the resulting paper be?
(Use the fact that $\log_{10} 2$ is approximately 0.3.)

67. An archaeologist claims that a bone he has discovered is seventeen thousand years old. If this were true, how much of the original amount of carbon 14 would you expect to remain in the bone?

68. The earth's population is now 6 billion and is increasing at the rate of 2% a year. If it continues at this rate, how long will it take to double? What would be the population in 100 years time?
(Use the fact that $\log_{10} 2 \approx 0.3$ and $\log_{10} 1.02 \approx 0.0086$.)

69. When a ship docks, its rope is thrown to the quay, wound round a bollard, and then held by a sailor. If θ is the angle in radians that the rope turns through when wound round the bollard, then a sailor exerting a tension of M kilograms on the rope can hold a tension of T kilograms from the ship, where

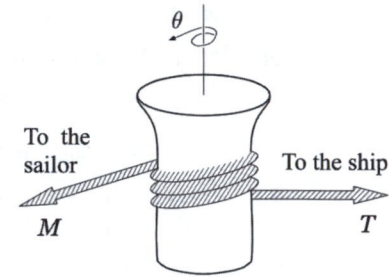

$$T = Me^{\mu\theta}.$$

The constant μ is the coefficient of friction between the rope and the bollard. How many times would the sailor have to wrap the rope around the bollard to hold a tension of 160 tonnes if $\mu = 0.5$ and the sailor could exert a tension of 20 kilograms?

70. If $1000 is placed in a savings bank at 10% interest, compounded daily, how much will be in the account after one year?

71. If X is a finite set, show that the following conditions on a function $f : X \to X$ are equivalent.

    (a) f is an injection.

    (b) f is a surjection.

    (c) f is a bijection.

72–79. If $\rho = \begin{pmatrix} 1\ 2\ 3\ 4 \\ 2\ 1\ 4\ 3 \end{pmatrix}$, $\sigma = \begin{pmatrix} 1\ 2\ 3\ 4 \\ 2\ 3\ 4\ 1 \end{pmatrix}$, and $\tau = \begin{pmatrix} 1\ 2\ 3\ 4 \\ 3\ 2\ 4\ 1 \end{pmatrix}$ are permutations of the set $\mathbb{P}_4 = \{1, 2, 3, 4\}$, find the following permutations.

72. $\rho \circ \sigma$ **73.** $\sigma \circ \rho$

74. $\rho^2 = \rho \circ \rho$ **75.** $(\rho \circ \sigma) \circ \tau$

76. ρ^{-1} **77.** τ^{-1}

78. $\rho^{-1} \circ \tau^{-1}$ **79.** $(\tau \circ \rho)^{-1}$

80. If S is a finite set with more than two elements, show that there are permutations σ and τ of S such that $\sigma \circ \tau \neq \tau \circ \sigma$.

Problem Set 6

81–82. For each of the following functional equations satisfied by the function $f : \mathbb{R} \to \mathbb{R}$, find

(a) $f(0)$

(b) $f(-x)$ in terms of $f(x)$

(c) $f(n)$, where $n \in \mathbb{P}$

(d) $f(n)$, where $n \in \mathbb{Z}$

(e) $f(r)$, where $r \in \mathbb{Q}$.

81. $f(x + y) = f(x) + f(y)$ and $f(1) = p$
82. $f(x + y) = f(x) \cdot f(y)$ and $f(1) = q$

83. Give one example of a function $f : \mathbb{R} \to \mathbb{R}$ satisfying $f(x + y) = f(x) \cdot f(y)$ with $f(1) = 2$.

84. Give two examples of functions $f : \mathbb{R} \to \mathbb{R}$ satisfying $f(x + y) = f(x) \cdot f(y)$.

85. Solve the functional equation

$$f(xy) \;=\; f(x) + f(y)$$

when f is a function from $\mathbb{R}$ to $\mathbb{R}$.

86. If $f(x + y) = f(x) \cdot f(y)$ and f is a bijection, show that its inverse satisfies the functional equation

$$f^{-1}(xy) \;=\; f^{-1}(x) + f^{-1}(y).$$

87. Find all real functions f, of a real variable, of the form

$$f(x) = \frac{ax+b}{cx+d}, \quad \text{where } a, b, c, d \in \mathbb{R},$$

such that $f(f(x)) = x$ for all x except when $cx + d = 0$.

88. Find a function $f : \mathbb{R} \to \mathbb{R}$, different from any of those of the previous example, for which $f \circ f = 1_{\mathbb{R}}$.

[Noncontinuous functions are allowed, though there are continuous examples.]

*89–92. In Section 3.3 we introduced the idea of a relation. Just as the concept of a function $f : X \to Y$ was made precise by defining it to be a special subset of $X \times Y$, we can give a precise definition of a relation. A **relation**, R between the set X and the set Y is any subset of $X \times Y$. If (x, y) is in this subset, we say that x is related to y and we write xRy. If (x, y) is not in the subset, then x is not related to y and we write $x \not\!R y$. A function from X to Y is therefore a special kind of relation between X and Y.*

89. What are all the elements of the subset $\mathbb{P}_4 \times \mathbb{P}_4$ that define the relation greater than?

90. Which subset of $\mathbb{P}_6 \times \mathbb{P}_6$ defines the relation divides? Is this relation a function from $\mathbb{P}_6$ to $\mathbb{P}_6$?

91. Sketch the subset of $\mathbb{R} \times \mathbb{R}$ that defines the equals relation. Do the same for the less than or equals relation. Are either of these relations functions from $\mathbb{R}$ to $\mathbb{R}$?

92. Sketch the subset of $\mathbb{R} \times \mathbb{R}$ that defines the relation R, where xRy if and only if $x^2 + y^2 = 4$. Is this relation a function from $\mathbb{R}$ to $\mathbb{R}$?

93. If the telephone directory defines a bijection from the set of listed subscribers to the set of listed numbers, how would you find the inverse of a given number?

94. Let ℓ and m be lines and P a point in a plane. We try to define a function f, from the points of ℓ to the points of m, by taking any point X on ℓ, and letting the line PX intersect m in the point $f(X)$. Under what geometric conditions will f be a bijection?

95. Does the function $f : \mathbb{R} \to \mathbb{R}$ defined by

$$f(x) = x^3 + x + 1$$

have an inverse? If so, can you find an equation for the inverse function?

96. How many functions are there from $\mathbb{P}_r$ to $\mathbb{P}_n$?

97. How many injections are there from $\mathbb{P}_r$ to $\mathbb{P}_n$?

98. **(a)** Find the number of surjections from a 3-element set to a 2-element set.

   (b) How would you tackle the problem of finding the number of surjections from $\mathbb{P}_r$ to $\mathbb{P}_n$ in general?

   [There is no simple formula. *Hint:* If f is a surjection and a_i is the number of elements of $\mathbb{P}_r$ that f sends to $i \in \mathbb{P}_n$, then $a_1 + a_2 + \cdots + a_n = r$ and $a_i \geq 1$ for all $i, 1 \leq i \leq n$.]

99. Prove that the function $f : X \to Y$ is injective if and only if it satisfies the following condition. "For any set T, and functions $g : T \to X$ and $h : T \to X$, $f \circ g = f \circ h$ implies that $g = h$."

100. Prove that the function $f : X \to Y$ is surjective if and only if it satisfies the following condition. "For any set Z, and functions $g : Y \to Z$ and $h : Y \to Z$, $g \circ f = h \circ f$ implies $g = h$."

   [*Hint:* To prove f is surjective, take $Z = \{z_1, z_2\}$ with $g(y) = z_1$ for all $y \in Y$, and $h(y) = z_1$ for $y \in f(X)$ but $h(y) = z_2$ for $y \notin f(X)$.]

101. If X is any set, finite or infinite, and $\mathcal{P}(X)$ is the set of all subsets of X, then show that there is no surjection from X to $\mathcal{P}(X)$.

   [This shows that the cardinality of $\mathcal{P}(X)$ is always larger than that of X. This gives a method for constructing infinite sets of larger and larger cardinality. Starting with the set $\mathbb{P}$, we obtain $\mathcal{P}(\mathbb{P})$, $\mathcal{P}(\mathcal{P}(\mathbb{P}))$, etc., each of which has larger cardinality than the previous set. It can be shown that $\#\mathcal{P}(\mathbb{P}) = \#\mathbb{R}$. However, it is an extremely difficult problem to determine whether there is an infinite set whose cardinality is strictly greater than $\mathbb{P}$ but strictly less than $\mathbb{R}$. The *continuum hypothesis* states that there is no such set. Many interesting conclusions can be derived from this hypothesis, but it has been shown that the continuum hypothesis cannot be proven, and it cannot be disproven!]

102–105. *Define addition of infinite cardinals by*

$$\#A + \#B \;=\; \#(A \cup B), \quad \text{if } A \cap B = \emptyset$$

and multiplication by

$$\#A \cdot \#B \;=\; \#(A \times B).$$

For finite sets, these definitions agree with the usual notions of addition and multiplication.

102. Show that $\#\mathbb{P} \cdot \#\mathbb{P} = \#\mathbb{P}$.

103. Show that $\#\mathbb{P} + \#\mathbb{P} = \#\mathbb{P}$.

104. Show that $\#\mathbb{P} + \#\mathbb{P}_n = \#\mathbb{P}$.

105. Show that $\#\mathbb{R} + \#\mathbb{P} = \#\mathbb{R}$.

106. (*For Discussion*) A man takes an hour walk. After half an hour two mosquitoes land on him, and he immediately manages to kill one of them. After three quarters of an hour, two more mosquitoes land on him, and he kills one of the three on him. In general, at time $\left(1 - \frac{1}{2^n}\right)$ hours, two mosquitoes land on him. He kills one of those on him, leaving n still alive. How many mosquitoes will be on him when he finishes his walk?

107. A man wishes to invest the same amount of money each year and to have $1000 after 10 years. How much will he have to invest each year if his money earns 10% compounded annually?

108. A mortgage of $20,000 is to be paid off in 10 years by equal payments at the end of each year. If the interest rate is 10% compounded annually, what payments must be made each year?

109. If b is a real number greater than 1, show that b^x is a monotone increasing function, for $x \in \mathbb{Q}$, by proving

   (i) $b^m > b^n$ if $m > n$ and $m, n \in \mathbb{P}$

  (ii) $b^m > b^n$ if $m > n$ and $m, n \in \mathbb{Z}$

 (iii) $a^m > b^m$ if $a > b > 1$ and $m \in \mathbb{P}$

 (iv) $a^m > b^m$ if $a > b > 1$ and m is a positive rational

  (v) $b^m > b^n$ if $m > n$ and $m, n \in \mathbb{Q}$.

110. Let X and Y be real intervals. Prove that if $f : X \to Y$ is a monotone increasing function, then f is an injection.

111. It can be shown that the value of the infinite series

$$\sum_{r=0}^{\infty} \frac{1}{r!} \;=\; e.$$

(a) Prove that, if $m > n \geq 1$, then

$$\frac{1}{(n+1)!} + \frac{1}{(n+2)!} + \cdots + \frac{1}{m!} < \frac{1}{n!}.$$

(b) Calculate e, correct to 4 decimal places, using the above infinite sum.

112. Use the following idea to prove that e is irrational. Suppose that $e = p/q$, where $p, q \in \mathbb{P}$. Multiply the equation

$$\frac{p}{q} \;=\; \sum_{r=0}^{\infty} \frac{1}{r!}$$

by $q!$ and use the fact that

$$\left[\frac{1}{(q+1)} + \frac{1}{(q+1)(q+2)} + \cdots\right] < \left[\frac{1}{2} + \frac{1}{2 \cdot 2} + \cdots\right] \;=\; 1.$$

113–116. *A permutation that interchanges just two elements i and j is called a* **transposition** *and is often denoted by (ij).*

113. What is the inverse of the transposition (ij)?

114. Write the permutation $\begin{pmatrix} 1\ 2\ 3\ 4 \\ 3\ 4\ 1\ 2 \end{pmatrix}$ as a composition of transpositions.

115. Write the permutation $\begin{pmatrix} 1\ 2\ 3 \\ 2\ 3\ 1 \end{pmatrix}$ as a composition of transpositions.

116. Write the permutation $\begin{pmatrix} 1\ 2\ 3\ 4 \\ 2\ 3\ 4\ 1 \end{pmatrix}$ as a composition of transpositions.

117. Let p be a prime and r be a positive integer relatively prime to $p-1$. Prove that $f : \mathbb{Z}_p \to \mathbb{Z}_p$, defined by $f[x] = [x^r]$, is a bijection.

118. For which values of $k \in \mathbb{Z}_7$ is $f_k : \mathbb{Z}_7 \to \mathbb{Z}_7$, defined by $f_k([x]) = [x^4 + kx]$, a bijection? Find the inverse function, f_k^{-1}, for each bijection.

119. If a and b are fixed integers, when is the function $f : \mathbb{Z} \times \mathbb{Z} \to \mathbb{Z}$, defined by $f(x, y) = ax + by$, a surjection?

CHAPTER 7

An Introduction to Cryptography

7.1 CRYPTOGRAPHY

Cryptography is the study of sending messages in a secret or hidden form so that only those people authorized to receive the message will be able to read it. *Crypto* is derived from the Greek word which means "hidden." *Cryptanalysis*, on the other hand, is the science of breaking such cryptographic messages. Cryptography and cryptanalysis are collectively referred to as *cryptology*.

A typical situation is illustrated in the diagram below. The message is to be sent across some communication channel, such as the Internet, a telephone line, wireless, by mail, or through a digital storage device. We assume that this channel is not secure and that an eavesdropper could read any messages in the channel. The sender *encrypts* the message to make it unreadable by an eavesdropper. The receiver then has to *decrypt* the encoded message to recover the original message. For example, you would want to use such a system if you send your credit card number over the Internet or if you want to make a secure backup of some data that only you could read.

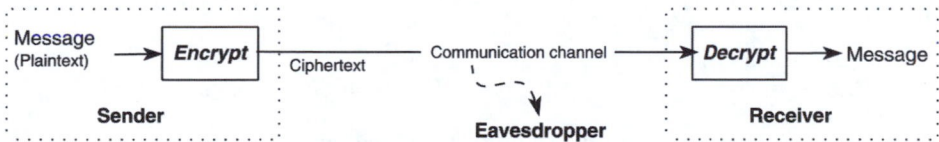

An early user of cryptography was Julius Caesar. Messages sent to his troops were disguised by the following simple method. A letter of the alphabet is replaced by another letter of the alphabet, which is three letters say, to the right in the natural ordering (we are assuming here that the letters of the alphabet are on a circle, so that A follows Z). Hence the word

ATTACK   would become   DWWDFN.

The message being sent, in this case "ATTACK," is usually referred to as the *plaintext*, and the encrypted message, "DWWDFN," is referred to as the *ciphertext*. The above cipher is commonly referred to as a *Caesar cipher* and is, of course, very easily *broken*. That is, the plaintext can be deduced from the ciphertext without knowing how the replacement of letters was done.

You might like to try your hand at cryptanalyzing the following piece of ciphertext. A standard practice is to suppress blanks in the plaintext and block the ciphertext into a fixed number of characters.

```
PDAEZ   AWXAD   EJZPD   EOXKK   GEOPD   WPWHH   PDAQO
ABQHP   KLEYO   EJWYK   HHACA   WHCAX   NWYKQ   NOASE
HHBKH   HKSBN   KIPDA   MQAOP   PKOKH   RALKH   UJKIE
WHAMQ   WPEKJ   OKRAN   RWNEK   QOJQI   XANOU   OPAIO
```

In a Caesar cipher we apply the same shift to each letter in the plaintext message. We could complicate the situation a bit more by using more than one translation and applying them in some systematic fashion. In order to describe this method let's assign each letter of the alphabet the integers 1 through 26, respectively. So A will be 1, B will be 2, and so on. Select a word, say MATH, as the *key* for the cipher. Suppose the plaintext of the message is

THE TIME HAS COME.

To put this into a ciphertext (or to encrypt it), we will add the key word letter by letter to the plaintext (repeating the key word as often as necessary) and reduce modulo 26. For example,

THE	TIME	HAS	COME	$\Longleftarrow$	*Plaintext*
MAT	HMAT	HMA	THMA	$\Longleftarrow$	*Key*
GIY	BVNY	PNT	WWZF	$\Longleftarrow$	*Ciphertext*

The ciphertext is formed as follows:

$$T + M = 20 + 13 = 33 \equiv 7 \pmod{26}$$

and since G is 7 we have that
$$T + M = G.$$

The ciphertext would likely be transmitted as

GIYBV NYPNT WWZF

so that the length of the individual words in the plaintext would not be given away by the ciphertext. Notice that every fourth letter of the plaintext is being translated by the same amount. That is, every fourth letter is being acted upon by the same Caesar cipher and, in total, four distinct Caesar ciphers are being combined here.

In order for two people to communicate using a Caesar cipher, they must both know the key. If either party divulges it, the cipher scheme becomes useless. We would then say that the cipher scheme has been *compromised*. Of course, it may be possible for an opponent to break the scheme without either of the legitimate users giving away the key. An eavesdropping opponent might be able to deduce the key by simply looking at enough ciphertext. When the key has few letters in it this is quite likely, and hence very long keys are needed.

A further way to reduce this problem is to change the keys regularly. This poses problems for legitimate users since they must somehow have a mechanism to exchange very long keys securely.

The Caesar cipher is a special case of what is called a *substitution cipher*. To describe a substitution cipher we let $\mathcal{A}$ be the set of letters of the alphabet and we let

$$f : \mathcal{A} \longrightarrow \mathcal{A}$$

be a permutation on $\mathcal{A}$. If we associate the letters of the alphabet with the integers $1, 2, \ldots, 26$ or to be more precise, the congruence classes $[1], [2], \ldots, [26]$, then the permutation describing the Caesar cipher is

$$f(x) \equiv x + 3 \pmod{26}.$$

There are 25 simple Caesar ciphers (plus one trivial one), and altogether 26! different permutations that could be used as substitution ciphers. One type of substitution cipher that makes use of some of the algebraic ideas developed earlier in the book is as follows. Suppose we select an integer a, such that the $\gcd(a, 26) = 1$. Then define

$$f(x) \equiv ax \pmod{26}.$$

It is not difficult to check that this function permutes the congruence classes of the integers modulo 26. We made use of this fact in the proof of Fermat's Little Theorem.

7.2   PRIVATE-KEY CRYPTOGRAPHY

A *private-key system* is a method for data encryption (and decryption) that requires the parties who communicate to share a common key. The substitution ciphers of the previous section are examples of private-key systems. It may be useful to illustrate one more example of a private-key system.

Let's assume that our data consists of binary strings; that is, sequences of 0's and 1's. This is how data is commonly represented in computers and telecommunication networks. Our method of encryption will be as follows.

A positive integer n is selected. A key will be a sequence of n 0's and 1's. Hence there will be 2^n possible keys. We will assume that the data to be encrypted is a sequence whose length is a multiple of n. (This is not a severe restriction since we can always pad out a data sequence by adding 0's to the end of it.) Partition the data sequence into blocks of length n. We will encrypt each block of data by adding the key to each block of data digit by digit using addition modulo 2, so that the resulting ciphertext digit is also a 0 or a 1. We illustrate this with an example.

Example 7.21.   Encrypt the data sequence (or plaintext)

$$0\ 0\ 1\ 1\ 0\ 0\ 1\ 0\ 0\ 0\ 0\ 1\ 1\ 0\ 1\ 0\ 1\ 0\ 1\ 0\ 0$$

using blocks of length $n = 7$ and the key $1\ 0\ 1\ 1\ 0\ 1\ 1$.

Solution.   Partition the data sequence into the three sequences

$$M_1 \ = \ 0011001 \qquad M_2 \ = \ 0000110 \qquad M_3 \ = \ 1010100.$$

Add the key $K = 1011011$ to M_1, using addition modulo 2 (*not* addition of binary numbers), digit by digit, to get the ciphertext C_1 as follows:

$$\begin{aligned} M_1 &= 0011001 \\ K &= 1011011 \\ C_1 &= 1000010. \end{aligned}$$

For example, adding the third digits of M_1 and K, we obtain $1 + 1 = 0$, and there are no carry digits. Then do the same to the other blocks, by adding K to M_2 and M_3, to get the ciphertext   100001010111010001111. $\square$

To decrypt the ciphertext, we need only add the same key to the blocks of the ciphertext, digit by digit modulo 2, and the plaintext message is recovered.

In practice the number n would be selected to be much larger, for otherwise a cryptanalyst could simply try all possible keys, and see which one gives a meaningful sequence. In the example above, there would be only $2^7 = 128$ keys to try, which would be an easy task using a computer.

Private-key systems have advantages and disadvantages. One very practical advantage is that they can be designed and built to handle very large volumes of data quickly, because the digit addition modulo 2 can be performed extremely quickly using exclusive-or gates (which were mentioned in Exercise 14 of Chapter 1). For example, some of the commercially successful private-key schemes can handle data in the range of gigabits (10^9 0's or 1's) per second.

In 1979, the National Security Agency (NSA) of the U.S. government endorsed the private-key scheme known as the Data Encryption Standard (DES) that used 56-bit keys. The NSA is most likely the largest code-making and code-breaking organization and among the largest employers of mathematicians in the world. Although endorsed for U.S. government use, the DES was adopted by the commercial community and became the most widely used cryptographic algorithm in history. However, by 1999 special-purpose hardware could crack the code in approximately 22 hours by trying all the 2^{56} DES keys. This spurred a movement to what is referred to as triple DES or 3DES, which uses three iterations of the DES algorithm with two distinct 56-bit keys, resulting in 2^{112} possible keys. Triple DES provided sufficient security for commercial uses.

In 2002, the U.S. government endorsed another private-key scheme called the Advanced Encryption Standard (AES) as a replacement for the aging DES and 3DES. There are three key sizes that the AES supports: 128, 192, 256 bits. These provide different levels of security, though it takes longer to encrypt and decrypt the data using the larger key sizes.

Perhaps the biggest disadvantage of these systems is the fact that two users of the scheme must somehow exchange a common key in a secure fashion before any data is encrypted. These keys should also be changed relatively often for other cryptographic reasons. To illustrate why this is a problem, consider a network of 500 users, where each user may want to communicate with each other user in private, using a common private-key scheme. Each pair of users should use a different key, which requires the exchange, in some secure way, of $\binom{500}{2} = \frac{(500)(499)}{2} = 124,750$ keys. If this key exchange needs to be done once a day, or even once a month, then there is a major problem to overcome. In the next section we will show how to surmount this difficulty.

7.3 PUBLIC-KEY CRYPTOGRAPHY

In 1976, a radically new idea was introduced into cryptography. Very simply, the idea was as follows. Suppose that we could find a method for data encryption and decryption where the key for encryption was different from the key for decryption and that the knowledge of one of these keys would not allow one to find the other. Why is this of interest? To answer this, suppose that two people A and B wish to communicate in private, and that A has produced an encryption key k_1 and a decryption key k_2. A sends B the encryption key k_1. B can now encrypt data using this key k_1 and send it to A, who decrypts it using the key k_2. An opponent might intercept the transmission of the key k_1 and hence would be able to send encrypted messages to A. However, the point is that only A can decrypt them. If A wants to send messages to B, then B will also have to construct an encryption key ℓ_1 and a decryption key ℓ_2, and then send A the key ℓ_1. Therefore, we no longer care whether or not opponents intercept the encryption keys, and we might just as well make them public knowledge. Hence, the term **public-key cryptosystem** has been used to describe this method. The encryption key k_1 is called A's **public key**, while the decryption key k_2 is called A's **private key**.

If it were possible to realize such a scheme, then our problem of managing keys is greatly reduced. Referring back to our 500-user network, if each user produces a public encryption key and a private decryption key, then all the encryption keys can be placed in a central public file accessible to all members of the network. If user A wishes to communicate with another user B, then A simply looks up B's encryption key in the public file, encrypts the data, and sends it to B. Only B knows the secret decrypting key, and hence only B can recover the original data. This has reduced the number of keys to be transmitted in the system from 124,750 to 500.

Conceptually, the idea of a public-key system is very attractive, but is it possible to realize it? The answer is a resounding yes! Several ways to realize public-key schemes give rise to what we consider are the most exciting applications of classical algebra to be developed in recent years.

Although the preceding description of the basic concept of public-key cryptography would appear to solve the key management problem, there are other practical issues to be resolved. Suppose A wishes to send an encrypted message to B. A requests B's public key from the public directory. The key received by A is purportedly that of B, but what if an adversary, Eve, intercepts B's key, replaces it with her own public key, and forwards it on to A? Using this key, A encrypts a message that she believes only B can read. Of course, since A used Eve's key, only Eve is able to read the message. Eve could now re-encrypt the message she reads from A, using B's public key, and then forward it to B. A and B would be none the wiser. Such a deception is commonly referred to as a "man-in-the-middle attack'" (even if Eve is the culprit). This shows the necessity of authenticating public keys. When A receives a public key purportedly belonging to B, she should be able to verify the authenticity of it. There are methods to deal with authentication, but they will not be covered here.

It should be mentioned that many realizations of public-key cryptography have been developed since public-key cryptography was discovered in 1976 by Diffie, Hellman, and Merkle. Some are only of theoretical interest, some have different and perhaps better properties than others, while only a few are practical and commercially viable. In the next section one of the first public-key schemes, which is both practical and commercially viable, will be described. This scheme has an additional property called a digital signature. This feature is extremely important for authentication and non-repudiation. Non-repudiation refers to the feature where A produces a digital signature for a message, and at any future date A cannot (even if she wants to) deny that she signed the message.

7.4   THE RSA SCHEME

In 1977, shortly after the idea of a public-key system was proposed, three mathematicians, Ron Rivest, Adi Shamir, and Len Adleman, gave a concrete example of a public-key cryptosystem. In honor of its discoverers the method is commonly referred to as the *RSA scheme*. We describe how it works.

We are going to think of our data as being integers. If our data is simply English text, then we might let

$$\text{blank} \leftrightarrow 00, \quad \text{A} \leftrightarrow 01, \quad \text{B} \leftrightarrow 02, \quad \text{C} \leftrightarrow 03, \quad \ldots \quad \text{Z} \leftrightarrow 26$$

and a piece of text such as MATH would be thought of as the integer 13012008.

A user, called Ursula, who wishes to participate in the network must first produce two keys: one public key for encryption, and one private key for decryption. To use the RSA scheme, Ursula begins by selecting two large distinct prime numbers p and q and then multiplies them together to obtain the integer $n = pq$. Let $\phi(n) = (p-1)(q-1)$. This is the Euler phi function from Section 3.7, since

$$\phi(n) \;=\; \phi(pq) \;=\; \phi(p)\phi(q) \;=\; (p-1)(q-1).$$

The reason why this number $(p-1)(q-1)$ is important is because of Corollary 3.73, which is a generalization of Fermat's Little Theorem to a product of two primes. Ursula now selects an integer $e > 1$ such that $\gcd(e, \phi(n)) = 1$. In practice this is easy to do, since a randomly selected e will have a fairly high probability of being relatively prime to $\phi(n)$, if n is large enough. This integer e will turn out to be part of the encryption key. Ursula now solves the linear Diophantine equation

$$ed \equiv 1 \pmod{\phi(n)}$$

for d. Recall from Theorem 3.54 that since $\gcd(e, \phi(n)) = 1$, there is precisely one congruence class modulo $\phi(n)$ that satisfies this congruence and hence exactly one integer between 0 and $\phi(n)$ that satisfies it. Take d to be the integer in this range. Ursula now makes the integers e and n public knowledge and can destroy p and q, if she desires. (We will see a little later on that there may be reasons to keep p and q.) The pair of integers (e, n) is Ursula's *public key* that is used for encryption. The pair of integers (d, n) is her *private key* that she uses for decryption. She must keep d secret and not divulge it to anybody.

Now suppose that Sue is in the network and wants to send a message M to Ursula. Recall that messages are integers. We restrict our messages even further by requiring that they be in the range 0 to $n-1$. This is not a severe restriction, as the following example illustrates.

If $n = 12319$ and we want to send the message MATH which is 13012008 then we would form two messages, one being MA and the other TH to get two integers 1301 and 2008 respectively, that have fewer than five digits. So in general we simply block off the digits of the entire message into messages which are smaller than n.

To encrypt the message M, the sender Sue obtains Ursula's public encryption key (e, n). Sue now exponentiates M to the power e to get M^e and divides by n to get a remainder C. This can be expressed as

$$M^e \;=\; qn + C, \quad \text{where } 0 \le C < n$$

for some integer quotient q. In congruence notation this is

$$M^e \equiv C \pmod{n}, \quad \text{where } 0 \le C < n.$$

The sender Sue takes C as the ciphertext to transmit to Ursula.

Decryption is the inverse of encryption. Ursula decrypts the received ciphertext C in the same way that Sue encrypted the message, except that Ursula uses her private key (d, n). Ursula computes C^d and divides it by n to get a remainder R. As above, this can be expressed as

$$C^d \equiv R \pmod{n}, \quad \text{where } 0 \le R < n.$$

We claim that this decrypted message R is the same as the original message M.

The following proof is independent of Section 3.7 on the Euler phi function, though it does essentially include a proof of Corollary 3.73.

Theorem 7.41.   *Let p and q be distinct primes, $n = pq$, and e and d be positive integers such that $ed \equiv 1 \,(\mathrm{mod}\ (p-1)(q-1))$. If $0 \le M < n$ and*

$$
\begin{aligned}
M^e &\equiv C \pmod{n} \\
C^d &\equiv R \pmod{n}, \quad \text{where } 0 \le R < n,
\end{aligned}
$$

then $R = M$.

Proof.   The proof is a nice application of Fermat's Little Theorem 3.42 and the Chinese Remainder Theorem 3.62. Since $ed \equiv 1 \,(\mathrm{mod}\ p - 1)(q - 1))$, there exists a positive integer k such that

$$ed = 1 + k(p-1)(q-1).$$

Now

$$
\begin{aligned}
R &\equiv C^d \pmod{n} \\
&\equiv (M^e)^d \pmod{n} \\
&\equiv M^{ed} \pmod{n} \\
&\equiv M^{1+k(p-1)(q-1)} \pmod{n} \\
&\equiv M M^{k(p-1)(q-1)} \pmod{n}.
\end{aligned}
$$

We show that the right side is congruent to M by splitting the congruence into a congruence modulo p and a congruence modulo q.

For the congruence modulo p, consider two cases depending on whether $p \nmid M$ or $p | M$.

Case 1.   If $p \nmid M$, then Fermat's Little Theorem 3.42 implies that

$$M^{p-1} \equiv 1 \pmod{p}.$$

Hence

$$
\begin{aligned}
M^{k(p-1)(q-1)} &\equiv (M^{p-1})^{k(q-1)} \pmod{p} \\
&\equiv 1^{k(q-1)} \equiv 1 \pmod{p}.
\end{aligned}
$$

Multiplying both sides of this congruence by M gives

$$MM^{k(p-1)(q-1)} \equiv M \pmod{p}.$$

Case 2.   If $p|M$, then $M \equiv 0 \pmod{p}$. Hence

$$MM^{k(p-1)(q-1)} \equiv 0 \pmod{p}$$
$$MM^{k(p-1)(q-1)} \equiv M \pmod{p}.$$

In both Case 1 and Case 2 we have $MM^{k(p-1)(q-1)} \equiv M \pmod{p}$.

An analogous argument shows that

$$MM^{k(p-1)(q-1)} \equiv M \pmod{q}.$$

We therefore have the two simultaneous congruences

$$MM^{k(p-1)(q-1)} \equiv M \pmod{p}$$
$$MM^{k(p-1)(q-1)} \equiv M \pmod{q}.$$

Since p and q are distinct primes, the Chinese Remainder Theorem 3.62 implies that the simultaneous congruences are equivalent to the congruence

$$MM^{k(p-1)(q-1)} \equiv M \pmod{n},$$

where $n = pq$. Hence $R \equiv M \pmod{n}$ and, since R and M are both integers between 0 and $n-1$, it follows that $R = M$.

Hence the receiver has decrypted the received message R to obtain the original message M. $\qquad\square$

Note that we have just proved that

$$M^{k\phi(n)+1} \equiv M \pmod{n} \quad \text{for all } M \text{ and all positive } k.$$

Encryption, using the public key (e, n), can be considered as the function

$$f : \mathbb{Z}_n \longrightarrow \mathbb{Z}_n \quad \text{defined by} \quad f([M]) = [M^e],$$

and decryption, using the private key (d, n), can be considered as the function

$$g : \mathbb{Z}_n \longrightarrow \mathbb{Z}_n \quad \text{defined by} \quad g([C]) = [C^d].$$

The above theorem shows that f and g are inverse functions.

RSA Cryptographic Scheme 7.42.   *The RSA scheme for encrypting and decrypting messages is summarized in the following steps. The message to be sent is represented by a large integer M.*

SELECTING THE PUBLIC AND PRIVATE KEYS

1. Select large distinct prime numbers p and q, and form $n = pq$.
2. Select an integer e such that $\gcd(e, (p-1)(q-1)) = 1$ and $1 < e < (p-1)(q-1)$.
3. Solve the congruence

$$ed \equiv 1 \quad (\mathrm{mod}\ (p-1)(q-1))$$

   for an integer d, where $1 < d < (p-1)(q-1)$.
4. The public encryption key is (e, n).
5. The private decryption key is (d, n).

SENDING ENCRYPTED MESSAGES TO A USER

1. Obtain the user's public key (e, n).
2. Make sure that the message M is an integer such that $0 \le M < n$.
3. Compute
$$M^e \equiv C \quad (\mathrm{mod}\ n), \quad \text{where } 0 \le C < n.$$
4. Transmit the integer C.

DECRYPTING RECEIVED MESSAGES

1. Use your private key (d, n).
2. Receive the integer C, where $0 \le C < n$.
3. Compute
$$C^d \equiv R \quad (\mathrm{mod}\ n), \quad \text{where } 0 \le R < n.$$
4. R is the original message.

Example 7.43. Find a set of keys in the RSA system using the primes $p = 631$ and $q = 421$.

Solution. The product of the primes is $n = pq = 265651$ and

$$\phi(n) \;=\; (p-1)(q-1) \;=\; 264600.$$

Selecting $e = 11$, we check that $\gcd(e, \phi(n)) = 1$. Solving

$$11d \equiv 1 \pmod{264600}$$

by the Extended Euclidean Algorithm 2.25 gives $d = 216491$.

Hence the public key is the integer pair $(e, n) = (11, 265651)$ and the private key is the integer pair $(d, n) = (216491, 265651)$.

Check. $ed = 11(216491) = 2381401 = 9(264600) + 1 \equiv 1 \pmod{264600}$ $\qquad\square$

Example 7.44. Send the message MATHEMATICS to the user with public key $(e, n) = (11, 265651)$, using the RSA system.

Solution. If we group the letters of the message in threes, then each group gives rise to an integer that is smaller than n and hence is suitable as a message block for the RSA system. In order that each block has three letters in it, we add a blank to the last block. The blocks and their associated integers are

$$
\begin{array}{ccccc}
\text{MAT} & \longleftrightarrow & 130120 & = & M_1 \\
\text{HEM} & \longleftrightarrow & 080513 & = & M_2 \\
\text{ATI} & \longleftrightarrow & 012009 & = & M_3 \\
\text{CS\_} & \longleftrightarrow & 031900 & = & M_4.
\end{array}
$$

We now encrypt the message blocks individually using the public key. We can either use a computer algebra system, such as *Maple* or *Mathematica*, to compute the eleventh power of the congruence class, or use the Square and Multiply Algorithm 7.46. We show how to calculate M_1^{11} in Example 7.47.

$$
\begin{array}{rcll}
M_1^{11} & \equiv & 163629 & \pmod{265651} \\
M_2^{11} & \equiv & 263790 & \pmod{265651} \\
M_3^{11} & \equiv & 177426 & \pmod{265651} \\
M_4^{11} & \equiv & 169750 & \pmod{265651}
\end{array}
$$

Hence the ciphertext is   163629   263790   177426   169750. $\qquad\square$

Example 7.45. Decrypt the ciphertext $C = 163629$ using the RSA system with the private key $(d, n) = (216491, 265651)$.

Solution.   We have to compute $C^d \pmod{n}$. Using a computer algebra system or the Square and Multiply Algorithm 7.46, we find

$$C^d \;=\; 163629^{216491} \;\equiv\; 130120 \pmod{265651}.$$

Hence the integer message must be 130120. This corresponds to the letters MAT in the first block of the message from Example 7.44. $\square$

We claimed at the beginning of this section that the RSA system is a public-key cryptosystem. We will be justified in saying this if we can be convinced that a knowledge of the public integers e and n does not reveal the integer d that is used in the private key. An opponent knows the public key (e, n) and so knows the integer n that is the product of two large primes p and q. If it were possible to determine p and q from n, then it would be a simple matter to compute d from e. (After all, Ursula was able to deduce d given e, p and q.)

The security of the RSA system relies on the fact that it is very difficult to factor a large integer into its prime factors. A factoring algorithm, such as the sieve of Eratosthenes given in Theorem 2.55, where we divide by all the primes up to $\sqrt{n}$, is hopelessly inadequate when the integer n is large.

In practice, the primes p and q are chosen to be about 300 decimal digits, so that n is about a 600-digit number. At present, nobody knows how to factor such a large number. Let us do a few calculations to get a feeling for the magnitude of the numbers involved.

Suppose that we are given $n = pq$, where p and q are both primes with about 300 decimal digits each, so that n is approximately 10^{600}. How long would the sieve of Eratosthenes take to factor n? There is a famous theorem, called the Prime Number Theorem, that gives a very good approximation to the number of prime numbers in the interval from 1 to x. It says that there are about $x/\log_e x$ primes in this interval. Hence there are about $\sqrt{n}/\log_e \sqrt{n}$ primes from 1 to $\sqrt{n}$. The sieve of Eratosthenes would require us to do about $\sqrt{n}/\log_e \sqrt{n}$ divisions to find the prime factors of n. If $n \approx 10^{600}$, then $\sqrt{n} \approx 10^{300}$, and the number of divisions is about 10^{297}. There are less than 4×10^7 seconds in a year, so if we could do one trillion divisions every second, we could do at most 4×10^{19} divisions per year of computing. Hence, it would take us approximately 10^{277} years to factor n. Since the solar system is only expected to last another 10^{10} or so years, we can forget about this approach to factoring n.

There are better algorithms known for factoring integers than simply dividing by all primes less than $\sqrt{n}$. But when n is about 10^{600}, these are still infeasible for finding the factors of n. The reason for choosing n as the product of two primes is that integers of this type are among the hardest to factor.

It is believed that as long as n cannot be factored, then it is impossible to find the private decrypting key d from a knowledge of n. Under this assumption, the RSA system is a public-key scheme. The reason we say "believed" is that no proof of this fact has as yet been given. The problem has been studied by many outstanding mathematicians around the world.

The RSA scheme is the basis for a commercially viable security mechanism. In 1982, the inventors of the RSA scheme founded the company RSA Security Inc. to market their idea, and it sells an extensive range of security products that use the scheme.

The RSA algorithm is now widely deployed, and the reader may be familiar with some of the following products and protocols that incorporate the RSA scheme.

- Pretty Good Privacy (PGP), the encryption program that is widely used by individuals on the Web, since it free for noncommercial use.

- Secure Sockets Layer (SSL), the standard Internet protocol for secure communications. For example, this is used to encrypt information, such as a credit card number or personal information, and send it securely to a server, via HTTP over SSL (called HTTPS).

- Lotus Notes, a messaging, calendaring, scheduling, and collaborative application from IBM.

- Smart card payment schemes using the EMV protocol set up by Europay, MasterCard, and Visa.

What is the largest number that can be factored in practice? In 2003, a team of researchers in Germany and several other countries managed to factor the number RSA-576, a challenge number proposed by RSA Laboratories, the research arm of RSA Security Inc. The number RSA-576 has 576 bits (i.e., binary digits), or 174 decimal digits, and is the product of two 87-decimal-digit numbers. This was the first of eight RSA challenge numbers, ranging in size from 576 bits to 2048 bits (617 decimal digits), to be factored in a challenge that started in 2001.

Even though we cannot factor 600-digit numbers, there are methods for determining whether a 600-digit number is composite of not. There are also efficient methods for finding very large prime numbers. For example, in 2003, Michael Shafer, a chemical engineering student at Michigan State University, found a record new prime

$$2^{20,996,011} - 1.$$

It took just over two years to find, using a distributed network of more than 200,000 computers. Such a prime of the form $2^p - 1$ is called a Mersenne Prime. This prime is 6,320,430 digits long and is only the 40th Mersenne prime to have been found. Shafer was one of 60,000 volunteers who contributed spare processing power to the Great Internet Mersenne Prime Search (GIMPS), a service started by George Woltman in 1996.

As we have seen, in order for the RSA scheme to be secure, the keys must be very large integers. This poses problems as far as computations are concerned. How can we compute M^e and M^d modulo n when the numbers have 600 digits? Could you calculate 263790^{216491} or even just 130120^{11}, modulo 265651, on your calculator? You could calculate M^{11} by computing M, M^2, $M^3, \ldots, M^{10}$, M^{11} modulo n, but this would be impossible, even for a computer, if the exponent e had 200 digits.

It is possible to perform these exponentiations using the *square and multiply algorithm*. For example, suppose the exponent is $e = 331$. First find the binary representation of e as

$$e = (101001011)_2 = 2^8 + 2^6 + 2^3 + 2^1 + 2^0.$$

Then, using the law of exponents, we have

$$M^e = M^{331} = M^{2^8} M^{2^6} M^{2^3} M^{2^1} M^{2^0}.$$

We can compute

$$M, \ M^2, \ M^{2^2}, \ M^{2^3}, \ M^{2^4}, \ M^{2^5}, \ M^{2^6}, \ M^{2^7}, \ M^{2^8} \quad (\text{mod } n)$$

by successively squaring the previous term, since

$$\begin{aligned}
M^{2^3} &= M^8 &= (M^4)^2 \\
M^{2^4} &= M^{16} &= (M^8)^2 \\
&\vdots \\
M^{2^8} &= M^{256} &= (M^{2^7})^2.
\end{aligned}$$

We can then multiply the appropriate terms together to obtain M^e modulo n. For $e = 331$, this method would use 12 multiplications to compute M^{331} modulo n.

Square and Multiply Algorithm 7.46.   *The Square and Multiply Algorithm for exponentiating a large power of an integer M modulo n is summarized in the following steps.*

THE SQUARE AND MULTIPLY ALGORITHM

1. To compute M^e modulo n for large e, first write e in binary as

$$e = (r_t \ldots r_2 r_1 r_0)_2, \quad \text{where each } r_i = 0 \text{ or } 1.$$

2. Compute

$$M, \ M^2, \ M^4, \ M^8, \ \ldots, \ M^{2^{t-1}}, \ M^{2^t} \quad (\text{mod } n)$$

by squaring the previous term in the sequence.

3. Then multiply the appropriate terms together, modulo n, to obtain

$$M^e \equiv \prod_{r_i=1} M^{2^i} \quad (\text{mod } n).$$

Step 2 requires t modular multiplications, and step 3 will take at most t modular multiplications. Therefore, M^e can be computed in at most $2t$ modular multiplications, where $t = \log_2(e)$. For a 600-digit integer e this requires about 4200 multiplications. Computer chips have been designed and built to do precisely what we have described. For a randomly chosen e, it takes less than a tenth of a second to compute M^e modulo n for n approximately 10^{600}.

Example 7.47. Use the square and multiply algorithm to compute

$$130120^{11} \pmod{265651}.$$

Solution.   The binary representation of the exponent is $11 = (1011)_2 = 8+2+1$. Hence $M^{11} = M^8 M^2 M$. Now calculate the powers of M. Since the modulus is 265651, and $265651^2 \approx 7 \times 10^{10}$, you can only do this on your calculator if it is accurate to 11 places.

$$
\begin{aligned}
M &\equiv 130120 \pmod{265651} \\
M^2 &\equiv 213566 \pmod{265651} \\
M^4 \equiv (M^2)^2 &\equiv 19213 \pmod{265651} \\
M^8 \equiv (M^4)^2 &\equiv 150130 \pmod{265651} \\
M^{11} &\equiv M^8 M^2 M \pmod{265651} \\
&\equiv M^8(253763) \pmod{265651} \\
&\equiv 163629 \pmod{265651}
\end{aligned}
$$

This is the first block of the ciphertext in Example 7.44. $\square$

At the beginning of this section we indicated that Ursula might not want to destroy the primes p and q. In our discussion above, we did not use the primes p and q after the keys were selected. However, p and q can be used to speed up the decryption significantly. To decrypt the ciphertext C, Ursula must compute C^d (mod n). Since $n = pq$ and $\gcd(p,q) = 1$, Ursula can compute the congruences

$$
\begin{aligned}
C^d &\pmod{p} \\
C^d &\pmod{q}
\end{aligned}
$$

and recombine them using the Chinese Remainder Theorem, to obtain $C^d \pmod{pq}$. Note that this only works for decryption, because the person encrypting the message does not know the primes p and q but only knows their product n.

Example 7.48.   Decrypt the second block, 263790, of the ciphertext sent in Example 7.44 by computing

$$263790^{216491} \pmod{265651}$$

using the Chinese Remainder Theorem.

Solution.   We know that $n = 265651$ factors as $n = pq = (631)(421)$, so we shall compute the exponential modulo 631 and 421 and then use the Chinese Remainder Theorem.

Now $263790 \equiv 32 \pmod{631}$. Since 631 is prime, Fermat's Little Theorem 3.42 implies that

$$a^{630} \equiv 1 \pmod{631}$$
$$a^{630+k} \equiv a^k \pmod{631}$$

and we only need to know the exponent modulo 630. Now $216491 \equiv 401 \pmod{630}$, so

$$263790^{216491} \equiv 32^{401} \pmod{631}$$
$$\equiv 376 \pmod{631},$$

where the computation can be done by the square and multiply algorithm on any calculator.

Similarly, since $263790 \equiv 244 \pmod{421}$ and $216491 \equiv 191 \pmod{420}$,

$$263790^{216491} \equiv 244^{191} \pmod{421}$$
$$\equiv 102 \pmod{421}.$$

If $R = 263790^{216491}$, then

$$R \equiv 376 \pmod{631}$$
$$R \equiv 102 \pmod{421}.$$

The Chinese Remainder Theorem 3.62 guarantees that these congruences have a unique solution modulo n, and it can be computed to be

$$R \equiv 080513 \pmod{265651}$$

This is the second block of the original message in Example 7.44, corresponding to the letters HEM. □

Fermat's Little Theorem 3.42 can be used in conjunction with the square and multiply algorithm to test whether a large number is prime. If $\gcd(a, q) = 1$, the contrapositive of Fermat's Little Theorem is that if $a^{q-1} \not\equiv 1 \pmod{q}$, then q is not prime. Hence you could prove that q was not prime, by finding such an integer a, even though you had no factorization of q.

For example, we shall show that $q = 323$ is not prime, without factoring it. Let us try a simple value of a, say $a = 2$, that is relatively prime to 323. To compute 2^{322} modulo 323, we have

$$
\begin{aligned}
2^8 &\equiv 256, \\
2^{16} &\equiv 256^2 \equiv 290, \\
2^{32} &\equiv 290^2 \equiv 120, \\
2^{64} &\equiv 120^2 \equiv 188, \\
2^{128} &\equiv 188^2 \equiv 137, \\
2^{256} &\equiv 137^2 \equiv 35 \quad (\mathrm{mod}\ 323).
\end{aligned}
$$

Since $322 = 256 + 64 + 2 = (101000010)_2$,

$$
\begin{aligned}
2^{322} &\equiv 2^{256} \cdot 2^{64} \cdot 2^2 \\
&\equiv 35 \cdot 188 \cdot 4 \\
&\equiv 157 \\
&\not\equiv 1 \quad (\mathrm{mod}\ 323)
\end{aligned}
$$

and we have shown that 323 is composite without factoring it. Of course, for this small example it would be quicker to factor it. If we had chosen $a = 18$ instead of 2, then we would get $18^{322} \equiv 1 \ (\mathrm{mod}\ 323)$, but this tells us nothing about the primality of 323. Note that there are a few composite numbers q for which $a^{q-1} \equiv 1 \ (\mathrm{mod}\ q)$ for all a relatively prime to q.

One of the truly remarkable features of the public-key system is its ability to allow a user to sign a message electronically by providing a *digital signature*. Suppose that Ursula is in a network, and she wants to authorize her stockbroker Sue to buy 100000 shares of Dome Petroleum. Let u be Ursula's encryption function defined on integers and based on her public key (e, n). Ursula is the only person who can compute the inverse function u^{-1} because it requires her private key (d, n). Let M represent the message to be sent. Ursula uses her private key to compute $u^{-1}(M)$ and transmits this to Sue. Sue wants to guarantee that this transmission is from Ursula, so she obtains Ursula's public key and computes $u(u^{-1}(M))$. If the result is not gibberish, then this proves that the message must have been sent by Ursula, since she is the only person who can produce u^{-1} with her private key.

Notice that under this scenario, the RSA scheme has not been used for secrecy. A person intercepting the message $u^{-1}(M)$ could obtain u and compute $u(u^{-1}(M))$. If Ursula would like to send the message M privately, then she would use Sue's encryption function s, based on Sue's public key (e', n'). Ursula would send

$$
C = s(u^{-1}(M))
$$

to Sue. Sue would then compute

$$
u(s^{-1}(C)) = u(s^{-1}(s(u^{-1}(M))))
$$

to obtain M. Digital signatures are one of the most attractive features of this system.

Exercise Set 7

1. The following is known to be a simple substitution cipher. Break the code.

PCTPG	ANJHT	GDURG	NEIDV	GPEWN	LPHYJ	AXJHR	PTHPG
BTHHP	VTHHT	CIIDW	XHIGD	DEHLT	GTSXH	VJXHT	SQNIW
TUDAA	DLXCV	HXBEA	TBTIW	DS			

2. The following is known to be a simple substitution cipher with a key length of 3. The following table may be of some help in cryptanalyzing the cipher. It gives the frequency distribution of letters in the English language.

Letter	A	B	C	D	E	F	G	H	I	J	K	L	M
%	8.0	1.5	3.0	4.0	13.0	2.0	1.5	6.0	6.5	0.5	0.4	3.5	3.0

Letter	N	O	P	Q	R	S	T	U	V	W	X	Y	Z
%	7.0	8.0	2.0	0.2	6.5	6.0	9.0	3.0	1.0	1.5	0.5	2.0	0.2

VRMGX	LGKDQ	EZVYA	QVDGD	PGQMP	OZCVI	NQMDB	IKMWT
ZWNIV	QWQCV	MSEIV	SWPVQ	GCIVD	PGRMC	BBQPB	JOJTK
VERWH	WIVRM	OKBKM	AEKTN	OLEVI	UCQEK	TCVOG	LZCMT
CCAKM	INKTI	OJTKM	PMWOZ	IUCMU	WWUDW	HDPGK	TIOJT
KLKCK	QFMTO	LRBQQ	BBQDP	GXQPO	BGOVV	RKGXB	WBGCB
WWXLC	LWWDM	KQPVO	MPRCP	NZGNQ	VGIUZ	ZQFMF	DPCDI
RBMES	AGCWN	EBKYV	VYBJO	OGXMT	KTRYT	AXWOS	INOYW
KBKYV	YKAKW	XQCAK	LTGKV	FDPKC	EQBSN	OIFDW	VRMFO
DGVWR	WMPDW	HGPCD	QUUVQ	GVCCU	QNMTX	INQMD	BI

3–6. For each of the following values of p, q, and e, find the public key (e, n), and the associated private key (d, n) of an RSA scheme.

3. $p = 17$, $q = 19$, $e = 25$
5. $p = 97$, $q = 107$, $e = 5$

4. $p = 59$, $q = 67$, $e = 1003$
6. $p = 211$, $q = 241$, $e = 65$

7–10. For each of the following public keys (e, n), determine the associated private key (d, n) of an RSA scheme.

7. $(e, n) = (5, 7663)$
9. $(e, n) = (1277, 47083)$

8. $(e, n) = (197, 6283)$
10. $(e, n) = (100937, 295927)$

11. Given $n = pq$, $p > q$, and $\phi(n) = (p - 1)(q - 1)$, prove that

$$p + q = n - \phi(n) + 1 \quad \text{and} \quad p - q = \sqrt{(p + q)^2 - 4n}.$$

12–15. Each integer n is the product of two primes p and q, and the Euler phi function $\phi(n) = (p - 1)(q - 1)$. Determine the prime factors p and q.

12. $n = 19837$, $\phi(n) = 19516$
14. $n = 71531$, $\phi(n) = 70992$

13. $n = 6887$, $\phi(n) = 6720$
15. $n = 2751121$, $\phi(n) = 2747700$

16. (a) Prove that the encryption function $f : \mathbb{Z}_n \to \mathbb{Z}_n$, for the RSA system defined by $f[x] = [x^e]$, is a bijection.

   (b) Find the permutation of $\mathbb{Z}_{15}$ defined by the bijection $f : \mathbb{Z}_{15} \to \mathbb{Z}_{15}$ with $f[x] = [x^7]$.

17–20. Encrypt each message M, using the RSA public key (e, n).

17. $M = 47$, $(e, n) = (5, 119)$ **18.** $M = 10$, $(e, n) = (7, 143)$
19. $M = 2425$, $(e, n) = (17, 28459)$ **20.** $M = 21421$, $(e, n) = (13, 101617)$

21–24. Decrypt each received ciphertext C, using the RSA private key (d, n).

21. $C = 32$, $(d, n) = (77, 119)$ **22.** $C = 99$, $(d, n) = (103, 143)$
23. $C = 7415$, $(d, n) = (263, 13261)$ **24.** $C = 1701$, $(d, n) = (519, 2773)$

25–26. Find each congruence class by the square and multiply algorithm.

25. $873^{193} \pmod{1000}$ **26.** $567^{81} \pmod{1024}$

27–30. Encrypt each message M, using the RSA public key (e, n) and the square and multiply algorithm.

27. $M = 1240$, $(e, n) = (17, 4757)$ **28.** $M = 2041$, $(e, n) = (13, 3599)$
29. $M = 2607$, $(e, n) = (21, 12193)$ **30.** $M = 1425$, $(e, n) = (19, 12091)$

31–33. For each exponent e, determine the number of modular multiplications to encrypt an RSA message using the square and multiply algorithm.

31. $e = 92487$ **32.** $e = 1247683$
33. $e = 524289$ **34.** $e = 46321$

35–37. Use the Chinese Remainder Theorem to decrypt each received ciphertext C using the RSA private key (d, n), where $n = pq$.

35. $C = 762$, $d = 899$, $p = 31$, $q = 37$
36. $C = 1120$, $d = 5051$, $p = 79$, $q = 131$
37. $C = 113261$, $d = 9809$, $p = 367$, $q = 401$

38. Let $(e, n) = (1837, 9379)$ be the public encryption key for an RSA system, and let $(d, n) = (5, 9379)$ be the corresponding private decryption key. Decrypt the following received message blocks, where the plaintext has been grouped into message blocks of two letters per block, using the equivalence A $\leftrightarrow$ 01, and so on. You will find it useful to take advantage of the prime factorization of n as $83 \cdot 113$.

$$2485 \quad 1169 \quad 1981 \quad 2897$$

39. Let (e, n) be the public encryption key for an RSA system. Suppose that it takes 10^{-4} seconds to do one modular multiplication and that e has 200 digits with 100 ones in its binary representation. Assuming that modular multiplication is the only time-consuming operation, determine the time required to encrypt a message.

40. Anne has a public key $(e_A, n_A) = (7, 8453)$ and private key $(d_A, n_A) = (7087, 8453)$. Bill has public key $(e_B, n_B) = (1837, 9379)$ and private key $(d_B, n_B) = (5, 9379)$. Anne sends to Bill a signed message encrypted under Bill's public key. The ciphertext comes in two enciphered blocks.

$$5752 \quad 7155$$

Find the message sent by Anne.

41. Suppose that Anne's encryption function is a based on the public key (e_A, n_A), and her decryption function is a^{-1} based on the private key (d_A, n_A). Bill's encryption function is b based on the public key (e_B, n_B), and his decryption function is b^{-1} based on the private key (d_B, n_B). Suppose also that Anne wants to send a digitally signed message M in private to Bill, by sending him $b(a^{-1}(M))$. Anne would first compute $M^{d_A} \equiv C \pmod{n_A}$ with $0 \le C < n_A$. Anne would then compute $C^{e_B} \pmod{n_B}$ provided C is such that $0 \le C < n_B$. This will be true if $n_A < n_B$. If $n_A > n_B$, this may be false and complications arise. Describe several ways to overcome this problem.

Problem Set 7

42. Which elements are fixed under the function

$$f : \mathbb{Z}_{77} \to \mathbb{Z}_{77} \quad \text{defined by} \quad f[x] = [x^7] \,?$$

That is, for which $[x] \in \mathbb{Z}_{77}$ is $f[x] = [x]$?

43. Two people A and B communicate using an RSA system for privacy. An opponent finds out that the messages being passed between the two are limited to a set of 100 messages and the opponent has a list of the messages. Describe a method by which the opponent can read the messages passing between A and B, without ever having to factor n. Can you devise a way for A and B to alter their RSA system slightly so as to avoid such an attack?

44. Suppose that an opponent discovers a nonzero message M that is not relatively prime to the modulus $n = pq$ of an RSA system.

   (a) Show that the opponent can factor n and, hence, break the system.

   (b) If the opponent selects a message at random, determine the probability that the message M is not relatively prime to n.

   (c) If both p and q are larger than 10^{100}, show that the probability in (b) is less than 10^{-99}.

45. We would like to distribute some information to s people so that if any two of the s people combine their information they can deduce the secret positive integer k, but no person alone can do so. Select a prime number p larger than s and k. Then select a polynomial $f(x) = ax + k \in \mathbb{Z}_p[x]$ with $a \neq 0$. Compute pairs $(i, f(i))$, $1 \leq i \leq s$, and distribute these to the s people. Prove that this scheme has the desired properties.

46. We would like to distribute some information to s people so that if any three of the s people combine their information, they can deduce the secret positive integer k, but any fewer than three cannot. Select a prime number p larger than s and k. Then select a polynomial $f(x) = ax^2 + bx + k \in \mathbb{Z}_p[x]$ with $a \neq 0$. Compute pairs $(i, f(i))$, $1 \leq i \leq s$, and distribute these to the s people. Prove that this scheme has the desired properties.

47. **(a)** Generalize the previous two problems so that any t of the s people have enough information to deduce a secret number k but any fewer than t can not.

(b) Six people receive the information pairs $(1,10)$, $(2,6)$, $(3,1)$, $(4,1)$, $(5,1)$, $(6,7)$ generated in this way using $p = 11$. It is known that any four of the six have enough information to deduce the secret integer k. Find k.

48. **(a)** Let $n = pq$, where p and q are primes. Prove that if $p - q$ is known, then n can be factored. (Exercise 11 will help.)

(b) Describe a way to break an RSA system with modulus $n = pq$ if $p - q$ is not too large. (This problem illustrates an important point. When constructing an RSA system, one must pick the primes so that their difference is large.)

49. Use the algorithm developed in the previous problem to break the RSA system having modulus $n = 26,850,099,599$.

50. Jane announces that her public RSA key pair is $(2743, 9797)$. Determine Jane's private key, and decrypt the message 3940 that is sent to her.

51. Compute 8^{132} and 9^{132} modulo 133. Do these calculations tell you anything about the primality of 133?

CHAPTER 8

Complex Numbers

We now come to the final extension of our number systems. The real number system allowed us to find roots of any *positive* number. The complex number system will be constructed from the real numbers so as to contain the roots of *any* number, positive or negative. In fact, any polynomial equation of the form

$$a_n x^n + a_{n-1} x^{n-1} + \cdots + a_1 x + a_0 = 0$$

will have a solution in the complex numbers. This result is known as the Fundamental Theorem of Algebra.

Taking the positive integers, $\mathbb{P}$, as a starting point, our number system was built up by first introducing zero and the negative integers, to obtain all the integers $\mathbb{Z}$. Because of the inability to solve some equations of the form $ax = b$ in $\mathbb{Z}$, the integers were extended to the rational number system, $\mathbb{Q}$. Every linear equation can be solved in $\mathbb{Q}$, but many quadratic and higher-order equations cannot. The real number system, $\mathbb{R}$, enabled us to solve equations of the form $x^n = a$ for positive a, and now the complex number system, $\mathbb{C}$, will allow us to solve such equations for negative a. Since any polynomial equation, even with complex coefficients, has a solution in the complex numbers, there will be no need, and no obvious way, to extend our number system any further.

Complex numbers appear fairly natural when seen from a modern mathematical viewpoint. However there was great suspicion about the square root of negative numbers when they were introduced to solve cubic and quartic equations in the 1500s. They were called impossible, or imaginary, numbers. At that time, even negative numbers were not really considered as numbers and were discarded when they appeared as roots of equations. Complex numbers remained mysterious until the 1800s, when they were given a geometric interpretation as points in the plane and an algebraic interpretation as pairs of real numbers. Soon after this, physicists and engineers were finding complex numbers essential for work in alternating-current electricity, radio communications, and later in quantum mechanics. We introduce complex numbers in Section 8.2 and consider their geometric interpretation as points in the complex plane in Section 8.3.

8.1 QUADRATIC EQUATIONS

Before we construct the complex numbers, we shall derive the well-known formula for solving quadratic equations with real coefficients.

Quadratic Formula 8.11. *If $a, b, c \in \mathbb{R}$, $a \neq 0$ and $b^2 - 4ac \geq 0$, then the quadratic equation*

$$ax^2 + bx + c = 0$$

has the solution

$$x = \frac{-b \pm \sqrt{b^2 - 4ac}}{2a}.$$

Proof. Since $a \neq 0$, we can divide by it to obtain

$$x^2 + \frac{b}{a}x + \frac{c}{a} = 0.$$

Now *complete the square* in x; that is, add a constant to the terms containing x so that they become the square of a linear expression in x. We have

$$x^2 + \frac{b}{a}x + \left(\frac{b}{2a}\right)^2 - \left(\frac{b}{2a}\right)^2 + \frac{c}{a} = 0$$

$$\left(x + \frac{b}{2a}\right)^2 - \frac{b^2}{4a^2} + \frac{c}{a} = 0$$

$$\left(x + \frac{b}{2a}\right)^2 = \frac{b^2 - 4ac}{4a^2}.$$

If $b^2 - 4ac \geq 0$, the right side is positive and has a real square root. Hence

$$x + \frac{b}{2a} = \pm\sqrt{\frac{b^2 - 4ac}{4a^2}}$$

$$= \pm\frac{\sqrt{b^2 - 4ac}}{2a}$$

since $\sqrt{4a^2} = 2|a|$. If $a < 0$, then the $\pm$ sign becomes $\mp$, which gives the same formula. The solution to the quadratic is therefore

$$x = \frac{-b \pm \sqrt{b^2 - 4ac}}{2a}. \qquad \square$$

If $b^2 - 4ac > 0$, then there are two distinct solutions to the quadratic equation. If $b^2 - 4ac = 0$, then there is just one solution. If $b^2 - 4ac < 0$, then there is no real solution.

However, the complex numbers will allow us to find square roots of negative numbers, and then the above result will show that we can also solve the quadratic equation when $b^2 - 4ac < 0$. In fact, the quadratic formula holds even if a, b, and c are complex.

8.2   COMPLEX NUMBERS

A complex number will be defined as a pair of real numbers $(x, y) \in \mathbb{R} \times \mathbb{R}$, but we shall write it as $x + yi$, where i will be a new symbol. We shall define arithmetic on these complex numbers so that the symbol i has the property that $i^2 = -1$.

Definition.   A **complex number** z in *standard form* is an expression of the form $x + yi$, where $x, y \in \mathbb{R}$. The set of all complex numbers is denoted by

$$\mathbb{C} \;=\; \{x + yi \mid x, y \in \mathbb{R}\}.$$

If $z = x + yi$ is a complex number, we shall also write it as $x + iy$. The real number x is called the *real part* of z and is denoted by $\mathrm{Re}(z)$. The real number y is called the *imaginary part* of z and is denoted by $\mathrm{Im}(z)$.

For example, the real part of the complex number $5 - 4i$ is 5, while the imaginary part is -4. If the real part of a complex number is zero, then that complex number is called *purely imaginary*. For example, $0 + \sqrt{3}i = \sqrt{3}i$ is purely imaginary. A complex number with zero imaginary part will be treated as a real number. For example, $3 + 0i$ will just be the real number 3. Denote $1i$ by i.

Electrical engineers normally use j, instead of the symbol i, because they have historically used i to represent the current.

Addition and Multiplication of Complex Numbers 8.21.

$$(a + bi) + (c + di) \;=\; (a + c) + (b + d)i$$
$$(a + bi) \cdot (c + di) \;=\; (ac - bd) + (ad + cb)i$$

These operations can be treated just as the usual addition and multiplication of algebraic expressions, where i is handled as an algebraic symbol with the property that whenever i^2 occurs, it is replaced by -1.

For example,

$$(5 + 4i) + (3 + i) \;=\; (5 + 3) + (4 + 1)i \;=\; 8 + 5i$$
$$(5 + 4i) \cdot (3 + i) \;=\; 5 \cdot 3 + 4 \cdot 3i + 5 \cdot 1i + 4i^2 \;=\; 15 + 17i - 4 \;=\; 11 + 17i$$
$$(2 - 3i) \cdot \left(\frac{1}{2}i\right) \;=\; i - \frac{3}{2}i^2 \;=\; \frac{3}{2} + i$$
$$(1 + i)^2 \;=\; 1 + 2i + i^2 \;=\; 1 + 2i - 1 \;=\; 2i.$$

Consider the operations on two complex numbers with zero imaginary part.

$$(a + 0i) + (c + 0i) \;=\; (a + c) + 0i$$
$$(a + 0i) \cdot (c + 0i) \;=\; ac + 0i$$

These are exactly the same as the operations on the real numbers a and c. We shall therefore identify a complex number with zero imaginary part, $x + 0i$, with the real number x. This means that the real numbers $\mathbb{R}$ are a subset of the complex numbers $\mathbb{C}$, and the complex number system is now an extension of the real number system.

Now consider the complex number i. This is not a real number, because its imaginary part is not zero. When we square this number, we obtain

$$(0+i)^2 \;=\; (0+i)\cdot(0+i) \;=\; -1+0i,$$

which is the real number -1. Hence i is one complex square root of -1; another complex square root of -1 is $-i$.

If a is any positive real number, then

$$(\sqrt{a}i)^2 \;=\; -a+0i \;=\; -a$$

and we have achieved our objective of extending the number system to include square roots of all negative numbers.

Example 8.22.   Let $z=4+i$ and $w=-3+2i$ be complex numbers. Find (i) $z+w$,   (ii) $z-w$,   (iii) z^2w,   (iv) the real and imaginary parts of w^4.

Solution. (i) $z+w \;=\; (4+i)+(-3+2i) \;=\; 1+3i$

(ii) $z-w \;=\; (4+i)-(-3+2i) \;=\; 7-i$

(iii)

$$
\begin{aligned}
z^2w &= (4+i)^2(-3+2i) \;=\; (16+8i+i^2)(-3+2i)\\
&= (15+8i)(-3+2i) \;=\; -45-24i+30i+16i^2\\
&= -61+6i
\end{aligned}
$$

(iv)

$$
\begin{aligned}
w^4 &= (-3+2i)^4\\
&= (-3)^4+4(-3)^3(2i)+\frac{4\cdot 3}{1\cdot 2}(-3)^2(2i)^2+4(-3)(2i)^3+(2i)^4\\
&= 81-216i+216i^2-96i^3+16i^4 \;=\; 81-216i-216+96i+16\\
&= -119-120i
\end{aligned}
$$

The real part of w^4 is therefore -119, and the imaginary part -120. $\square$

Notice in the above calculation that, since $i^2=-1$,

$$i^4 \;=\; (i^2)^2 \;=\; (-1)^2 \;=\; +1.$$

In fact, all the different powers of i are as follows.

Proposition 8.23.

$$i^n = \begin{cases} 1 & \text{if } n \equiv 0 \pmod 4 \\ i & \text{if } n \equiv 1 \pmod 4 \\ -1 & \text{if } n \equiv 2 \pmod 4 \\ -i & \text{if } n \equiv 3 \pmod 4 \end{cases}$$

Proof.   We shall first show that $i^n = i^m$ if $n \equiv m \pmod 4$. If $m = n + 4k$, where $k \in \mathbb{Z}$, then

$$i^m = i^{n+4k} = i^n \cdot i^{4k} = i^n (i^4)^k = i^n \cdot 1^k = i^n.$$

The result now follows because $i^0 = 1$, $i^1 = i$, $i^2 = -1$, and $i^3 = i \cdot i^2 = -i$. $\square$

Example 8.24.   Calculate $(2+i)^6$.

Solution.   By the Binomial Theorem,

$$\begin{aligned} (2+i)^6 &= 2^6 + 6 \cdot 2^5 i + \frac{6 \cdot 5}{1 \cdot 2} \cdot 2^4 i^2 + \frac{6 \cdot 5 \cdot 4}{1 \cdot 2 \cdot 3} \cdot 2^3 i^3 + \frac{6 \cdot 5}{1 \cdot 2} \cdot 2^2 i^4 + 6 \cdot 2 i^5 + i^6 \\ &= 64 + 192i - 240 - 160i + 60 + 12i - 1 \\ &= -117 + 44i. \end{aligned}$$

$\square$

The operations on complex numbers have the following properties, which we would expect of an addition and multiplication.

Proposition 8.25.   *Let $z, w, t \in \mathbb{C}$ and let $z = x + yi$ and $w = u + vi$, where $x, y, u, v \in \mathbb{R}$. Then*

(i) *$z = w$ if and only if $x = u$ and $y = v$;*
   in other words, two complex numbers are equal if and only if their real parts are equal and their imaginary parts are equal.

(ii) *$(z + w) + t = z + (w + t)$ (associativity of addition).*

(iii) *$z + w = w + z$ (commutativity of addition).*

(iv) *The number $0 = 0 + 0i \in \mathbb{C}$ is such that $0 + z = z$ (existence of a zero).*

(v) *The number $-z = -x - yi \in \mathbb{C}$ is such that $z + (-z) = 0$*
   (existence of negatives).

(vi) *$(z \cdot w) \cdot t = z \cdot (w \cdot t)$ (associativity of multiplication).*

(vii) *$z \cdot w = w \cdot z$ (commutativity of multiplication).*

(viii) *The number $1 = 1 + 0i \in \mathbb{C}$ is such that $1 \cdot z = z$ (existence of a unit).*

(ix) *If $z \neq 0$, the element*

$$z^{-1} = \frac{1}{z} = \frac{x - yi}{x^2 + y^2}$$

is the inverse of z and satisfies $z \cdot z^{-1} = 1$ (existence of inverses).

(x) $z \cdot (w + t) = z \cdot w + z \cdot t$ *(distributive law).*

Proof. *(i)*   The complex numbers $z = x + yi$ and $w = u + vi$ represent the ordered pairs of real numbers (x, y) and (u, v), respectively. These two ordered pairs are equal if and only if their first elements are equal and their second elements are equal; that is $z = w$, if and only if $x = u$ and $y = v$.

(ii)–(x)   All these parts of the proposition can be proved by calculating each expression directly, using 8.21. We shall just perform this calculation for one of the parts and leave the remainder for the reader.

(ix)   The complex number $z = x + yi$ is zero if and only if $x = 0$ and $y = 0$. Hence if $z \neq 0$, then either x or y is nonzero, and it follows that $x^2 + y^2 > 0$. Therefore,

$$\frac{x - yi}{x^2 + y^2} = \frac{x}{x^2 + y^2} - \frac{yi}{x^2 + y^2} \in \mathbb{C}$$

$$\left(\frac{x - yi}{x^2 + y^2}\right)(x + yi) = \frac{x^2 + xyi - xyi + y^2}{x^2 + y^2} = 1.$$

Hence the complex number $\dfrac{x - yi}{x^2 + y^2}$ is the inverse of $x + yi$. $\square$

The properties (ii)–(x) show that the complex number system forms a *field*. Other examples of fields that we have encountered are $\mathbb{R}$, the rational numbers $\mathbb{Q}$, and $\mathbb{Z}_p$, when p is a prime.

One property of the real numbers that does not carry over to the complex numbers is the order relation, which is the ability to say that one number is bigger than another. If we could compare the sizes of all the complex numbers then either $i > 0$ or $i < 0$. In either case, the properties of any ordering would imply that $i^2 > 0$,

Frank and Ernest

PLEASE TAKE A NUMBER

and so $-1 > 0$. This contradicts our order relation in $\mathbb{R}$ and shows that there is no natural order relation in $\mathbb{C}$. It is therefore nonsense to talk about inequalities of complex numbers. However, we can talk about inequalities involving the real or imaginary parts of complex numbers and, as the next section will show, moduli of complex numbers. This is because, for any complex number, its real part, its imaginary part and its modulus are all real numbers.

The breakdown of the complex numbers into different subsets is illustrated in the following diagram.

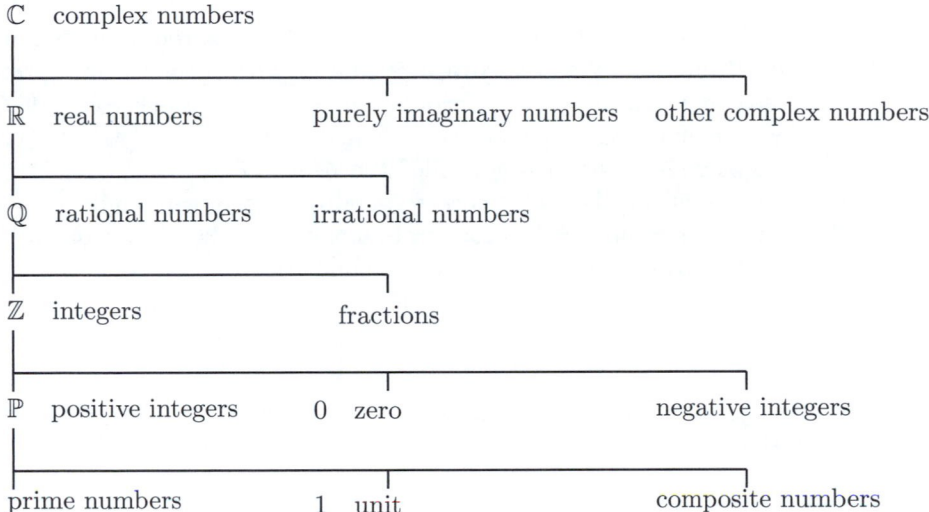

8.3   THE COMPLEX PLANE

The real numbers have a geometric representation as points of a number line. We shall now give a geometric interpretation of the complex numbers and of its various operations, such as addition and multiplication.

The complex number $x + iy$ is really just an ordered pair of real numbers $(x, y) \in \mathbb{R} \times \mathbb{R} = \mathbb{R}^2$, and the elements of $\mathbb{R}^2$ can naturally be interpreted as points of a plane. This allows us to associate to each complex number a point on a plane.

Choose rectangular coordinates in the plane, and label one axis the **real axis** and the other the **imaginary axis**. The geometric representation of the complex number $z = x + iy$ is the point (x, y) in the plane, whose coordinate along the real axis is the real part of z and whose coordinate along the imaginary axis is the imaginary part of z.

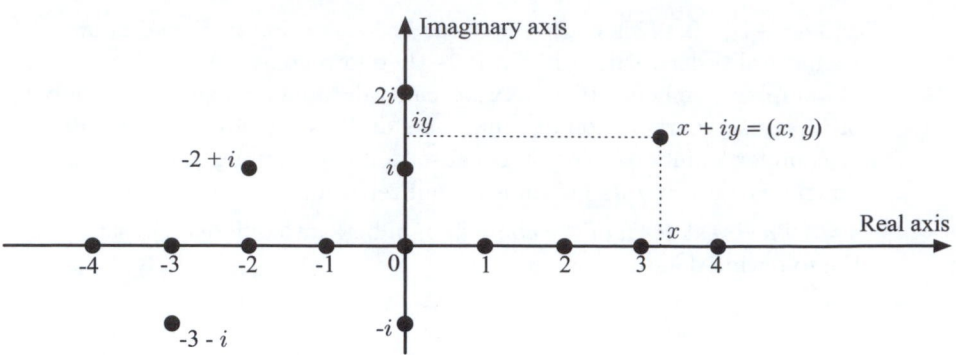

This defines a one-to-one correspondence between the complex numbers and the plane, which is called the **complex plane**. This geometric interpretation was discovered independently by the Norwegian surveyor Caspar Wessel (1745–1818), the Swiss bookkeeper Jean-Robert Argand (1768–1822) and Gauss (1777–1855). The complex plane is sometimes called the *Argand diagram.*

The points on the real axis correspond to all the real numbers $x + 0i$. Hence the real axis is a copy of the usual real number line. The points on the imaginary axis correspond to the purely imaginary numbers $0 + iy$.

If $z = x + iy$ and $w = u + iv$ are two complex numbers, their sum is the complex number

$$z + w = (x + u) + i(y + v).$$

In the complex plane, the point $z + w$ corresponds to the fourth vertex of the parallelogram, whose other vertices are the points z, 0, and w.

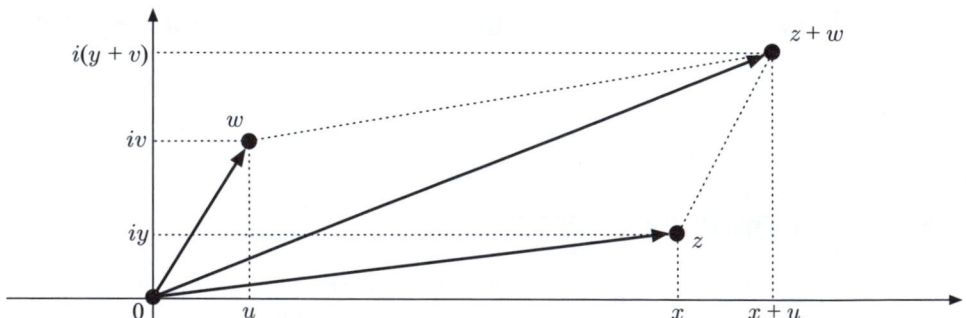

This addition is precisely the *parallelogram law of addition* for vectors. Because of this equivalence with vectors, we sometimes draw a directed arrow from the origin 0 to the point corresponding to a complex number. The vector $z + w$ is the diagonal from 0 to $z + w$ in the parallelogram $z, 0, w, z + w$.

Definition.   The **modulus** or **absolute value** of the complex number $z = x + iy$ is the nonnegative real number

$$|z| = |x + iy| = \sqrt{x^2 + y^2}.$$

If $z = x + i0$, then this complex modulus

$$|z| = \sqrt{x^2} = |x|,$$

the usual absolute value of the real number x.

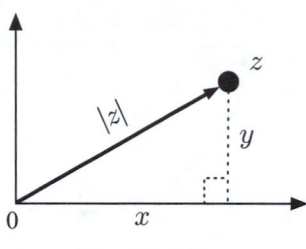

The Modulus

The modulus $|z| = \sqrt{x^2 + y^2}$ of a complex number $z = x + iy$ is just the distance of the point z from the origin in the complex plane. This is the length of the vector from 0 to z.

The distance from the point $w = u + iv$ to $z = x + iy$ is

$$|z - w| = \sqrt{(x - u)^2 + (y - v)^2}.$$

The modulus provides a crude method for comparing sizes of complex numbers. But note that many different complex numbers have the same modulus. For example,

$$\begin{aligned}
|1| &= |-1| \\
&= |i| = |-i| \\
&= \left|\frac{1}{\sqrt{2}} + \frac{i}{\sqrt{2}}\right| \\
&= \left|\frac{1}{2} - \frac{\sqrt{3}i}{2}\right| \\
&= 1.
\end{aligned}$$

Example 8.31.   Sketch the points in the complex plane that have modulus 1.

Solution.   If $z = x + iy$ and $|z| = 1$, then

$$\sqrt{x^2 + y^2} = 1$$

and

$$x^2 + y^2 = 1.$$

Therefore, all the points with modulus 1 lie on a circle of radius 1 and center at the origin.

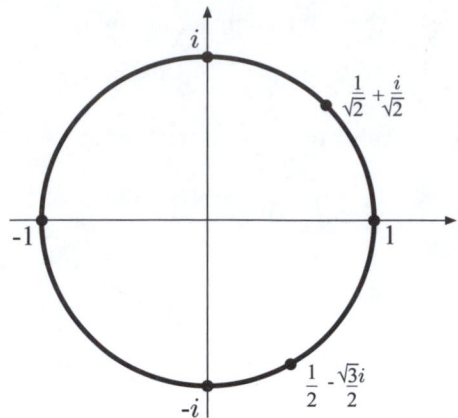

This shows that there is an infinite number of points with modulus 1; among them are the points $1, -1, i, -i, \frac{1}{\sqrt{2}} + \frac{i}{\sqrt{2}}$, and $\frac{1}{2} - \frac{\sqrt{3}i}{2}$. □

Definition.   The **complex conjugate** of $z = x + iy$ is the complex number

$$\overline{z} = x - iy$$

that is obtained by changing the sign of the imaginary part.

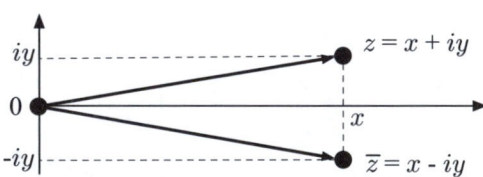

In the complex plane, the complex conjugate, $\overline{z} = x - iy$, is just the reflection of $z = x + iy$ in the real axis. If z is real, then z lies on the real axis, and $\overline{z} = z$.

Example 8.32.   If $z = 2 + i$, plot the points z, $\overline{z}$, and iz in the complex plane.

Solution.   If $z = 2 + i$, then

$$\overline{z} = 2 - i$$

and

$$iz = 2i + i^2$$
$$= -1 + 2i.$$

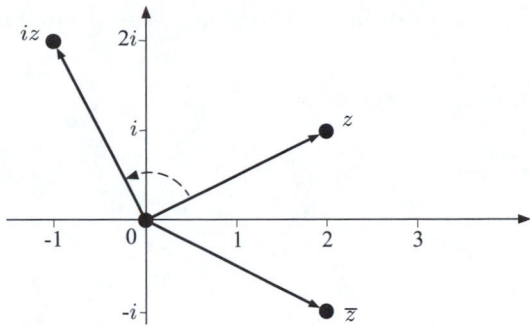

It should be noted that multiplication by i has the effect of rotating the complex number counter clockwise through an angle $\pi/2$. This fact is true for all complex numbers, because if $z = x + iy$, then $iz = -y + ix$, and the fact that z and iz are at right angles can be seen from the diagram. However, to interpret multiplication of complex numbers in general, we need to represent points in the complex plane by polar coordinates, which we introduce in Section 8.5.

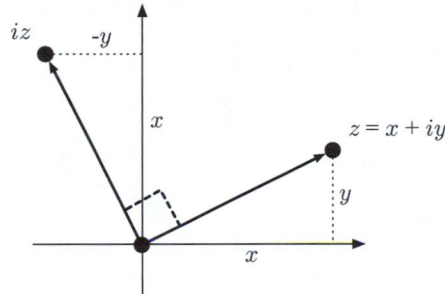

8.4   PROPERTIES OF COMPLEX NUMBERS

If $x + iy$ is any complex number, then $(x + iy)(x - iy) = x^2 + y^2$, and this is always a real number. That is, a complex number times its conjugate is always real. This property is very useful in the division of complex numbers.

The simplest way to calculate the inverse $1/z$ of any nonzero complex number is to multiply numerator and denominator by its conjugate $\bar{z}$. For example, the inverse of $4 + 5i$ is

$$
\begin{aligned}
\frac{1}{4 + 5i} &= \frac{4 - 5i}{(4 + 5i)(4 - 5i)} \\
&= \frac{4 - 5i}{16 + 25} \\
&= \frac{4}{41} - \frac{5i}{41}.
\end{aligned}
$$

The same method can be used to remove any denominator of complex numbers.

Example 8.41. Calculate $\dfrac{5 + 6i}{(7 - i)(-i)}$.

Solution.

$$
\begin{aligned}
\frac{5 + 6i}{(7 - i)(-i)} &= \frac{5 + 6i}{-1 - 7i} \\
&= \frac{(5 + 6i)(-1 + 7i)}{(-1 - 7i)(-1 + 7i)} \\
&= \frac{-5 + 35i - 6i + 42i^2}{1 + 49} \\
&= \frac{-47 + 29i}{50} \\
&= \frac{-47}{50} + \frac{29}{50}i
\end{aligned}
$$

$\square$

Proposition 8.42.   *If z and w are complex numbers, then*

(i) $\overline{z + w} = \overline{z} + \overline{w}$.

(ii) $\overline{zw} = \overline{z}\,\overline{w}$.

(iii) $\overline{\overline{z}} = z$.

(iv) $z\overline{z} = |z|^2$.

(v) $z + \overline{z}$ *is twice the real part of z,*

(vi) $z - \overline{z}$ *is $2i$ times the imaginary part of z.*

Proof.   Let $z = x + iy$ and $w = u + iv$, where $x, y, u, v \in \mathbb{R}$.

(i) $\quad \overline{z + w} \;=\; \overline{(x + u + i(y + v))} \;=\; x + u - i(y + v)$
$\qquad\qquad\; = \; x - iy + u - iv \;=\; \overline{z} + \overline{w}$

(ii) $\qquad \overline{zw} \;=\; \overline{(xu - yv + i(xv + yu))} \;=\; xu - yv - i(xv + yu)$
$\qquad\qquad\; = \; (x - iy)(u - iv) \;=\; \overline{z}\,\overline{w}$

(iii) $\qquad \overline{\overline{z}} \;=\; \overline{x - iy} \;=\; x + iy \;=\; z$

(iv) $\qquad z\overline{z} \;=\; (x + iy)(x - iy) \;=\; x^2 + y^2 \;=\; |z|^2$

(v) $\qquad z + \overline{z} \;=\; x + iy + x - iy \;=\; 2x$

(vi) $\qquad z - \overline{z} \;=\; x + iy - (x - iy) \;=\; 2iy$

$\square$

Corollary 8.43.   *If z is a nonzero complex number, then* $\dfrac{1}{z} = \dfrac{\overline{z}}{|z|^2}$.

Proof.   This follows directly from part (iv) of the preceding result.   □

Proposition 8.44.   *If z and w are complex numbers, then*

   (i) $|z| = 0$ if and only if $z = 0$.

   (ii) $|\bar{z}| = |z|$.

   (iii) $|zw| = |z||w|$.

   (iv) $|z + w| \leq |z| + |w|$ (the triangle inequality).

Proof.   Let $z = x + iy$ and $w = u + iv$, where $x, y, u, v \in \mathbb{R}$.

(i) $|z| = \sqrt{x^2 + y^2}$, which is zero if and only if $x^2 + y^2 = 0$. This happens if and only if $x = y = 0$.

(ii) $|\bar{z}| = |x - iy| = \sqrt{x^2 + (-y)^2} = \sqrt{x^2 + y^2} = |z|$.

(iii)   By Proposition 8.42 (iv),

$$|zw|^2 = (zw)(\overline{zw}) = zw\bar{z}\,\bar{w} = (z\bar{z})(w\bar{w}) = |z|^2|w|^2 = (|z||w|)^2.$$

Since the modulus is a nonnegative real number, it follows that $|zw| = |z||w|$.

(iv)   By Proposition 8.42,

$$\begin{aligned}
|z + w|^2 &= (z + w)(\overline{z + w}) = (z + w)(\bar{z} + \bar{w}) \\
&= z\bar{z} + z\bar{w} + \bar{z}w + w\bar{w} \\
&= |z|^2 + |w|^2 + (z\bar{w} + \bar{z}\,\bar{w}).
\end{aligned}$$

Hence $(|z| + |w|)^2 - |z + w|^2 = 2|z||w| - (z\bar{w} + \bar{z}\,\bar{w}) = 2|z\bar{w}| - (z\bar{w} + (\overline{z\bar{w}}))$. By Proposition 8.42(v), $z\bar{w} + (\overline{z\bar{w}})$ is twice the real part of $z\bar{w}$, and this is always less than or equal to twice the modulus of $z\bar{w}$. Therefore,

$$(|z| + |w|)^2 \geq |z + w|^2$$

and, since the moduli are nonnegative real numbers,

$$|z| + |w| \geq |z + w|.$$ □

The triangle inequality in Proposition 8.44(iv) has a simple geometric interpretation. In the parallelogram with vertices z, 0, w, $z + w$, the distance from 0 to w is $|w|$, and this is the same as the distance from z to $z + w$. Hence, the sides of the triangle with vertices $z, 0$ and $z + w$ have lengths $|z|, |z + w|$ and $|w|$. The inequality $|z + w| \leq |z| + |w|$ expresses the fact that the length of one side of a triangle cannot exceed the sum of the lengths of the other two sides.

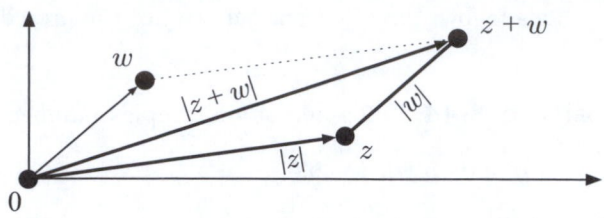

The Triangle Inequality

Example 8.45. When does the equality $|z + w| = |z| + |w|$ hold?

Solution. From the above discussion of the triangle z, 0, $z + w$ in the complex plane, it follows that $|z + w| = |z| + |w|$ if and only if z lies on the line segment from 0 to $z + w$ (and lies between 0 and $z + w$). This happens if and only if the parallelogram $z, 0, w, z+w$ collapses so that w and z lie on a line through the origin and are both on the same side of the origin. Hence either w or z is zero or there exists a positive real number k such that $w = kz$.

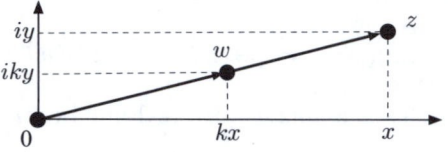

Therefore

$$|z + w| \;=\; |z| + |w|$$

if and only if $z = 0$ or $w = 0$, or $w = kz$ for some positive real number k. ☐

The Quadratic Formula 8.11 solves the quadratic equation

$$ax^2 + bx + c \;=\; 0$$

with real coefficients, when $b^2 - 4ac \geq 0$. If $b^2 - 4ac < 0$, put $b^2 - 4ac = -d$, where $d > 0$; then $i\sqrt{d}$ and $-i\sqrt{d}$ are both complex numbers whose square is $b^2 - 4ac$. The Quadratic Formula can still be used to yield the two complex solutions

$$
\begin{aligned}
x \;&=\; \frac{-b \pm \sqrt{b^2 - 4ac}}{2a} \;=\; \frac{-b \pm i\sqrt{d}}{2a} \\[2mm]
&=\; \frac{-b}{2a} + \frac{i\sqrt{d}}{2a} \quad \text{or} \quad \frac{-b}{2a} - \frac{i\sqrt{d}}{2a}.
\end{aligned}
$$

Notice that these two solutions are complex conjugates of each other.

Hence, if $a, b, c \in \mathbb{R}$, the quadratic equation

$$ax^2 + bx + c \;=\; 0$$

has the following types of solutions depending on the value of the *discriminant* $b^2 - 4ac$.

- If $b^2 - 4ac > 0$, there are two real and distinct solutions.
- If $b^2 - 4ac = 0$, there is one real solution.
- If $b^2 - 4ac < 0$, there are two complex conjugate solutions.

Example 8.46. Solve the equation $2x^2 - 3x + 5 = 0$ for $x \in \mathbb{C}$.

Solution. By the Quadratic Formula we have

$$ x = \frac{3 \pm \sqrt{9 - 40}}{4} = \frac{3 \pm \sqrt{-31}}{4} = \frac{3 \pm i\sqrt{31}}{4}. \qquad \square$$

Example 8.47. Solve the equation $x^2 = -1$.

Solution. We can either solve this by the Quadratic Formula to obtain

$$ x = \pm\sqrt{-1} = \pm i $$

or we can factor $x^2 + 1$ as $(x + i)(x - i)$ and obtain the same solutions. $\qquad \square$

Notice that $\pm\sqrt{-1}$ is well defined as $\pm i$, but $\sqrt{-1}$ alone can lead to trouble, since it is not always clear which complex square root it refers to. See Problem 120.

8.5 POLAR REPRESENTATION

The pair of real numbers (x, y), in the rectangular *Cartesian coordinate system system* with origin O corresponds to the point P in the plane, where the projections of OP onto the two axes are x and y. The *polar coordinate system* is another way of associating to each point in the plane a pair of real numbers. To form the polar coordinate system, we start with the origin O and take the x-axis as polar axis. A point P has polar coordinates (r, θ) if r is the distance from O to P and θ is the angle, in radians measured counter clockwise, between the polar axis and the line OP.

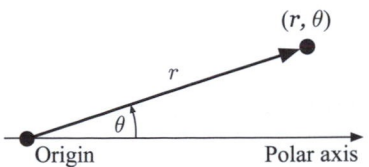
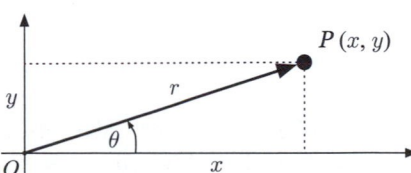

For example, the point with Cartesian coordinates $(1, 1)$ has polar coordinates $(\sqrt{2}, \pi/4)$. The point with Cartesian coordinates $(-2, 0)$ has polar coordinates $(2, \pi)$.

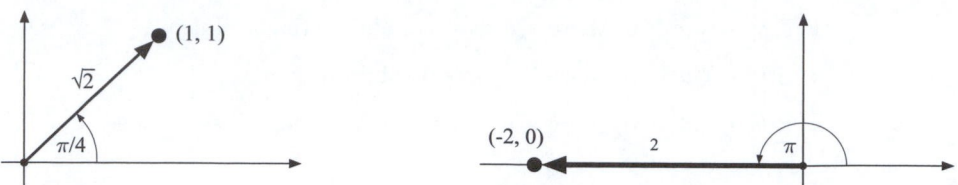

If (r, θ) are the polar coordinates of a point, then r is always a positive or zero real number, while θ can be any real number, positive or negative (a negative angle is measured clockwise). However, unlike Cartesian coordinates, polar coordinates are not unique. For example, the point with polar coordinates $(\sqrt{2}, \pi/4)$ is also represented by the polar coordinates $(\sqrt{2}, 9\pi/4)$ and by $(\sqrt{2}, -7\pi/4)$.

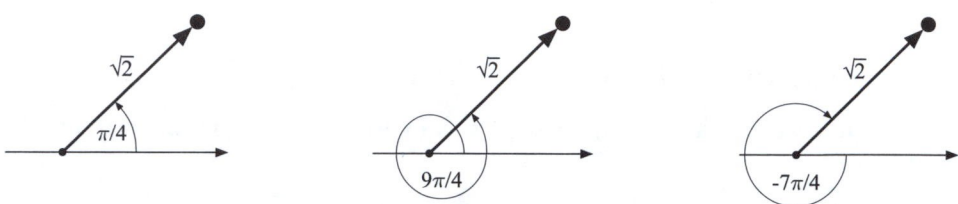

The polar coordinates (r, θ) and $(r, \theta + 2k\pi)$ represent the same point, for any $k \in \mathbb{Z}$. Furthermore, the polar coordinates $(0, \theta)$ represent the origin for all real values of the angle θ.

By applying trigonometry to the right-angled triangle in the following diagram, we can convert the polar coordinates (r, θ) into the Cartesian coordinates (x, y) by means of the following relations.

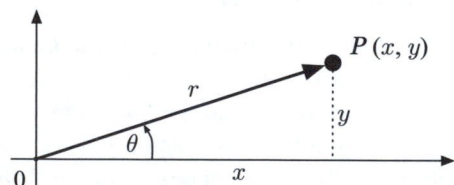

Convert from Polar to Cartesian Coordinates 8.51.

$$x = r \cos \theta$$
$$y = r \sin \theta$$

Conversely, a point whose Cartesian coordinates are (x, y) has the polar coordinates (r, θ) where

$$r = \sqrt{x^2 + y^2}$$

and θ is an angle such that

$$\cos \theta = \frac{x}{r}, \quad \sin \theta = \frac{y}{r} \quad \text{and} \quad \tan \theta = \frac{y}{x}.$$

If $r = 0$, then θ can be any angle. There is a slight complication in finding the exact angle θ; it is either $\mathrm{Tan}^{-1}(y/x)$ or $\pi + \mathrm{Tan}^{-1}(y/x)$, depending on the signs of x and y. The safest way to find the angle is to plot the required point in the plane to see which quadrant it lies in. Then θ can be found once $\mathrm{Tan}^{-1}(y/x)$ or $\mathrm{Sin}^{-1}(y/r)$ or $\mathrm{Cos}^{-1}(y/r)$ is known.

Let $z = x + iy$ be any complex number and let (r, θ) be the polar coordinates of the point whose Cartesian coordinates are (x, y). Then the complex number z can be written in the form $z = r\cos\theta + ir\sin\theta$.

Definition.   The **polar form** of the complex number z is

$$z \;=\; r(\cos\theta + i\sin\theta).$$

The real number r is just the *modulus* of the complex number z. The angle θ is called the **argument** or *amplitude* of z. The expression $\cos\theta + i\sin\theta$ is often abbreviated to cis θ, so that the polar form becomes $z = r$ cis θ.

Example 8.52.   Convert the complex numbers i, $-1 + i$, $\sqrt{3} - 3i$, and -4 to polar form.

Solution.   Since $|i| = 1$ and the argument of i is $\frac{\pi}{2}$, it follows that the polar form is

$$i \;=\; 1\left(\cos\frac{\pi}{2} + i\sin\frac{\pi}{2}\right).$$

Now $|-1 + i| = \sqrt{1^2 + 1^2} = \sqrt{2}$ and it can be seen from the diagram that the argument of $-1 + i$ is $\frac{3\pi}{4}$. Hence

$$-1 + i \;=\; \sqrt{2}\left(\cos\frac{3\pi}{4} + i\sin\frac{3\pi}{4}\right).$$

We have $|\sqrt{3} - 3i| = \sqrt{3 + 9} = \sqrt{12} = 2\sqrt{3}$. If θ is the argument of $\sqrt{3} - 3i$, then $\sin\theta = \frac{-3}{2\sqrt{3}} = -\frac{\sqrt{3}}{2}$. It is well known that $\mathrm{Sin}^{-1}\frac{\sqrt{3}}{2} = \frac{\pi}{3}$ (see the Appendix if it is not well known to you) and hence it follows from the diagram that $\theta = -\frac{\pi}{3}$ or $\frac{5\pi}{3}$. Therefore,

$$\sqrt{3} - 3i \;=\; \sqrt{3}\left(\cos\frac{5\pi}{3} + i\sin\frac{5\pi}{3}\right).$$

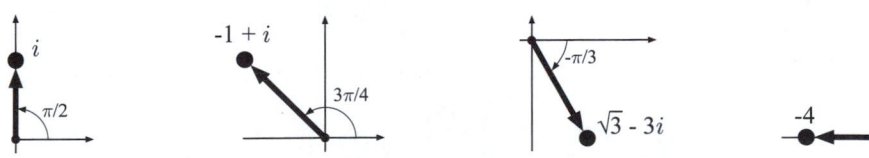

Since $|-4| = 4$ and the argument of -4 is π,

$$-4 \;=\; 4(\cos\pi + i\sin\pi). \qquad \square$$

One reason for converting complex numbers into polar form is that multiplication then takes on a rather elementary form.

Theorem 8.53.   If $z_1 = r_1(\cos\theta_1 + i\sin\theta_1)$ and $z_2 = r_2(\cos\theta_2 + i\sin\theta_2)$ are two complex numbers in polar form, then

$$z_1 z_2 = r_1 r_2[\cos(\theta_1 + \theta_2) + i\sin(\theta_1 + \theta_2)].$$

Proof.

$$\begin{aligned}
z_1 z_2 &= r_1(\cos\theta_1 + i\sin\theta_1)\cdot r_2(\cos\theta_2 + i\sin\theta_2)\\
&= r_1 r_2[\cos\theta_1\cos\theta_2 - \sin\theta_1\sin\theta_2 + i(\cos\theta_1\sin\theta_2 + \sin\theta_1\cos\theta_2)]\\
&= r_1 r_2[\cos(\theta_1 + \theta_2) + i\sin(\theta_1 + \theta_2)]
\end{aligned}$$

using the formulas for the sine and cosine of the sum of two angles.   □

Hence, to multiply two complex numbers, we multiply their moduli and *add* their arguments.

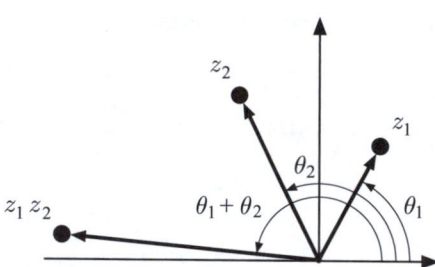

If the complex number z has modulus 1 and argument θ, then multiplication by z corresponds to a rotation through an angle θ. In particular, multiplication by i corresponds to a rotation through an angle $\frac{\pi}{2}$.

Example 8.54.   Multiply the complex number $z_1 = 2\left(\cos\dfrac{\pi}{4} + i\sin\dfrac{\pi}{4}\right)$ by the complex number $z_2 = \dfrac{1}{\sqrt{2}}\left(\cos\dfrac{7\pi}{6} + i\sin\dfrac{7\pi}{6}\right)$.

Solution.

$$\begin{aligned}
z_1 z_2 &= 2\cdot\frac{1}{\sqrt{2}}\left[\cos\left(\frac{\pi}{4} + \frac{7\pi}{6}\right) + i\sin\left(\frac{\pi}{4} + \frac{7\pi}{6}\right)\right]\\
&= \sqrt{2}\left(\cos\frac{17\pi}{12} + i\sin\frac{17\pi}{12}\right)
\end{aligned}$$

□

8.6   DE MOIVRE'S THEOREM

The previous section showed that the result of multiplying two complex numbers of unit modulus together is another number of unit modulus, whose argument is the sum of their arguments. By repeatedly applying this result we obtain the following important theorem named after the mathematician Abraham de Moivre (1667–1754), who was born in France but worked in England.

De Moivre's Theorem 8.61.   *For any real number θ and integer n,*

$$(\cos\theta + i\sin\theta)^n \;=\; \cos n\theta + i\sin n\theta.$$

Proof.   We shall prove the three cases (i) $n > 0$, (ii) $n = 0$, and (iii) $n < 0$ separately.

(i)   We shall use induction to prove the result when $n > 0$.
If $n = 1$, the result is clearly true.
Now suppose that the result is true for $n = k$, where $k \in \mathbb{P}$. Then

$$
\begin{aligned}
(\cos\theta + i\sin\theta)^{k+1} \;&=\; (\cos\theta + i\sin\theta)^k(\cos\theta + i\sin\theta) \\
&=\; (\cos k\theta + i\sin k\theta)(\cos\theta + i\sin\theta) \\
&=\; \cos(k+1)\theta + i\sin(k+1)\theta, \quad \text{by Theorem 8.53.}
\end{aligned}
$$

Hence the result is true for $n = k + 1$ if it is true for $n = k$. It follows from Mathematical Induction that the theorem is true for all $n \in \mathbb{P}$.

(ii)   If $n = 0$, $(\cos\theta + i\sin\theta)^0 = 1$, by the standard convention that $z^0 = 1$ for every nonzero number z. Note that $\cos\theta + i\sin\theta$ cannot be zero, because it always has modulus 1. Hence the result is true if $n = 0$, since

$$\cos 0\theta + i\sin 0\theta \;=\; 1 + i0 \;=\; 1.$$

(iii)   Suppose that n is a negative integer, say $n = -m$, where $m \in \mathbb{P}$. Then

$$
\begin{aligned}
(\cos\theta + i\sin\theta)^n \;&=\; (\cos\theta + i\sin\theta)^{-m} \\
&=\; \frac{1}{(\cos\theta + i\sin\theta)^m} \\
&=\; \frac{1}{(\cos m\theta + i\sin m\theta)}, \quad \text{by part (i)} \\
&=\; \cos m\theta - i\sin m\theta, \quad \text{by Corollary 8.43} \\
&=\; \cos(-m\theta) + i\sin(-m\theta) \\
&=\; \cos n\theta + i\sin n\theta.
\end{aligned}
$$

Hence the theorem is true for all integer values of n. $\Box$

Corollary 8.62.   *If $z = r(\cos\theta + i\sin\theta)$ then, for any integer n,*

$$z^n = r^n(\cos n\theta + i\sin n\theta).$$

Proof.   This follows immediately from the previous theorem.   □

Example 8.63.   Calculate $(\sqrt{3} + i)^{11}$.

Solution.   We could expand this by the Binomial Theorem. However, it is easier to convert the number to polar form and use De Moivre's Theorem.

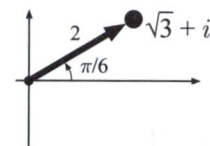 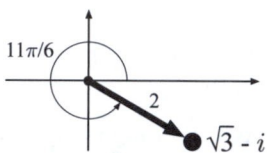

We have

$$\sqrt{3} + i = 2\left(\cos\frac{\pi}{6} + i\sin\frac{\pi}{6}\right)$$

and so

$$(\sqrt{3} + i)^{11} = 2^{11}\left(\cos\frac{11\pi}{6} + i\sin\frac{11\pi}{6}\right)$$

$$= 2^{11}\left(\frac{\sqrt{3} - i}{2}\right)$$

$$= 1024\sqrt{3} - 1024i.$$ □

Example 8.64.   Show that

$$\cos 3\theta = 4\cos^3\theta - 3\cos\theta$$
$$\sin 3\theta = 3\sin\theta - 4\sin^3\theta.$$

Solution.   By De Moivre's Theorem,

$$\cos 3\theta + i\sin 3\theta = (\cos\theta + i\sin\theta)^3$$

and by the Binomial Theorem,

$$(\cos\theta + i\sin\theta)^3 = \cos^3\theta + 3i\cos^2\theta\sin\theta - 3\cos\theta\sin^2\theta - i\sin^3\theta.$$

Hence

$$\cos 3\theta + i\sin 3\theta = \cos^3\theta - 3\cos\theta\sin^2\theta + i(3\cos^2\theta\sin\theta - \sin^3\theta).$$

By Proposition 8.25 (i), two complex numbers are equal if and only if their real parts are equal and their imaginary parts are equal. Therefore, equating real parts, we have

$$\cos 3\theta \ = \ \cos^3 \theta - 3 \cos \theta \sin^2 \theta$$

and equating imaginary parts we have

$$\sin 3\theta \ = \ 3 \cos^2 \theta \sin \theta - \sin^3 \theta.$$

By using the relation $\sin^2 \theta + \cos^2 \theta = 1$, which is true for all θ, we obtain

$$\begin{aligned}
\cos 3\theta \ &= \ \cos^3 \theta - 3 \cos \theta (1 - \cos^2 \theta) \\
&= \ 4 \cos^3 \theta - 3 \cos \theta. \\
\sin 3\theta \ &= \ 3(1 - \sin^2 \theta) \sin \theta - \sin^3 \theta \\
&= \ 3 \sin \theta - 4 \sin^3 \theta.
\end{aligned} \qquad \Box$$

The multiplication of complex numbers of unit modulus by adding their arguments is reminiscent of the multiplication of powers by adding their exponents. This can be utilized in the following way. A knowledge of elementary calculus will show that the real exponential function $y = e^{kx}$ is the solution to the differential equation $\frac{dy}{dx} = ky$ with the initial condition that $y = 1$, when $x = 0$. Assuming that we can differentiate complex functions, by differentiating their real and imaginary parts, we have

$$\frac{d}{d\theta}(\cos \theta + i \sin \theta) \ = \ -\sin \theta + i \cos \theta \ = \ i(\cos \theta + i \sin \theta)$$

and $\cos \theta + i \sin \theta = 1$ when $\theta = 0$. Hence it is plausible to define the **complex exponential function** by

$$e^{i\theta} \ = \ \cos \theta + i \sin \theta.$$

This complex exponential obeys the usual laws of exponents because

$$\begin{aligned}
e^{i\theta} \cdot e^{i\phi} \ &= \ e^{i(\theta + \phi)} \qquad &\text{by Theorem 8.53} \\
(e^{i\theta})^n \ &= \ e^{in\theta} \qquad &\text{by De Moivre's Theorem.}
\end{aligned}$$

The polar form of a complex number z can now be written as

$$z \ = \ re^{i\theta},$$

where $r = |z|$ and θ is the argument of z.

Since $e^{i\pi} = \cos \pi + i \sin \pi = -1 + i0$, we obtain the famous equation, due to Euler, that connects the numbers π, e, i, 1, and 0, namely

$$e^{i\pi} + 1 \ = \ 0.$$

We have so far only defined the complex exponential function with purely imaginary exponents. However, if $z = x + iy$ we can define

$$e^z \;=\; e^{x+iy} \;=\; e^x \cdot e^{iy} \;=\; e^x(\cos y + i \sin y).$$

The equation $e^{i\theta} = \cos\theta + i\sin\theta$ expresses the exponential function in terms of trigonometric functions. We can use this to express the trigonometric functions in terms of exponential functions as follows. Since

$$e^{-i\theta} \;=\; \cos(-\theta) + i\sin(-\theta) \;=\; \cos\theta - i\sin\theta,$$

by adding and subtracting the equations

$$e^{i\theta} \;=\; \cos\theta + i\sin\theta$$
$$e^{-i\theta} \;=\; \cos\theta - i\sin\theta$$

we obtain the formula for the cosine

$$\cos\theta \;=\; \frac{e^{i\theta} + e^{-i\theta}}{2}$$

and for the sine

$$\sin\theta \;=\; \frac{e^{i\theta} - e^{-i\theta}}{2i}.$$

8.7   ROOTS OF COMPLEX NUMBERS

Our reason for introducing complex numbers was to be able to find the roots of any number. We have already shown that complex numbers can be used to find the square roots of any negative real number. Using De Moivre's Theorem, we shall now show how to find the nth roots of any number, positive, negative or complex. In other words, we will show how to solve the equation

$$z^n \;=\; a \quad \text{for any } a \in \mathbb{C}.$$

Example 8.71.   Find all the complex fourth roots of -16.

Solution.   We have to solve the equation $z^4 = -16$. Let $z = r(\cos\theta + i\sin\theta)$ so that

$$z^4 \;=\; r^4(\cos 4\theta + i\sin 4\theta).$$

In polar form, $-16 = 16(\cos\pi + i\sin\pi)$. Hence, if $z^4 = -16$, then

$$r^4(\cos 4\theta + i\sin 4\theta) \;=\; 16(\cos\pi + i\sin\pi).$$

This equation is an equality between two complex numbers. Therefore, by Proposition 8.25 (i), their moduli must be equal, and their arguments must define the same angle. Hence $r^4 = 16$ and, since the modulus is a nonnegative real number, $r = 2$. The arguments 4θ and π will define the same angle if and only if they differ by an integer multiple of 2π. Therefore,

$$4\theta = \pi + 2k\pi, \qquad \text{where } k \in \mathbb{Z}$$
$$\theta = \frac{\pi}{4} + k\frac{\pi}{2}, \qquad \text{where } k \in \mathbb{Z}.$$

The solutions to the equation $z^4 = -16$ are

$$z = 2\left[\cos\left(\frac{\pi}{4} + k\frac{\pi}{2}\right) + i\sin\left(\frac{\pi}{4} + k\frac{\pi}{2}\right)\right], \qquad \text{where } k \in \mathbb{Z}.$$

When $k = 0$, this gives $z = 2(\cos\frac{\pi}{4} + i\sin\frac{\pi}{4}) = 2(\frac{1}{\sqrt{2}} + \frac{i}{\sqrt{2}}) = \sqrt{2} + i\sqrt{2}$.

When $k = 1$, $z = 2(\cos\frac{3\pi}{4} + i\sin\frac{3\pi}{4}) = 2(\frac{-1}{\sqrt{2}} + \frac{i}{\sqrt{2}}) = -\sqrt{2} + i\sqrt{2}$.

When $k = 2$, $z = 2(\cos\frac{5\pi}{4} + i\sin\frac{5\pi}{4}) = 2(\frac{-1}{\sqrt{2}} - \frac{i}{\sqrt{2}}) = -\sqrt{2} - i\sqrt{2}$.

When $k = 3$, $z = 2(\cos\frac{7\pi}{4} + i\sin\frac{7\pi}{4}) = 2(\frac{1}{\sqrt{2}} - \frac{i}{\sqrt{2}}) = \sqrt{2} - i\sqrt{2}$.

When $k = 4$, $z = 2(\cos\frac{9\pi}{4} + i\sin\frac{9\pi}{4}) = 2(\cos\frac{\pi}{4} + i\sin\frac{\pi}{4})$, and this gives the same value as was obtained by putting $k = 0$.

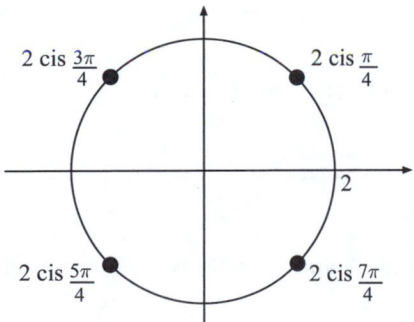

In fact, there are only four different fourth roots because $(\frac{\pi}{4} + k\frac{\pi}{2})$ and $(\frac{\pi}{4} + \ell\frac{\pi}{2})$ define the same angle if and only if

$$k \equiv \ell \pmod 4.$$

Therefore, the fourth roots of -16 are

$$\sqrt{2} + i\sqrt{2}, \quad -\sqrt{2} + i\sqrt{2}, \quad -\sqrt{2} - i\sqrt{2}, \quad \text{and} \quad \sqrt{2} - i\sqrt{2} \qquad \square$$

The method used in the preceding example can be used in general to prove the following result.

Theorem 8.72.   If $r(\cos\theta + i\sin\theta)$ is the polar form of a complex number, then all its complex nth roots are

$$\sqrt[n]{r}\left[\cos\left(\frac{\theta + 2k\pi}{n}\right) + i\sin\left(\frac{\theta + 2k\pi}{n}\right)\right] \qquad \text{for } k = 0, 1, 2, 3, \ldots, n-1.$$

The modulus $\sqrt[n]{r}$ is the unique real nonnegative nth root of r.

This result shows that any nonzero complex (or real) number has exactly n different complex nth roots.

Proof.   Let $w = s(\cos\phi + i\sin\phi)$ be an nth root of $r(\cos\theta + i\sin\theta)$. Then

$$w^n \;=\; s^n(\cos n\phi + i\sin n\phi) \;=\; r(\cos\theta + i\sin\theta).$$

Equating moduli, we obtain $s^n = r$, and so $s = \sqrt[n]{r}$. If $r \neq 0$, the arguments must define the same angle and hence must differ by an integer multiple of 2π. Therefore,

$$n\phi \;=\; \theta + 2k\pi$$
$$\phi \;=\; \frac{\theta + 2k\pi}{n}, \qquad \text{where } k \in \mathbb{Z}.$$

Hence the nth roots of $r(\cos\theta + i\sin\theta)$ are

$$\sqrt[n]{r}\left[\cos\left(\frac{\theta + 2k\pi}{n}\right) + i\sin\left(\frac{\theta + 2k\pi}{n}\right)\right] \qquad \text{for } k \in \mathbb{Z}.$$

The integers k_1 and k_2 will give the same solution:

$$\text{iff} \qquad \frac{\theta + 2k_1\pi}{n} - \frac{\theta + 2k_2\pi}{n} = 2\ell\pi \qquad \text{for some } \ell \in \mathbb{Z}$$

$$\text{iff} \qquad \frac{2\pi(k_1 - k_2)}{n} = 2\ell\pi \qquad \text{for some } \ell \in \mathbb{Z}$$

$$\text{iff} \qquad k_1 - k_2 = n\ell \qquad \text{for some } \ell \in \mathbb{Z}$$

$$\text{iff} \qquad k_1 \equiv k_2 \pmod{n}.$$

Hence, if $r \neq 0$, the number $r(\cos\theta + i\sin\theta)$ has exactly n distinct complex nth roots, namely when $k = 0, 1, 2, 3, \ldots, n-1$. □

Note that all the nth roots have the same modulus and so, in the complex plane, they all lie on a circle center the origin, radius $\sqrt[n]{r}$. All their arguments differ by multiples of $2\pi/n$ from each other, and so the n different nth roots lie at the vertices of a regular n-gon, whose center is the origin.

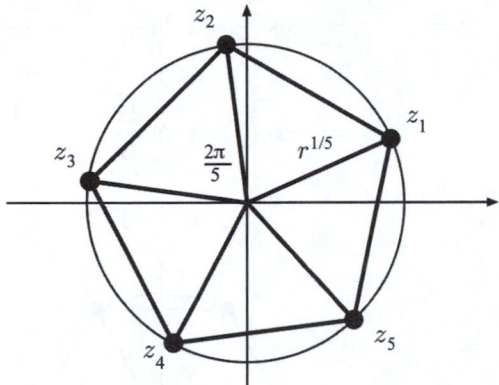

The preceding diagram illustrates a typical example of the five fifth roots z_1, z_2, z_3, z_4, and z_5 of a complex number.

It is clear that if we could find one nth root of a complex number, the other $n-1$ could easily be sketched in the complex plane.

Theorem 8.72 can be viewed as an extension of De Moivre's Theorem to rational exponents. It is true that *one* of the values of $[r(\cos\theta + i\sin\theta)]^{1/n}$ is

$$r^{1/n}\left(\cos\frac{\theta}{n} + i\sin\frac{\theta}{n}\right).$$

It is also straightforward to show that *one* of the values of $[r(\cos\theta + i\sin\theta)]^{p/q}$ is

$$r^{p/q}\left(\cos\frac{p\theta}{q} + i\sin\frac{p\theta}{q}\right).$$

Example 8.73.   Find all the sixth roots of unity.

Solution.   A root of unity is a complex number that has an integer power that is 1. We have to solve the equation $z^6 = 1$. We can write 1, in polar form, as $1(\cos 0 + i\sin 0)$. Hence, by Theorem 8.72, the sixth roots of unity are

$$
\begin{aligned}
z_k &= \sqrt[6]{1}\left(\cos\frac{2k\pi}{6} + i\sin\frac{2k\pi}{6}\right) \\
&= \cos\frac{k\pi}{3} + i\sin\frac{k\pi}{3} \qquad \text{for } k = 0, 1, 2, 3, 4, 5.
\end{aligned}
$$

The different roots are

$$
\begin{aligned}
z_0 &= \cos 0 + i \sin 0 & &= 1 \\
z_1 &= \cos \frac{\pi}{3} + i \sin \frac{\pi}{3} & &= \frac{1}{2} + \frac{\sqrt{3}i}{2} \\
z_2 &= \cos \frac{2\pi}{3} + i \sin \frac{2\pi}{3} & &= \frac{-1}{2} + \frac{\sqrt{3}i}{2} \\
z_3 &= \cos \pi + i \sin \pi & &= -1 \\
z_4 &= \cos \frac{4\pi}{3} + i \sin \frac{4\pi}{3} & &= \frac{-1}{2} - \frac{\sqrt{3}i}{2} \\
z_5 &= \cos \frac{5\pi}{3} + i \sin \frac{5\pi}{3} & &= \frac{1}{2} - \frac{\sqrt{3}i}{2}.
\end{aligned}
$$

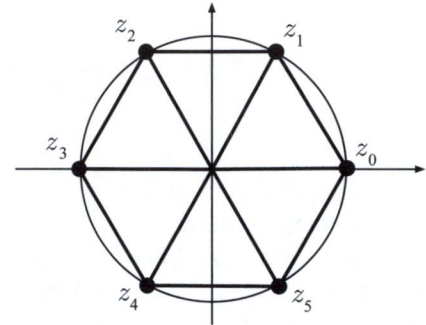

Hence the six sixth roots of unity are ± 1 and $\dfrac{\pm 1}{2} \pm \dfrac{\sqrt{3}i}{2}$. $\square$

Example 8.74.   Find all the square roots of $-2i$.

Solution.   In polar form

$$
-2i = 2\left(\cos \frac{3\pi}{2} + i \sin \frac{3\pi}{2} \right).
$$

Hence by Theorem 8.72, its square roots are

$$
\sqrt{2}\left[\cos\left(\frac{3\pi}{4} + k\pi \right) + i \sin\left(\frac{3\pi}{4} + k\pi \right) \right]
$$

for $k = 0$ and 1. That is,

$$
\begin{aligned}
\sqrt{2}\left(\cos \frac{3\pi}{4} + i \sin \frac{3\pi}{4} \right) &= \sqrt{2}\left(-\frac{1}{\sqrt{2}} + \frac{i}{\sqrt{2}} \right) &= -1 + i \\
\sqrt{2}\left(\cos \frac{7\pi}{4} + i \sin \frac{7\pi}{4} \right) &= \sqrt{2}\left(\frac{1}{\sqrt{2}} - \frac{i}{\sqrt{2}} \right) &= 1 - i.
\end{aligned}
$$

The two square roots of $-2i$ are therefore $\pm(1-i)$.   □

Example 8.75. Solve the quadratic equation

$$iz^2 - 2(1+i)z + 1 \;=\; 0 \quad \text{for } z \in \mathbb{C}.$$

Solution.   The proof of the Quadratic Formula 8.11 will still be valid if the coefficients of the quadratic equation are complex. Therefore, we can use the formula to solve our quadratic equation to obtain

$$
\begin{aligned}
z &= \frac{2(1+i) \pm [4(1+i)^2 - 4i]^{1/2}}{2i} \\[2mm]
&= \frac{1+i \pm [(1+i)^2 - i]^{1/2}}{i},
\end{aligned}
$$

where $\pm[(1+i)^2 - i]^{1/2}$ are the two complex square roots of $[(1+i)^2 - i]$. Now

$$
\begin{aligned}
(1+i)^2 - i &= 2i - i = i \\[2mm]
&= \cos\frac{\pi}{2} + i\sin\frac{\pi}{2}.
\end{aligned}
$$

Hence, the two square roots of i are

$$
\begin{aligned}
\cos\frac{\pi}{4} + i\sin\frac{\pi}{4} &= \frac{1}{\sqrt{2}} + \frac{i}{\sqrt{2}} \\[2mm]
\cos\frac{5\pi}{4} + i\sin\frac{5\pi}{4} &= -\frac{1}{\sqrt{2}} - \frac{i}{\sqrt{2}}.
\end{aligned}
$$

The solutions to the quadratic equation are therefore

$$z = \frac{1 + i \pm \left(\frac{1+i}{\sqrt{2}}\right)}{i} = \left(\frac{1+i}{i}\right)\left(1 \pm \frac{1}{\sqrt{2}}\right) = (-i+1)\left(\frac{2 \pm \sqrt{2}}{2}\right).$$

These two solutions are $\dfrac{(2+\sqrt{2})}{2}(1-i)$ and $\dfrac{(2-\sqrt{2})}{2}(1-i)$. $\qquad\square$

Example 8.76. Solve the equation

$$z^6 - z^3 - 2 = 0 \quad \text{for } z \in \mathbb{C}.$$

Solution. The above equation of the sixth degree is a quadratic in z^3, which factors as

$$(z^3 - 2)(z^3 + 1) = 0.$$

Hence $z^3 = 2$ or -1.

The cube roots of $2 = 2(\cos 0 + i \sin 0)$ are

$$\sqrt[3]{2}(\cos 0 + i \sin 0) = \sqrt[3]{2}$$

$$\sqrt[3]{2}\left(\cos\frac{2\pi}{3} + i \sin\frac{2\pi}{3}\right) = \sqrt[3]{2}\left(\frac{-1}{2} + \frac{\sqrt{3}i}{2}\right)$$

$$\sqrt[3]{2}\left(\cos\frac{4\pi}{3} + i \sin\frac{4\pi}{3}\right) = \sqrt[3]{2}\left(\frac{-1}{2} - \frac{\sqrt{3}i}{2}\right).$$

The cube roots of $-1 = 1(\cos\pi + i \sin\pi)$ are

$$\cos\frac{\pi}{3} + i \sin\frac{\pi}{3} = \frac{1}{2} + \frac{\sqrt{3}i}{2}$$

$$\cos\frac{3\pi}{3} + i \sin\frac{3\pi}{3} = -1$$

$$\cos\frac{5\pi}{3} + i \sin\frac{5\pi}{3} = \frac{1}{2} - \frac{\sqrt{3}i}{2}.$$

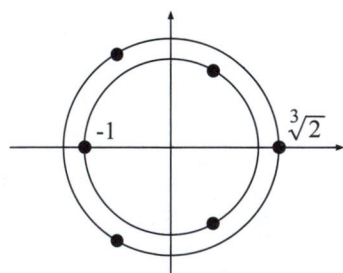

Hence the six solutions to the equation $z^6 - z^3 - 2 = 0$ are

$$z = -1, \quad \sqrt[3]{2}, \quad \frac{\sqrt[3]{2}(-1 \pm \sqrt{3}i)}{2}, \quad \text{and} \quad \frac{1 \pm \sqrt{3}i}{2}. \qquad \square$$

8.8   THE FUNDAMENTAL THEOREM OF ALGEBRA

The complex number system allows us to solve all quadratic equations of the form

$$a_2 z^2 + a_1 z + a_0 = 0,$$

where the coefficients a_0, a_1, and a_2 can be real or complex, and also to solve equations of the form

$$z^n - a_0 = 0,$$

where a_0 is real or complex. However, far more is true. Every polynomial equation with real or complex coefficients has a solution in the complex numbers. This result is so important that it is called the Fundamental Theorem of Algebra. The theorem only tells us that a solution *exists*. It does not tell us how to find the exact solution. In Chapter 9 we will investigate various methods for finding solutions to polynomial equations.

The first proof of this Fundamental Theorem of Algebra was given by Gauss in his doctoral thesis in 1799. However, none of the known proofs are easy, and they all require the use of concepts outside the realm of pure algebra. We now give an intuitive proof using concepts from topology. Topology is a branch of mathematics often known as "rubber-sheet geometry." The reason for this name should soon become clear.

Fundamental Theorem of Algebra 8.81.   *Every equation of the form*

$$a_n z^n + a_{n-1} z^{n-1} + \cdots + a_2 z^2 + a_1 z + a_0 = 0,$$

where $a_n, a_{n-1}, \ldots, a_2, a_1, a_0 \in \mathbb{C}$, $n \geq 1$ *and* $a_n \neq 0$, *has at least one solution in the complex numbers.*

Sketch Proof.   Since $a_n \neq 0$, we can divide by it, so that the coefficient of the highest power of z is 1. Let

$$f(z) = z^n + b_{n-1} z^{n-1} + \cdots + b_1 z + b_0,$$

where $b_i = a_i / a_n$.

This defines a function

$$f : \mathbb{C} \longrightarrow \mathbb{C}$$

from the complex plane to the complex plane. To show that the original equation has a solution, we have to show that there exists a complex number w such that $f(w) = 0$; in other words, we have to show that the origin of the complex plane lies in the image of the function f. We shall look at the image under f of the different circles centered at the origin in the domain; that is, the images of the complex numbers of a fixed modulus.

When the modulus of z is large, we shall show that the function $f(z)$ is dominated by its first term z^n. Let

$$g : \mathbb{C} \longrightarrow \mathbb{C}$$

be defined by the first term, so that $g(z) = z^n$. Then, for any complex number $z = r(\cos \theta + i \sin \theta)$,

$$g(z) \;=\; r^n(\cos \theta + i \sin \theta)^n \;=\; r^n(\cos n\theta + i \sin n\theta)$$

by De Moivre's Theorem.

Denote the circle of radius r by

$$L_r \;=\; \{z \mid |z| = r\}.$$

Therefore, if $z \in L_r$, $|z| = r$, $|g(z)| = |z^n| = r^n$ and furthermore, as z travels once round the circle L_r, $g(z) = z^n$ will travel n complete times around the circle of radius r^n.

Let us see how close $f(z)$ and $g(z)$ are for points on the circle L_r. If $z \in L_r$,

$$
\begin{aligned}
|f(z) - g(z)| \;&=\; |f(z) - z^n| \;=\; |b_{n-1}z^{n-1} + \cdots + b_1 z + b_0| \\[2mm]
&=\; |z|^n \cdot \left| \frac{b_{n-1}}{z} + \cdots + \frac{b_1}{z^{n-1}} + \frac{b_0}{z^n} \right| \qquad \text{by Proposition 8.44 (iii)} \\[2mm]
&\le\; |z|^n \left(\left| \frac{b_{n-1}}{z} \right| + \cdots + \left| \frac{b_1}{z^{n-1}} \right| + \left| \frac{b_0}{z^n} \right| \right) \qquad \text{by Proposition 8.44 (iv)} \\[2mm]
&<\; r^n \left(\frac{1}{2n} + \cdots + \frac{1}{2n} + \frac{1}{2n} \right)
\end{aligned}
$$

if r is larger than the fixed real number

$$R \;=\; \max\{2n|b_{n-1}|, (2n|b_{n-2}|)^{1/2}, \ldots, (2n|b_0|)^{1/n}\}.$$

Hence, if $z \in L_r$, where $r > R$, then

$$|f(z) - g(z)| < \frac{r^n}{2}.$$

The points $f(z)$ and $g(z)$ are therefore within $r^n/2$ of each other in the complex plane, whenever z lies on the circle L_r. Since $g(z)$ always lies on the circle of radius r^n, it follows that

$$\frac{r^n}{2} < |f(z)| < \frac{3r^n}{2} \quad \text{whenever } z \in L_r \text{ and } r > R.$$

Hence, if $r > R$, the image of L_r under f will be a curve wrapped n times around the origin and lying inside an annulus of outer radius $3r^n/2$ and inner radius $r^n/2$.

It can be shown that the function f is continuous and so the image $f(L_r)$ is a continuous curve.

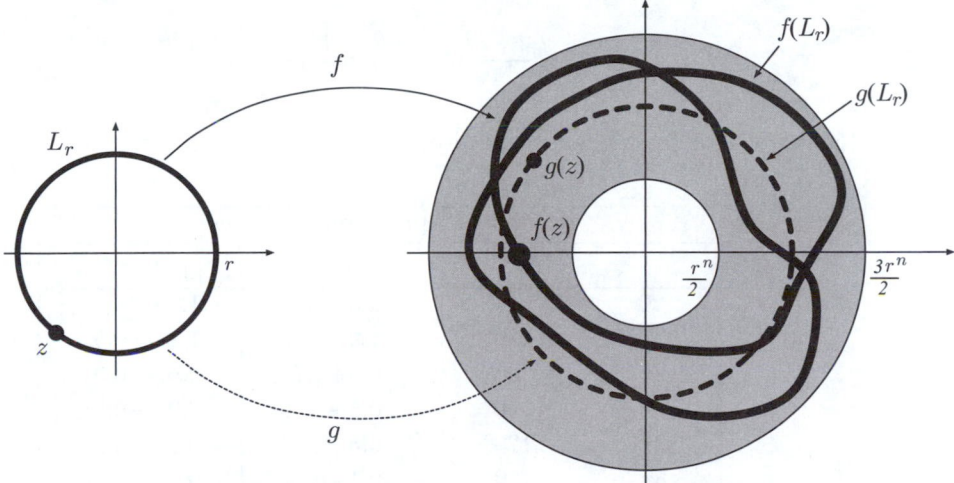

Now let r get smaller and smaller so that the circle L_r shrinks continuously toward the origin. Its image $f(L_r)$ must shrink continuously to the point $f(0) = b_0$. We can visualize this by imagining the curve $f(L_r)$ to be an elastic band in the complex plane. This elastic band is initially wrapped n times around the origin, and the band must be contracted onto the point b_0 without breaking it. At some point the band $f(L_r)$ must pass through the origin. Hence the origin lies in the image of f and there exists a complex number w such that $f(w) = 0$. $\qquad \square$

From the above argument, it can be seen that the solution w and, in fact all solutions, have modulus less than R.

Example 8.82. Let $f : \mathbb{C} \to \mathbb{C}$ be defined by

$$f(z) = z^2 + z + \frac{1}{2}.$$

Sketch the images, under f, of the circles $|z| = 1$ and $|z| = \frac{1}{2}$.

Solution.   Any point on the circle $|z| = 1$ can be written as $e^{i\theta} = \cos\theta + i\sin\theta$, and any point on $|z| = \frac{1}{2}$ can be written as $e^{i\theta}/2 = (\cos\theta + i\sin\theta)/2$. It follows from De Moivre's Theorem that

$$
\begin{aligned}
f(e^{i\theta}) &= (\cos\theta + i\sin\theta)^2 + (\cos\theta + i\sin\theta) + \frac{1}{2} \\
&= \left(\cos 2\theta + \cos\theta + \frac{1}{2}\right) + i(\sin 2\theta + \sin\theta)
\end{aligned}
$$

$$
\begin{aligned}
f\left(\frac{e^{i\theta}}{2}\right) &= \frac{(\cos\theta + i\sin\theta)^2}{4} + \frac{(\cos\theta + i\sin\theta)}{2} + \frac{1}{2} \\
&= \left(\frac{\cos 2\theta}{4} + \frac{\cos\theta}{2} + \frac{1}{2}\right) + i\left(\frac{\sin 2\theta}{4} + \frac{\sin\theta}{2}\right).
\end{aligned}
$$

Evaluate these functions for one revolution of θ.

Table of Approximate Values			
θ		$f(e^{i\theta})$	$f(0.5e^{i\theta})$
0	0°	2.5	1.25
$\pi/6$	30°	1.9+1.4i	1.1 + 0.5i
$\pi/3$	60°	0.5+1.7i	0.6 + 0.6i
$\pi/2$	90°	−0.5+ i	0.25+ 0.5i
$2\pi/3$	120°	−0.5	0.1 + 0.2i
$5\pi/6$	150°	0.1−0.4i	0.2 +0.03i
π	180°	0.5	0.25
$7\pi/6$	210°	0.1+0.4i	0.2 −0.03i
$4\pi/3$	240°	−0.5	0.1 − 0.2i
$3\pi/2$	270°	−0.5− i	0.25− 0.5i
$5\pi/3$	300°	0.5−1.7i	0.6 − 0.6i
$11\pi/6$	330°	1.9−1.4i	1.1 − 0.5i
2π	360°	2.5	1.25

Notice that the values for the function for θ and $2\pi - \theta$ are conjugates of each other, so that the curve is symmetric about the real axis. This follows because $f(\bar{z}) = \overline{f(z)}$ whenever f has real coefficients, a fact that will be proved in general in the Conjugate Roots Theorem 9.24.

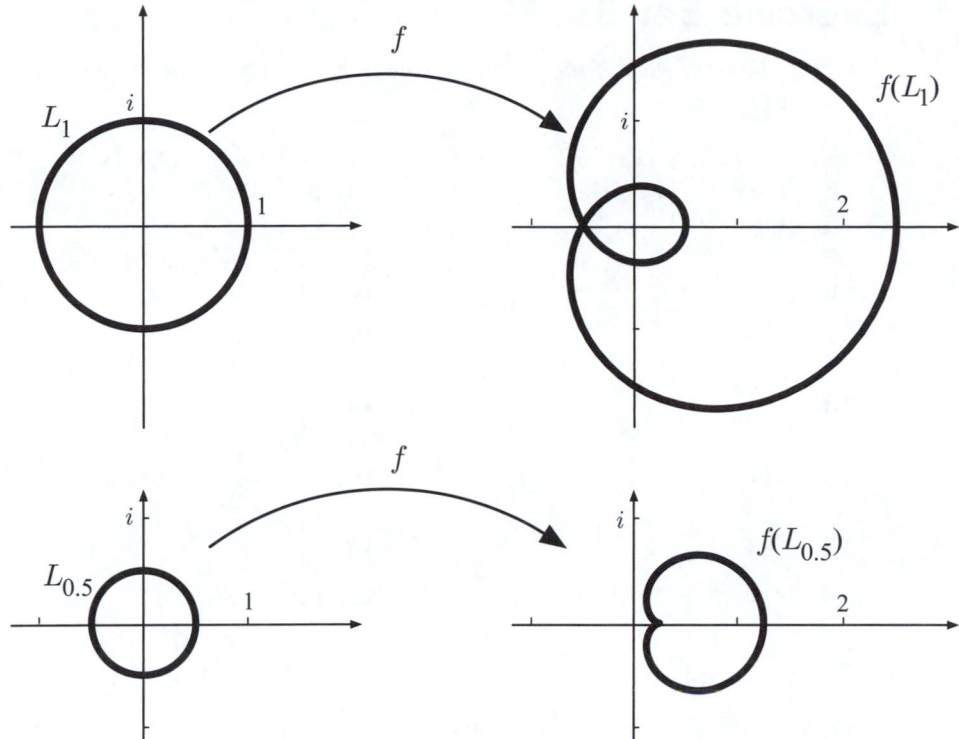

We see that as r decreases from 1 to $\frac{1}{2}$, $f(L_r)$ shrinks continuously from a curve that loops around the origin to a curve that is to the right of the imaginary axis. Hence it must pass through the origin for some $\frac{1}{2} < r < 1$. Therefore, $f(z)$ has a root with modulus between $\frac{1}{2}$ and 1.

Of course, in this example, we can solve the quadratic equation

$$z^2 + z + \frac{1}{2} = 0$$

by the Quadratic Formula 8.11 to obtain the two roots

$$z = \frac{(-1 \pm i)}{2}.$$

These both have moduli $1/\sqrt{2} \approx 0.71$ and so lie between the circles $|z| = 1$ and $|z| = \frac{1}{2}$. □

Exercise Set 8

1–22. Express each of the following complex numbers in the standard form $x + iy$, where $x, y \in \mathbb{R}$.

1. $(1 + i\sqrt{2}) + (1 - i\sqrt{2})$
3. $(7 + 6i) - (4 - 5i)$
5. $(1 + i\sqrt{2})(1 - i\sqrt{2})$
7. $(4 - i)(3 + i)$
9. $(3 - i)(1 + i)^2$

2. $(1 + i\sqrt{2}) - (1 - i\sqrt{2})$
4. $5i + (2 - 3i)$
6. $(2 + 3i)^2$
8. $(\sqrt{2} - i\sqrt{3})(\sqrt{2} + i\sqrt{3})$
10. $(4 - i)(3 - i)(2 - i)$

11. $\dfrac{1}{2 + i}$
12. $\dfrac{i}{3 - \sqrt{2}i}$

13. $\dfrac{7 - i}{4 + 3i}$
14. $\dfrac{1}{\sqrt{2}i}$

15. $\dfrac{1 + i}{1 - i}$
16. $\dfrac{1 + i\sqrt{2}}{1 - i\sqrt{2}}$

17. $1 + i + i^2 + i^3$
18. $\dfrac{1}{i^{65}}$

19. $(2 - i)^3$
20. $(\sqrt{5} - i\sqrt{3})^4$

21. $\dfrac{1}{(1 - 2i)(2 - i)}$
22. $\dfrac{(\sqrt{2} - i)^2}{(\sqrt{2} + i)(1 - \sqrt{2}i)}$

23–26. What are the real and imaginary parts of the following complex numbers?

23. $4 - 7i$
24. $-10 + \sqrt{2}i$
25. -3
26. $(6 - i)^2$

27–32. What is the complex conjugate and modulus of each of the following numbers?

27. $3 - 2i$
28. $1 + i\sqrt{5}$
29. $-i$
30. 2
31. $2 + \frac{i}{\sqrt{2}}$
32. $-\sqrt{3} - i\sqrt{7}$

33–37. Plot the following points in the complex plane.

33. $z = -i$, $w = 1 + i$, $z + w$, $2z$, and $3w$
34. $z = 1 - i$, $w = 2 + i$, $2z$, $z + w$, $z - w$, and $z - 2w$
35. $z = 2i$, z^2, z^3, and $1/z$
36. $z = 1 + i$, z^2, z^3, z^4, and $1/z$
37. $z = 3 + i$, iz, $i^2 z$, $i^3 z$, and $i^4 z$

38–41. Find polar coordinates for each of the following points that are given in Cartesian form.

38. $(2, 2)$
39. $(1, -1)$
40. $(-\sqrt{3}, 1)$
41. $(-1, -\sqrt{3})$

42–45. Each of the following pairs of numbers gives the polar coordinates of a point in the plane. Express each point in Cartesian coordinates.

42. $(1, \pi/2)$
44. $(7, 7\pi/6)$

43. $(\sqrt{3}, 3\pi/4)$
45. $(1, -2\pi/3)$

46–51. Express each of the following complex numbers in their polar form.

46. $-1 - i$
48. $\sqrt{3} - i$
50. 5

47. -8
49. $-2i$
51. $\frac{1}{\sqrt{3}} + \frac{i}{\sqrt{3}}$

52–55 Express each of the following numbers $r\operatorname{cis}\theta = r(\cos\theta + i\sin\theta)$ in the standard form $x + iy$ where $x, y \in \mathbb{R}$.

52. $\operatorname{cis}(3\pi/2)$
54. $\sqrt{3}\operatorname{cis}(4\pi/3)$

53. $4\operatorname{cis} 2\pi$
55. $2\operatorname{cis}(-3\pi/4)$

56–59. Find the modulus and argument of each of the following complex numbers.

56. $2 - 2i$
58. $\sqrt{2} + (3 - \sqrt{5})i$

57. $-3i$
59. $(1 + i)^{30}$

60. Show that $(1 + i^{2n})(1 + i^n)$ is either 0 or 4 for $n \in \mathbb{P}$. Describe the values of n that yield the various answers.

61. Solve the equation $\bar{z} = z^2$ for $z \in \mathbb{C}$.

62–65. Express the following complex numbers in standard form.

62. $(1 - i)^6$

63. $\left(\dfrac{\sqrt{3}}{2} + \dfrac{i}{2}\right)^{25}$

64. $(1 - \sqrt{3}i)^8$

65. $\dfrac{1}{(-1 - \sqrt{3}i)^{10}}$

66. Use De Moivre's Theorem to show that

$$\cos 2\theta = 2\cos^2\theta - 1$$
$$\sin 2\theta = 2\sin\theta\cos\theta.$$

67. **(a)** Use De Moivre's Theorem to show that

$$\cos 4\theta = 8\cos^4\theta - 8\cos^2\theta + 1$$
$$\sin 4\theta = 4\cos\theta(\sin\theta - 2\sin^3\theta).$$

(b) Calculate $\cos 4\theta$ if $\theta = \operatorname{Cos}^{-1}(1/\sqrt{3})$.

68. Find expressions for $\cos 5\theta$ and $\sin 5\theta$ in terms of $\cos\theta$ and $\sin\theta$. Calculate $\cos 5\theta$ if $\cos\theta = 0.1$.

69. Prove that $(1+i)^n = 2^{\frac{n}{2}}\left(\cos\dfrac{n\pi}{4} + i\sin\dfrac{n\pi}{4}\right)$.

70. Find all the cube roots of unity.

71. Find all the fourth roots of unity.

72–83. *Solve each of the following equations for* $\mathbf{z} \in \mathbb{C}$ *and plot your solutions on the complex plane.*

72. $z^3 = \dfrac{1 - i\sqrt{3}}{8}$

73. $z^4 + 1 = 0$

74. $z^3 = 9i$

75. $z^4 + 1 + i\sqrt{3} = 0$

76. $z^8 = 16$

77. $z^2 = i$

78. $2z^2 + 5 = 0$

79. $7z^2 - z + 4 = 0$

80. $z^2 - 5z + 6 = 0$

81. $4z^2 + iz - 1 = 0$

82. $z^2 - (2 + 2i)z - 1 + 2i = 0$

83. $iz^3 + 1 + i = 0$

84. If $w \in \mathbb{C}$ is a solution to the equation $ax^2 + bx + c = 0$, where $a, b, c \in \mathbb{R}$, show that the complex conjugate $\overline{w}$ is also a solution.

85. By substitution and calculation, show that each of the numbers $-1/2$, $1 + i\sqrt{5}$, and $1 - i\sqrt{5}$ are solutions to the equation

$$2z^3 - 3z^2 + 10z + 6 \;=\; 0.$$

86. **(a)** If $x_1, x_2 \in \mathbb{R}$, and $x_1^3 = x_2^3$, does it follow that $x_1 = x_2$?

   (b) If $z_1, z_2 \in \mathbb{C}$, and $z_1^3 = z_2^3$, does it follow that $z_1 = z_2$?

87. If $c + id = (a + ib)^n$, show that $c^2 + d^2 = (a^2 + b^2)^n$.

88. Find one value of $\sqrt[5]{\dfrac{1 - i}{\sqrt{3} + i}}$.

89–92 *Plot the solutions to the following equations in the complex plane.*

89. $z^5 = 1$

90. $z^4 = 16i$

91. $z^{16} + z = 0$

92. $z^2 = 2 - \sqrt{5}i$

93–103. *Shade each of the following regions of the complex plane.*

93. Real part of z less than or equal to 2

94. Imaginary part of z equal to -1

95. Imaginary part of z greater than -1

96. $|z| = 3$

97. $z\overline{z} = 9$

98. $|z - i| = 1$

99. $|z + 3i| \leq 2$

100. $z - \overline{z} = 1$

101. $(z + \overline{z})^2 = |z|^2 + |z - \overline{z}|^2$

102. $|z| \geq |\overline{z}|$

103. $2 < |z + 1| < 3$

104. Prove that, for all $z_1, z_2 \in \mathbb{C}$

$$|z_1 - z_2| \geq |z_1| - |z_2|.$$

What is the geometric interpretation of this inequality, and when does this equality hold?

105–108. Write the following numbers in the exponential form e^{x+iy}.

105. $-2i$ **106.** $-1 + i$

107. 3 **108.** $(\sqrt{3} - i)^{100}$

109. If w is a nonreal cube root of unity, prove that $w^2 + w + 1 = 0$.

110. If u is a nonreal sixth root of unity but not a cube root of unity, prove that $u^3 = -1$, and that $u^2 - u + 1 = 0$.

Problem Set 8

111.   **(a)** If $z = r \operatorname{cis} \theta$ and $w = s \operatorname{cis} \phi$, prove directly that

$$\frac{z}{w} = \frac{r}{s} \operatorname{cis}(\theta - \phi).$$

     (b) Evaluate $\dfrac{\sqrt{3} - 3i}{\sqrt{2} + \sqrt{6}i}$ using this result.

112. If $z, w \in \mathbb{C}$, prove that

$$|1 - z\overline{w}|^2 - |z - w|^2 = (1 - |z|^2)(1 - |w|^2).$$

113. If $z, w \in \mathbb{C}$, prove that

$$|z + w|^2 + |z - w|^2 = 2|z|^2 + 2|w|^2.$$

What is the geometric interpretation of this equality?

114. Shade the region of the complex plane for which

$$|z - 1| + |z + 1| \leq 4.$$

115. Shade the region of the complex plane for which

$$\frac{iz - 1}{z - i}$$

is real.

116–117. *Shade the following regions of the complex plane.*

116. $\left\{ \left. \dfrac{1}{z} \right| \; z \in \mathbb{C}, |z| = 1 \right\}$ **117.** $\{1 - z \mid z \in \mathbb{C}, |z| = 1\}$

118. (a) Prove that the sum of the fifth roots of unity is zero.
　　　(b) Generalize this result and prove your generalization.

119. Find all $z \in \mathbb{C}$ for which $\bar{z} = z^{n-1}$.

120. What is wrong with the following argument?

$$-1 \;=\; \sqrt{-1}\sqrt{-1} \;=\; \sqrt{-1}\sqrt{\frac{1}{-1}} \;=\; \frac{\sqrt{-1}}{\sqrt{-1}} \;=\; 1$$

121. (a) Prove that

$$\tan nt \;=\; \frac{\binom{n}{1}\tan t - \binom{n}{3}\tan^3 t + \binom{n}{5}\tan^5 t - \cdots}{1 - \binom{n}{2}\tan^2 t + \binom{n}{4}\tan^4 t - \cdots}.$$

　　　(b) Calculate $\tan 5t$ if $t = \mathrm{Tan}^{-1}\sqrt{2}$.

122. Find all the values of $(1+i)^{2/3}$.

123. Let ω be a complex cube root of unity, where $\omega \neq 1$. If b is one solution to $z^3 = a$, show that the other two solutions are $b\omega$ and $b\omega^2$.

124. (a) If $z \in \mathbb{C}$, prove by induction that $z^n + \dfrac{1}{z^n}$ can be written as a polynomial in $w = z + \dfrac{1}{z}$.

　　　(b) Prove that $\cos n\theta$ can be written as a polynomial in $\cos \theta$.

　　　(c) Is a similar result true for $\sin n\theta$?

125. If θ is not a multiple of 2π, show that

$$\sin \theta + \sin 2\theta + \sin 3\theta + \cdots + \sin n\theta \;=\; \sin \frac{(n+1)\theta}{2} \cdot \sin \frac{n\theta}{2} \left/ \sin \frac{\theta}{2} \right.$$

$$1 + \cos \theta + \cos 2\theta + \cdots + \cos n\theta \;=\; \sin \frac{(n+1)\theta}{2} \cdot \cos \frac{n\theta}{2} \left/ \sin \frac{\theta}{2} \right. .$$

126–130. *Find all the solutions to the following equations for $z \in \mathbb{C}$.*

126. $z^4 - 3z^2 + 4 = 0$ **127.** $z^4 - 30z^2 + 289 = 0$

128. $z^8 + z^4 + 1 = 0$ **129.** $z^4 + iz^2 + 2 = 0$

130. $(z+1)^n + (z-2)^n = 0$

131. If you know how to integrate exponential functions, evaluate the integrals

$$U \;=\; \int e^{ax} \cos bx \; dx$$

$$V \;=\; \int e^{ax} \sin bx \; dx$$

by computing

$$U + iV = \int e^{(a+ib)x} \, dx.$$

(Assume that the usual integration formulas hold when the constants are complex numbers.)

132. Sketch, by hand or by computer, the image of the two circles $|z| = 1$ and $|z| = 2$ under the complex function $f : \mathbb{C} \to \mathbb{C}$ defined by $f(z) = z^3 + z + 1$. What can you say about the complex solutions to the equation $z^3 + z + 1 = 0$?

133. Sketch, by computer, the image of the circle $|z| = 2$ under the complex function $f : \mathbb{C} \to \mathbb{C}$ defined by $f(z) = z^4 - 3z + 1$.

134. *(For discussion)*

   (a) Is the complex function $f : \mathbb{C} \to \mathbb{C}$, defined by $f(z) = e^z$, a bijection?

   (b) How would you define $\text{Log}_e z$ for $z \in \mathbb{C}$?

   (c) What is $\text{Log}_e i$?

   (d) What is i^i?

135. Use the fact that $\theta = 2\pi/5$ satisfies the equation $\cos 2\theta = \cos 3\theta$ to calculate $\cos(2\pi/5)$.

136. Find all the complex solutions to $z^9 + z^6 + z^3 + 1 = 0$.

137–142. The complex numbers with integer real and imaginary parts are called the Gaussian integers, *and they play the role of the integers in the complex numbers. All the Gaussian integers can be represented in the* complex base $b = -1 + i$ *using the binary digits $\{0, 1\}$ so that if $p, q \in \mathbb{Z}$, then*

$$p + iq \;=\; \sum_{r=0}^{k} a_r b^r, \quad \text{where } a_r \in \{0, 1\}.$$

For example, $1 - 2i = (101)_{-1+i}$, $2 = (1100)_{-1+i}$, and $-3 - 2i = (10101)_{-1+i}$. (See section 4.1 of The Art of Computer Programming, Volume 2: Seminumerical Algorithms *by Donald Knuth, Addison-Wesley, 1997.)*

137. What are the Cartesian forms of the complex numbers

$$(100)_{-1+i}, \;\; (10000)_{-1+i}, \;\; \text{and} \;\; (1111)_{-1+i} \;?$$

138. What are the Cartesian forms of the complex numbers

$$(11)_{-1+i}, \;\; (11100)_{-1+i}, \;\; \text{and} \;\; (111010001)_{-1+i} \;?$$

139. Find the base $-1 + i$ representation of $1 + i$ and 3.

140. Find the base $-1 + i$ representation of $-1 - i$ and $-3i$.

141. Just as every positive real number can be represented in base 10 by an infinite decimal expansion, every complex number can be represented in the base $b = -1 + i$ using negative powers of the base; for example

$$\begin{aligned}(0.11)_{-1+i} &= b^{-1} + b^{-2} &= -\tfrac{1}{2}, \\ (10.101)_{-1+i} &= b + b^{-1} + b^{-3} &= -\tfrac{5}{4} + \tfrac{1}{4}i.\end{aligned}$$

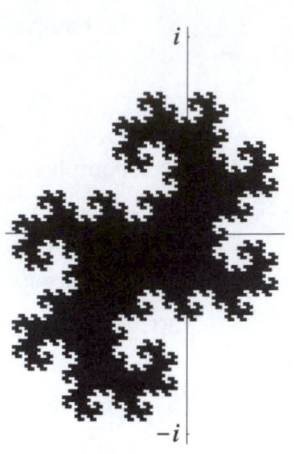

Write a computer program to display the complex numbers with zero integer part in the base $-1 + i$; that is, those with representations of the form

$$(0.a_1 \ldots a_s)_{-1+i} \quad \text{with } a_1, \ldots, a_s \in \{0, 1\}.$$

Increasing the value of s should produce a better approximation to the figure on the right.

Space-Filling Twin Dragon Curve

142. Only some of the Gaussian integers can be represented in the complex base $b = 1 + i$ using binary digits. For various values of k, sketch the Gaussian integers that can be written as $(a_k \cdots a_1 a_0)_{1+i}$, where $a_r \in \{0, 1\}$.

143. Consider the quadratic function $f(z) = z^2 + c$. If $|z| > |c| + 1$, show that $|f(z)| > |z|$.

144. For a fixed complex number c, the *filled-in Julia set* of the quadratic function $f : \mathbb{C} \to \mathbb{C}$, defined by

$$f(z) = z^2 + c,$$

consists of those complex numbers z whose iterates under f all remain bounded. Write a computer program to display an approximation to the filled-in Julia set. The parameters of the program should be the complex number c, the number n of times to iterate the function f, and the size of the grid in which the complex plane is dissected. From the previous problem, you can assume that the filled-in Julia set lies inside the circle $|z| \le |c| + 1$ and that the point z is *not* in the set if any one of the first n iterates of z under f satisfies

Filled-In Julia Set of $z^2 + 0.4 - 0.35i$

$$|f(f(\ldots f(f(z))\ldots))| > |c| + 1.$$

145. The *Mandelbrot set* consists of all the complex numbers c such that the images of 0 are bounded under all the iterates of $f(z) = z^2 + c$. It can be shown that $c \in \mathbb{C}$ is in the Mandelbrot set if and only if the filled-in Julia set of $f(z)$ is connected. Write a computer program to display an approximation to the Mandelbrot set. The parameters of the program should be the number n of times to iterate the function f and the size of the grid in which the complex plane is dissected. You can assume that the Mandelbrot set lies inside the circle $|z| \leq 2$ and that the point c is *not* in the set if any one of the first n iterates of 0 under f satisfies

$$|f(f(\ldots f(f(0))\ldots))| > 2,$$

that is, if any of the first n of the following are satisfied:

$$|c| > 2, \quad |c^2 + c| > 2, \quad |(c^2 + c)^2 + c| > 2, \quad |((c^2 + c)^2 + c)^2 + c| > 2 \ldots.$$

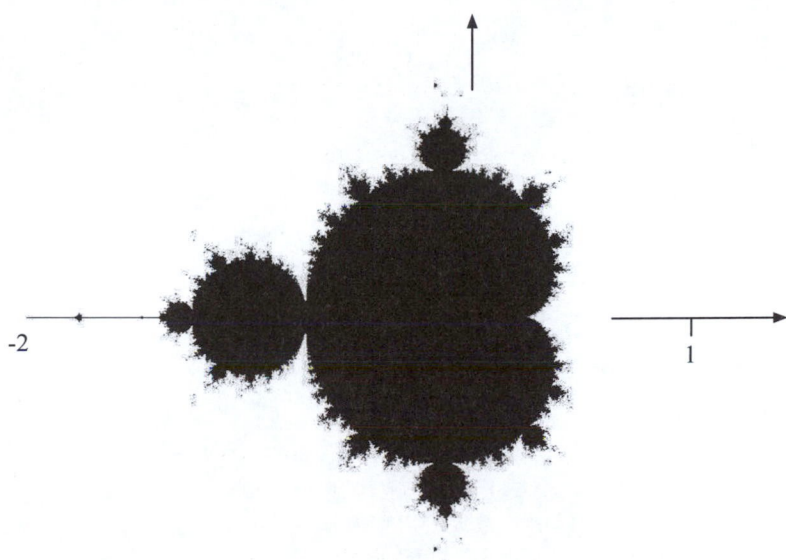

The Mandelbrot Set

CHAPTER 9

Polynomial Equations

In this chapter we discuss various methods for finding solutions to equations of the form

$$a_n x^n + a_{n-1} x^{n-1} + \cdots + a_1 x + a_0 = 0.$$

Such equations are called polynomial equations, and they form the most basic type of equation in mathematics.

There is no systematic method for solving all such equations. However, they can be solved in many special cases. We already know how to find all the complex solutions to equations of the form $a_2 x^2 + a_1 x + a_0 = 0$ and of the form $a_n x^n + a_0 = 0$.

Even if we cannot find an exact solution, we can often find the number of solutions and their approximate values.

Before we begin to actually solve polynomial equations, we must examine the polynomial expressions contained in the equations.

9.1 POLYNOMIALS AND FACTORING

Let $\mathbb{F}$ be a field, such as the rational numbers $\mathbb{Q}$, the real numbers $\mathbb{R}$, the complex numbers $\mathbb{C}$, or $\mathbb{Z}_p$, the integers modulo a prime p. An expression of the form

$$a_n x^n + a_{n-1} x^{n-1} + \cdots + a_1 x + a_0,$$

where $a_0, a_1, \ldots, a_n \in \mathbb{F}$, and $n \geq 0$ is called a *polynomial in x with coefficients from $\mathbb{F}$*. If $a_n \neq 0$, then the above polynomial is said to have *degree n*. The degree of a polynomial is the highest power of x that has a nonzero coefficient. If all the coefficients are zero, the polynomial is called the *zero polynomial*, and its degree is not defined.

For example, $2x^3 - \sqrt{3}x + 1$ is a polynomial in x of degree 3 with coefficients from $\mathbb{R}$, and $5z - 1 + i$ is a linear polynomial in z (that is, of degree 1) with coefficients from $\mathbb{C}$.

Polynomials of low degree are assigned the names given in the following table.

Degree	Name	General Example	
Undefined	Zero polynomial	0	
0	Constant polynomial	a_0	$(a_0 \neq 0)$
1	Linear polynomial	$a_1 x + a_0$	$(a_1 \neq 0)$
2	Quadratic polynomial	$a_2 x^2 + a_1 x + a_0$	$(a_2 \neq 0)$
3	Cubic polynomial	$a_3 x^3 + a_2 x^2 + a_1 x + a_0$	$(a_3 \neq 0)$
4	Quartic polynomial	$a_4 x^4 + a_3 x^3 + a_2 x^2 + a_1 x + a_0$	$(a_4 \neq 0)$
5	Quintic polynomial	$a_5 x^5 + a_4 x^4 + a_3 x^3 + a_2 x^2 + a_1 x + a_0$	$(a_5 \neq 0)$

The *set of all polynomials in x with coefficients from $\mathbb{F}$ is denoted by*

$$\mathbb{F}[x] = \{a_n x^n + \cdots + a_1 x + a_0 \mid a_i \in \mathbb{F}, n \geq 0\}.$$

A typical element in $\mathbb{F}[x]$ will often be denoted by $f(x)$, so that

$$f(x) = a_n x^n + \cdots + a_1 x + a_0 = \sum_{i=0}^{n} a_i x^i.$$

Two polynomials in $\mathbb{F}[x]$

$$
\begin{aligned}
f(x) &= a_n x^n + \cdots + a_1 x + a_0 \\
g(x) &= b_m x^m + \cdots + b_1 x + b_0
\end{aligned}
$$

are *equal* if they have the same degree and $a_i = b_i$ for all i.

A *polynomial equation* is an equation of the form

$$a_n x^n + a_{n-1} x^{n-1} + \cdots + a_1 x + a_0 = 0,$$

which will often be written as $f(x) = 0$, where $f(x) \in \mathbb{F}[x]$. An element $c \in \mathbb{F}$ is called a *root* of the polynomial $f(x)$ if $f(c) = 0$; in other words, a root of the polynomial $f(x)$ is a solution of the polynomial equation $f(x) = 0$.

Let $f(x) = \sum_{i=0}^{n} a_i x^i$ and $g(x) = \sum_{i=0}^{m} b_i x^i$ be two polynomials in $\mathbb{F}[x]$. These polynomials can be added, subtracted, and multiplied in the usual way that we deal with algebraic expressions. Their sum is

$$f(x) + g(x) = \sum_{i=0}^{\max(n, m)} (a_i + b_i) x^i,$$

where $a_i = 0$, if $i > n$, and $b_i = 0$, if $i > m$. Their difference is

$$f(x) - g(x) = \sum_{i=0}^{\max(n, m)} (a_i - b_i) x^i$$

and their product is

$$f(x) \cdot g(x) = \sum_{i=0}^{m+n} c_i x^i,$$

where

$$c_i = a_0 b_i + a_1 b_{i-1} + \cdots + a_{i-1} b_1 + a_i b_0 = \sum_{j=0}^{i} a_j b_{i-j}.$$

These are all polynomials in $\mathbb{F}[x]$ and, if we denote the degree of $f(x)$ by $\deg f(x)$, then, for different nonzero polynomials $f(x)$ and $g(x)$,

$$\begin{aligned}
\deg[f(x) \pm g(x)] &\leq \max\{\deg f(x), \deg g(x)\} \\
\deg[f(x) \cdot g(x)] &= \deg f(x) + \deg g(x).
\end{aligned}$$

For example, if $f(x) = 2x^2 - x + 4$, $g(x) = x^3 - 3$, and $h(x) = -2x^2 + x - 5$ are polynomials in $\mathbb{Q}[x]$. Then

$$\begin{aligned}
f(x) + g(x) &= x^3 + 2x^2 - x + 1 \\
f(x) + h(x) &= -1 \\
f(x)g(x) &= 2x^5 - x^4 + 4x^3 - 6x^2 + 3x - 12 \\
[f(x)]^2 &= 4x^4 - 4x^3 + 17x^2 - 8x + 16.
\end{aligned}$$

The set of polynomials $\mathbb{F}[x]$ is rather like the integers $\mathbb{Z}$, in that addition, subtraction, and multiplication can be performed in the set, but it is not always possible to divide one element by another nonzero element without leaving a remainder. For example, we can divide $x^2 - 4$ by $x + 2$ and obtain the quotient $x - 2$, because $x^2 - 4 = (x + 2)(x - 2)$. However, we cannot divide $x^2 - 4$ by x and obtain a polynomial as a result. (Remember, $x - 4x^{-1}$ is not a polynomial because polynomials do not contain negative powers of x.)

When faced with such a situation in the integers, we used the Division Algorithm to divide one integer by another and we allowed remainders. There is an analogous Division Algorithm for Polynomials. This asserts that the usual long division of polynomials can be carried out.

For example, in $\mathbb{Q}[x]$, let us divide the polynomial $f(x) = 3x^4 - 5x^3 - 2x^2 - 1$ by $g(x) = 3x^2 + x - 1$, using long division.

$$
\begin{array}{r}
x^2 - 2x + \frac{1}{3} \\
3x^2 + x - 1 \enclose{longdiv}{3x^4 - 5x^3 - 2x^2 + 0x - 1} \\
\underline{3x^4 + x^3 - x^2} \\
-6x^3 - x^2 + 0x \\
\underline{-6x^3 - 2x^2 + 2x} \\
x^2 - 2x - 1 \\
\underline{x^2 + \frac{1}{3}x - \frac{1}{3}} \\
-\frac{7}{3}x - \frac{2}{3}
\end{array}
$$

Hence when $f(x)$ is divided by $g(x)$, the quotient is $x^2 - 2x + \frac{1}{3}$ with remainder $-\frac{7}{3}x - \frac{2}{3}$. Therefore,

$$
f(x) = \left(x^2 - 2x + \frac{1}{3}\right) g(x) + \left(-\frac{7}{3}x - \frac{2}{3}\right).
$$

Note that the remainder in this case is not a constant, as might be expected. In general, the best we can do is to find a remainder whose degree is less than the degree of the divisor.

Division Algorithm for Polynomials 9.11.
If $\mathbb{F}$ is a field, and $f(x)$ and $g(x)$ are polynomials in $\mathbb{F}[x]$, where $g(x)$ is not the zero polynomial, then there exist unique polynomials $q(x), r(x) \in \mathbb{F}[x]$ such that

$$f(x) \;=\; q(x) \cdot g(x) + r(x), \quad \text{where } \deg r(x) < \deg g(x), \text{ or } r(x) = 0.$$

The polynomial $q(x)$ is called the **quotient**, and $r(x)$ is called the **remainder**, when $f(x)$ is divided by $g(x)$. If $r(x) = 0$, the zero polynomial, we say that $g(x)$ **divides** $f(x)$ or that $g(x)$ is a **factor** of $f(x)$ and write $g(x)|f(x)$.

Proof.   To prove the existence of $g(x)$ and $r(x)$, we proceed by induction on the degree of $f(x)$. If $f(x)$ is the zero polynomial, the result holds by taking $q(x)$ and $r(x)$ to be the zero polynomial. If $\deg g(x) = 0$, then $g(x)$ is a nonzero constant, say b_0 and

$$f(x) \;=\; \frac{f(x)}{b_0} \cdot g(x) + 0$$

so we can take $q(x) = f(x)/b_0$ and $r(x) = 0$.

If $\deg f(x) < \deg g(x)$, then, by taking $q(x)$ as the zero polynomial and $r(x)$ as $f(x)$, the assertion is true.

Consider the case when $\deg g(x) = m > 0$. We have just shown that the assertion is true if $f(x) = 0$ or if $\deg f(x) < m$. These provide the base cases for the induction, to prove the result when $\deg f(x) \geq m$.

Suppose the result is true for all polynomials $f(x)$ of degree less than k, for some $k \geq m$. Let

$$f(x) \;=\; a_k x^k + \cdots + a_1 x + a_0$$

be a polynomial of degree k, and let

$$g(x) \;=\; b_m x^m + \cdots + b_1 x + b_0, \quad \text{where } b_m \neq 0.$$

Now the polynomial $f(x) - \frac{a_k}{b_m} x^{k-m} g(x)$ has degree less than k and, by the induction hypothesis, can be written in the form $q_1(x) \cdot g(x) + r(x)$, where $\deg r(x) < \deg g(x)$ or $r(x) = 0$. Hence

$$f(x) \;=\; \left[\frac{a_k}{b_m} x^{k-m} + q_1(x) \right] g(x) + r(x),$$

which shows that the assertion is true for $\deg f(x) = k$, by leaving the remainder as $r(x)$ and taking

$$q(x) \;=\; \left[\frac{a_k}{b_m} x^{k-m} + q_1(x) \right].$$

The existence of the quotient $q(x)$ and remainder $r(x)$, for all polynomials $f(x)$ of degree $\geq m$, now follows by strong induction. The existence of $q(x)$ and $r(x)$ for polynomials $f(x)$ of degree $< m$, and for the special cases, in which $g(x)$ is a constant or $f(x)$ is the zero polynomial, has already been proven.

To show the uniqueness of $q(x)$ and $r(x)$, suppose that

$$f(x) \;=\; q_1(x) \cdot g(x) + r_1(x) \;=\; q_2(x) \cdot g(x) + r_2(x)$$

are two expressions for $f(x)$ satisfying the assertion. Then

$$[q_1(x) - q_2(x)]g(x) \;=\; r_2(x) - r_1(x),$$

where the right side is either the zero polynomial or has degree less than that of $g(x)$. If $q_1(x) \neq q_2(x)$, the left side would have degree greater than or equal to that of $g(x)$; this a contradiction. Hence $q_1(x) = q_2(x)$ and also $r_1(x) = r_2(x)$. $\square$

The quotient and remainder can always be found by long division. For example, if we divide $f(z) = z^4 - iz^3 + 2z^2 - z + 1 - 2i$ by $g(z) = z^2 - i$ in $\mathbb{C}[z]$, we obtain the following.

$$
\begin{array}{r}
z^2 - iz + 2 + i \\
\hline
z^2 + 0z - i \,\big)\; z^4 - iz^3 + 2z^2 - z + 1 - 2i \\
z^4 - iz^2 \\
\hline
-iz^3 + (2+i)z^2 - z \\
-iz^3 - z \\
\hline
(2+i)z^2 + 1 - 2i \\
(2+i)z^2 + 1 - 2i \\
\hline
0
\end{array}
$$

Since the remainder is zero, $g(z)$ divides $f(z)$ and

$$z^4 - iz^3 + 2z^2 - z + 1 - 2i \;=\; (z^2 - iz + 2 + i)(z^2 - i).$$

We can even apply long division to polynomials with coefficients in the finite field $\mathbb{Z}_p$, where p is a prime. Let us denote the elements of $\mathbb{Z}_5$ just by 0, 1, 2, 3, and 4, instead of by [0], [1], [2], [3], and [4]. We shall divide $f(x) = 4x^4 + 4x^3 + 1$ by $g(x) = 2x^2 + 3x + 1$ in $\mathbb{Z}_5[x]$. When carrying out this long division, keep in mind the rather peculiar properties of the field $\mathbb{Z}_5$; for example, $-2 = 3$ because we have $-2 \equiv 3 \pmod 5$, and 2 divides into 3 exactly 4 times because $3 \equiv 2 \cdot 4 \pmod 5$.

$$
\boxed{\text{In } \mathbb{Z}_5}
$$

$$
\begin{array}{r}
2x^2 + 4x + 3 \\
\hline
2x^2 + 3x + 1 \,\big)\; 4x^4 + 4x^3 + 0x^2 + 0x + 1 \\
4x^4 + x^3 + 2x^2 \\
\hline
3x^3 + 3x^2 + 0x \\
3x^3 + 2x^2 + 4x \\
\hline
x^2 + x + 1 \\
x^2 + 4x + 3 \\
\hline
2x + 3
\end{array}
$$

The quotient is $2x^2 + 4x + 3$ and the remainder is $2x + 3$; hence, in $\mathbb{Z}_5[x]$,

$$4x^4 + 4x^3 + 1 \ = \ (2x^2 + 4x + 3)(2x^2 + 3x + 1) + (2x + 3).$$

The reader will discover, after a little practice, that it is easy, and even fun, to manipulate polynomials over a finite field. One advantage is that, in long division, one never has any ugly fractions to deal with, as is often the case with polynomials over the rational, real, and complex fields.

When we divide by a *linear* polynomial, the remainder must be a constant. The value of this constant is given by the following well-known theorem.

Remainder Theorem 9.12. *The remainder when the polynomial $f(x)$ is divided by $(x - c)$ is $f(c)$.*

Proof. By the Division Algorithm, there exist unique polynomials $q(x)$ and $r(x)$ such that

$$f(x) \ = \ q(x) \cdot (x - c) + r(x),$$

where $\deg r(x) < 1$, or $r(x) = 0$. Therefore, the remainder $r(x)$ is a constant that could be zero and that we shall write as r_0. Hence

$$f(x) \ = \ q(x) \cdot (x - c) + r_0.$$

Substituting $x = c$ in this equation, we obtain $f(c) = r_0$, which completes the proof of the theorem. $\qquad\square$

Example 9.13. Find the remainder when the complex eighth-degree polynomial $f(z) = 5z^8 - 2iz^5 + (1 + i)z^3 - 2z + 1 - i$ is divided by $z + i$.

Solution. Applying the Remainder Theorem, we evaluate

$$
\begin{aligned}
f(-i) &= 5(-i)^8 &- 2i(-i)^5 &+ (1+i)(-i)^3 &- 2(-i) &+ 1 - i \\
&= 5 &- 2i(-i) &+ (1+i)(i) &+ 2i &+ 1 - i \\
&= 5 &- 2 &+ i - 1 &+ 2i &+ 1 - i \\
&= 3 + 2i
\end{aligned}
$$

and see that the remainder is $3 + 2i$. $\qquad\square$

The Remainder Theorem yields an important criterion for a linear polynomial to be a factor of another polynomial.

Factor Theorem 9.14. *The linear polynomial $(x - c)$ is a factor of the polynomial $f(x)$ if and only if $f(c) = 0$; in other words, if and only if c is a root of the polynomial $f(x)$.*

Proof. The polynomial $(x - c)$ is a factor of $f(x)$ if and only if the remainder when $f(x)$ is divided by $(x - c)$ is zero. By the Remainder Theorem 9.12, this remainder is $f(c)$. $\qquad\square$

Example 9.15. Is $x + 1$ a factor of $x^{12} + 1$, or $x^{13} + 1$?

Solution.   The polynomial $x + 1$ is a factor if and only if -1 is a root. Now

$$(-1)^{12} + 1 \;=\; 2, \quad \text{and} \quad (-1)^{13} + 1 \;=\; 0.$$

Hence $x + 1$ is a factor of $x^{13} + 1$ but not of $x^{12} + 1$. $\qquad\qquad \square$

Example 9.16.   Find a quartic polynomial whose roots are 3, -4, $2 - \sqrt{3}$, and $2 + \sqrt{3}$.

Solution.   Such a polynomial must have $(x - 3)$, $(x + 4)$, $(x - 2 + \sqrt{3})$, and $(x - 2 - \sqrt{3})$ as factors. Therefore, a suitable polynomial is

$$\begin{aligned}
f(x) \;&=\; (x - 3)(x + 4)(x - 2 + \sqrt{3})(x - 2 - \sqrt{3}) \\
&=\; (x^2 + x - 12)(x^2 - 4x + 1) \\
&=\; x^4 - 3x^3 - 15x^2 + 49x - 12.
\end{aligned}$$

$\qquad\qquad \square$

Theorem 9.17.   *A polynomial of degree n over the field $\mathbb{F}$ has at most n distinct roots in $\mathbb{F}$.*

Proof.   We shall prove this result by induction on the degree. A polynomial of degree 0 is a nonzero constant and so has no roots. A polynomial of degree 1, $a_1 x + a_0$, always has one root $-a_0/a_1$. Hence the result is true if $n \leq 1$.

Assume, as induction hypothesis, that every polynomial of degree k over $\mathbb{F}$ has at most k distinct roots in $\mathbb{F}$. Let $f(x) \in \mathbb{F}[x]$ have degree $k + 1$. If $f(x)$ has no roots in $\mathbb{F}$, then it certainly has at most $k + 1$ roots in $\mathbb{F}$. Otherwise, if $f(x)$ does have roots, let $c \in \mathbb{F}$ be one such root. By the Factor Theorem 9.14,

$$f(x) \;=\; q(x) \cdot (x - c)$$

for some $q(x) \in \mathbb{F}[x]$ of degree k. Any root of $f(x)$ must be a root of $q(x)$, or of $x - c$. By the induction hypothesis, $q(x)$ has at most k distinct roots, and $x - c$ has one root; hence $f(x)$ has at most $k + 1$ distinct roots in $\mathbb{F}$. $\qquad \square$

Any polynomial equation of degree n with rational, real, or complex solutions can therefore have at most n different solutions.

When we attempt to factor a polynomial into linear factors in order to find its roots, it may happen that some of the factors occur more than once. A root c is said to have *multiplicity* r if $(x - c)^r$ is the highest power of $(x - c)$ in the polynomial. If $r > 1$, c is called a *multiple root*; if $r = 1$, c is a *simple root*. For example, $x^6 + x^5 - 4x^4 - 2x^3 + 5x^2 + x - 2$ can be factored as $(x - 1)^3 (x + 1)^2 (x + 2)$, so 1 is a root of multiplicity 3, and -1 is a root of multiplicity 2, while -2 is a simple root.

When factoring the integers, we found that any integer could be written as a product of basic integers, called primes, that could not be factored any further. In an analogous way, a polynomial can be factored, over a given field, into a product of basic polynomials that will factor no further. We call these basic polynomials irreducible.

Definition.   A polynomial in $\mathbb{F}[x]$ of positive degree is called **reducible in** $\mathbb{F}[x]$ if it can be written as the product of two polynomials in $\mathbb{F}[x]$ of positive degree.

A polynomial in $\mathbb{F}[x]$ of positive degree is called **irreducible in** $\mathbb{F}[x]$ if it is not reducible; that is, if it cannot be written as a product of two polynomials in $\mathbb{F}[x]$ of positive degree.

For example, $x^2 - 1$ is reducible in $\mathbb{R}[x]$ because it factors as $(x - 1)(x + 1)$. However $x^2 + 1$ is irreducible in $\mathbb{R}[x]$; if $x^2 + 1$ were reducible in $\mathbb{R}[x]$ it would factor into two real linear factors and hence, by the Factor Theorem, would have a real root c such that $c^2 = -1$.

Reducibility depends very much on the given field. For example, $x^2 + 1$ is irreducible in $\mathbb{R}[x]$ but reducible in $\mathbb{C}[x]$ because it factors as $(x + i)(x - i)$.

In the Unique Factorization Theorem for Integers 2.54, we showed that the factorization of integers into primes is essentially unique. It is possible to prove in an analogous way, though we shall not do it here, that any nonzero polynomial in $\mathbb{F}[x]$ can be written as a product

$$a \cdot g_1(x) \cdot g_2(x) \cdots g_r(x),$$

where $a \in \mathbb{F}$, and each $g_i(x)$ is a monic irreducible polynomial in $\mathbb{F}[x]$. Furthermore, this expression is unique up to the order of the factors. A *monic polynomial* in x is one in which the coefficient of the highest power of x is 1. Any polynomial over a field can be written as a constant times a monic polynomial.

Reducible polynomials are analogous to composite numbers, and irreducible polynomials to prime numbers. While it is relatively straightforward to factor a small integer into primes, it is often very difficult, if not impossible, to factor a given polynomial into irreducible factors.

All polynomials of degree 1 are irreducible because they cannot be factored as the product of two polynomials of positive degree. It follows from the Factor Theorem that the polynomials of degree 1 are the only irreducible polynomials that have roots in $\mathbb{F}$. Hence any polynomial of degree higher than 1 that has a root in $\mathbb{F}$ must be reducible in $\mathbb{F}[x]$.

The absence of any roots is sufficient to show that a polynomial of degree 2 or 3 is irreducible, but not sufficient to show that polynomials of degree greater than 3 are irreducible.

Proposition 9.18.   *Let $f(x)$ be a polynomial of degree 2 or 3 in $\mathbb{F}[x]$. Then $f(x)$ is irreducible in $\mathbb{F}[x]$ if and only if $f(x)$ has no roots in the field $\mathbb{F}$.*

Proof.   It follows from the Factor Theorem that if $f(x)$ has a root in $\mathbb{F}$, then $f(x)$ is reducible in $\mathbb{F}[x]$.

If a quadratic polynomial is reducible in $\mathbb{F}[x]$, it must factor into two linear factors. If a cubic polynomial is reducible in $\mathbb{F}[x]$, it must factor into three linear factors or one linear factor and one quadratic factor. Hence, a polynomial of degree 2 or 3 that is reducible in $\mathbb{F}[x]$ must contain at least one linear factor and therefore must have a root in the field $\mathbb{F}$.

It follows that a polynomial of degree 2 or 3 that has no roots in $\mathbb{F}$ must be irreducible in $\mathbb{F}[x]$. □

To show that this result is not true for polynomials of degree 4 or higher, look at $f(x) = x^4 + 2x^2 + 1$ in $\mathbb{R}[x]$. This polynomial $f(x)$ has no real roots because $f(x) \geq 1$ for all $x \in \mathbb{R}$. However, it is reducible in $\mathbb{R}[x]$ because it factors as $f(x) = (x^2 + 1)(x^2 + 1)$.

9.2   COMPLEX ROOTS OF A POLYNOMIAL

The Fundamental Theorem of Algebra guarantees that any polynomial in $\mathbb{C}[x]$ has a root. This, together with the Factor Theorem, shows that any complex polynomial of degree n has n roots in the complex numbers. These n roots may not necessarily be distinct.

Theorem 9.21.   *Let $f(x) \in \mathbb{C}[x]$ be a polynomial of degree n. Then there are n complex numbers $c_1, c_2, \ldots, c_n$, not necessarily distinct, such that $f(x)$ factors into the product of n linear factors*

$$f(x) \quad = \quad a_n(x - c_1)(x - c_2) \cdots (x - c_n),$$

where a_n is the coefficient of x^n in $f(x)$.

Proof.   The proof will be by induction on n, the degree of the polynomial.

When $n = 0$, $f(x) = a_0$ and the result is true.

Suppose that the result is true for all polynomials of degree $n - 1$, and let $f(x) = a_n x^n + \cdots + a_0$ be a polynomial of degree n. By the Fundamental Theorem of Algebra 8.81, $f(x)$ has a root, say $c_n \in \mathbb{C}$. Therefore, by the Factor Theorem 9.14,

$$f(x) \quad = \quad g(x)(x - c_n).$$

By looking at the coefficients of highest degree, we see that the degree of the polynomial $g(x)$ must be $n - 1$, and the coefficient of x^{n-1} in $g(x)$ is a_n. By our induction hypothesis we can write

$$g(x) \quad = \quad a_n(x - c_1)(x - c_2) \cdots (x - c_{n-1}),$$

where $c_i \in \mathbb{C}$. Therefore,

$$f(x) \quad = \quad a_n(x - c_1)(x - c_2) \cdots (x - c_{n-1})(x - c_n).$$ □

Corollary 9.22.   *Any polynomial of degree n in $\mathbb{C}[x]$ has exactly n roots, where a root of multiplicity k counts as k roots.*

Proof.   If $f(x) \in \mathbb{C}[x]$, then, by the above theorem, we can write this as

$$f(x) = a_n(x - c_1)(x - c_2) \cdots (x - c_n),$$

where $c_1, c_2, \ldots, c_n$ are the n roots of $f(x)$ with a root of multiplicity k counting as k roots.   □

Corollary 9.23.   *A polynomial is irreducible in $\mathbb{C}[x]$ if and only if it has degree 1.*

Proof.   By Theorem 9.21, any polynomial of degree 2 or higher, factors into linear factors in $\mathbb{C}[x]$ and so is reducible. The corollary follows because all polynomials of degree 1 are irreducible.   □

There is one important theorem concerning the *complex* roots of a *real* polynomial. For example, consider the real quadratic polynomial $f(x) = x^2 - 4x + 5$. Its complex roots are

$$x = \frac{4 \pm \sqrt{16 - 20}}{2} = 2 \pm i.$$

The two roots are complex conjugates of each other. In fact, the nonreal complex roots of any real polynomial always occur in conjugate pairs.

Conjugate Roots Theorem 9.24.   *If $c \in \mathbb{C}$ is a root of a polynomial with real coefficients, then its complex conjugate $\bar{c}$ is also a root.*

Proof.   Let $f(x) = a_n x^n + a_{n-1} x^{n-1} + \cdots + a_0 \in \mathbb{R}[x]$.
   If c is a complex root of $f(x)$, then

$$f(c) = a_n c^n + a_{n-1} c^{n-1} + \cdots + a_0 = 0$$

and
$$f(\bar{c}) = a_n \bar{c}^n + a_{n-1} \bar{c}^{n-1} + \cdots + a_0.$$

Since each a_i is real, $a_i = \bar{a}_i$ and, using Proposition 8.42, we have

$$f(\bar{c}) = \bar{a}_n \bar{c}^n + \bar{a}_{n-1} \bar{c}^{n-1} + \cdots + \bar{a}_0 = \overline{a_n c^n} + \overline{a_{n-1} c^{n-1}} + \cdots + \bar{a}_0$$

$$= \overline{a_n c^n + a_{n-1} c^{n-1} + \cdots + a_0} = \overline{f(c)} = \bar{0}$$

$$= 0.$$

Hence $\bar{c}$ is also a root of $f(x)$.   □

Example 9.25.   If $-2 - \sqrt{2}i$ is given as a root of

$$f(x) = x^4 + 3x^3 + 4x^2 + 2x + 12,$$

find all its complex roots, and factor $f(x)$ as a product of linear factors in $\mathbb{C}[x]$.

Proof.   The polynomial $f(x)$ has real coefficients so, by the Conjugate Roots Theorem, if $-2 - \sqrt{2}i$ is one root, another must be its conjugate, $-2 + \sqrt{2}i$. Hence $(x + 2 + \sqrt{2}i)(x + 2 - \sqrt{2}i) = x^2 + 4x + 6$ is a factor of $f(x)$. Divide $f(x)$ by $x^2 + 4x + 6$ using long division.

$$
\begin{array}{r}
x^2 \quad - \quad x \quad + \quad 2 \\
x^2 + 4x + 6 \enclose{longdiv}{\; x^4 + 3x^3 + 4x^2 + 2x + 12} \\
x^4 + 4x^3 + 6x^2 \\
\hline
-\;x^3 - 2x^2 + 2x \\
-\;x^3 - 4x^2 - 6x \\
\hline
2x^2 + 8x + 12 \\
2x^2 + 8x + 12 \\
\hline
0
\end{array}
$$

Therefore, $f(x)$ is indeed divisible by $x^2 + 4x + 6$, and the information given, that $-2 - \sqrt{2}i$ is a root, is correct. Now

$$f(x) \;=\; (x^2 - x + 2)(x^2 + 4x + 6)$$

and, by the Quadratic Formula, the other two roots are $\frac{1 \pm i\sqrt{7}}{2}$.

The four complex roots of $f(x)$ are $-2 - i\sqrt{2}, -2 + i\sqrt{2}, \frac{1+i\sqrt{7}}{2}$, and $\frac{1-i\sqrt{7}}{2}$. Hence $f(x)$ factors into linear factors in $\mathbb{C}[x]$ as

$$f(x) \;=\; \left(x + 2 + i\sqrt{2}\right)\left(x + 2 - i\sqrt{2}\right)\left(x - \tfrac{1}{2} - i\tfrac{\sqrt{7}}{2}\right)\left(x - \tfrac{1}{2} + i\tfrac{\sqrt{7}}{2}\right). \qquad \square$$

The linear factorization of complex polynomials together with the Conjugate Roots Theorem allows us to determine all the irreducible polynomials in $\mathbb{R}[x]$.

Theorem 9.26.   *Let $f(x)$ be a polynomial of degree n in $\mathbb{R}[x]$. Then $f(x)$ can be factored in $\mathbb{R}[x]$ into irreducible factors as*

$$f(x) \;=\; a_n(x - c_1)(x - c_2) \cdots (x - c_r)q_1(x)q_2(x) \cdots q_s(x),$$

where a_n is the coefficient of x^n in $f(x)$, $q_i(x) = x^2 + k_i x + \ell_i$ with $k_i^2 - 4\ell_i < 0$, and $r + 2s = n$.

Proof.   By Theorem 9.21, $f(x)$ has n roots in $\mathbb{C}$, counting multiplicities. Let c_1, $c_2, \ldots, c_r$ be all the real roots, including multiplicities. By the Conjugate Roots Theorem 9.24, the other roots occur as complex conjugate pairs, say $d_1, \overline{d}_1, d_2, \overline{d}_2, \ldots, d_s, \overline{d}_s$, where $r + 2s = n$. Now

$$(x - d_i)(x - \overline{d}_i) \;=\; x^2 - (d_i + \overline{d}_i)x + d_i\overline{d}_i \;=\; x^2 + k_i x + \ell_i,$$

where $k_i = -(d_i + \overline{d}_i)$ and $\ell_i = d_i\overline{d}_i$. By Proposition 8.42, k_i and ℓ_i are real. The quadratic $x^2 + k_i x + \ell_i$ has no real roots; therefore, it is irreducible in $\mathbb{R}[x]$, and $k_i^2 - 4\ell_i < 0$.

Hence, if the coefficient of x^n in $f(x)$ is a_n,

$$\begin{aligned}
f(x) &= a_n(x - c_1)(x - c_2)\cdots(x - c_r)(x - d_1)(x - \bar{d}_1)\cdots(x - d_s)(x - \bar{d}_s) \\
&= a_n(x - c_1)(x - c_2)\cdots(x - c_r)q_1(x)\cdots q_s(x),
\end{aligned}$$

where $q_i(x) = x^2 + k_i x + \ell_i$, and $r + 2s = n$. $\qquad\square$

Corollary 9.27. *A polynomial is irreducible in $\mathbb{R}[x]$ if and only if it is linear, or it is a quadratic of the form $ax^2 + bx + c$ where $b^2 - 4ac < 0$.*

Proof. This follows directly from the previous theorem. $\qquad\square$

There is an analogous theorem to the Conjugate Roots Theorem for *certain irrational* roots of *rational* polynomials.

Theorem 9.28. *Let $f(x) \in \mathbb{Q}[x]$. If $a + b\sqrt{c}$ is a real root of $f(x)$, where $a, b, c \in \mathbb{Q}$, and $\sqrt{c}$ is irrational, then $a - b\sqrt{c}$ is also a real root.*

Proof. This theorem can be proved in a similar way to the Conjugate Roots Theorem 9.24, by defining the notion of conjugates for each fixed c, in the set of numbers $\{a + b\sqrt{c} \mid a, b \in \mathbb{Q}\}$, by $\overline{a + b\sqrt{c}} = a - b\sqrt{c}$ and establishing the analogue of Proposition 8.42 (i) and (ii). $\qquad\square$

Example 9.29. It is given that $1 - \sqrt{2}$ and $2 + \sqrt{3}i$ are roots of the polynomial

$$f(x) = x^6 - 6x^5 + 15x^4 - 16x^3 + 7x^2 - 10x - 7.$$

Find all the roots and factor $f(x)$ into irreducible polynomials in $\mathbb{C}[x]$, $\mathbb{R}[x]$, and $\mathbb{Q}[x]$.

Solution. Since $f(x)$ has rational coefficients, and $1 - \sqrt{2}$ and $2 + \sqrt{3}i$ are roots, it follows from Theorem 9.28 and the Conjugate Roots Theorem that $1 + \sqrt{2}$ and $2 - \sqrt{3}i$ are also roots. Therefore, a factor of $f(x)$ is

$$\begin{aligned}
(x - 1 + \sqrt{2})(x - 1 - \sqrt{2})(x - 2 - \sqrt{3}i)(x - 2 + \sqrt{3}i) \\
= (x^2 - 2x - 1)(x^2 - 4x + 7) \\
= x^4 - 6x^3 + 14x^2 - 10x - 7.
\end{aligned}$$

Divide $f(x)$ by this polynomial.

$$
\begin{array}{r}
x^2 \qquad\quad + \ 1 \\
x^4 - 6x^3 + 14x^2 - 10x - 7 \overline{\smash{\big)}\ x^6 - 6x^5 + 15x^4 - 16x^3 + \ 7x^2 - 10x - 7} \\
\underline{x^6 - 6x^5 + 14x^4 - 10x^3 - \ 7x^2} \\
x^4 - \ 6x^3 + 14x^2 - 10x - 7 \\
\underline{x^4 - \ 6x^3 + 14x^2 - 10x - 7} \\
0
\end{array}
$$

The quotient and other factor of $f(x)$ is $x^2 + 1$. Hence the roots of $f(x)$ are $1 \pm \sqrt{2}$, $\pm i$, and $2 \pm \sqrt{3}i$. In $\mathbb{C}[x]$, the polynomial factors into linear factors as

$$f(x) = (x - 1 + \sqrt{2})(x - 1 - \sqrt{2})(x - i)(x + i)(x - 2 - \sqrt{3}i)(x - 2 + \sqrt{3}i).$$

To factor the polynomial in $\mathbb{R}[x]$, we combine the pairs of factors corresponding to the pairs of complex conjugate roots to obtain

$$f(x) = (x - 1 + \sqrt{2})(x - 1 - \sqrt{2})(x^2 + 1)(x^2 - 4x + 7).$$

To factor the polynomial in $\mathbb{Q}[x]$, we combine the pair of factors containing square roots to obtain

$$f(x) = (x^2 - 2x - 1)(x^2 + 1)(x^2 - 4x + 7). \qquad \square$$

9.3   RATIONAL ROOTS OF A POLYNOMIAL

We can actually produce an algorithm for finding all the rational roots of a polynomial in $\mathbb{Q}[x]$.

This algorithm will only yield the rational roots of a polynomial. It will not find any irrational roots of a rational polynomial. For example, it will not find the roots $2 \pm \sqrt{3}$ of the polynomial $x^4 - 3x^3 - 15x^2 + 49x - 12$. Neither will the algorithm find rational roots of any real polynomial; it will be useless to find the rational roots of $x^3 - \sqrt{5}x + 4$.

First notice that the roots of a rational polynomial are the same as the solutions of a polynomial equation with *integer* coefficients. If $g(x) \in \mathbb{Q}[x]$, let k be the least common multiple of the denominators of the coefficients of $g(x)$. The roots of $g(x)$ are the same as the solutions to the equation $kg(x) = 0$, which has integer coefficients. The algorithm for finding the rational roots of an equation with integer coefficients is contained in the following theorem.

Rational Roots Theorem 9.31.   *Let $f(x) = a_n x^n + \cdots + a_0$ be a polynomial with integer coefficients. If p/q is a rational root, in its lowest terms, then $p|a_0$ and $q|a_n$.*

Therefore, in order to find the rational roots of $f(x)$, we need only examine a *finite* collection of rational numbers, those whose numerators divide the constant term and whose denominators divide the *leading coefficient* a_n. Note that the theorem only suggests those rational numbers that *may* be roots. It does not say that all, or even any, of these numbers are in fact roots.

Proof.   If p/q is a rational number in its lowest terms, then p and q have no factors in common. If p/q is a root of $f(x)$, then

$$a_n \left(\frac{p}{q}\right)^n + a_{n-1} \left(\frac{p}{q}\right)^{n-1} + \cdots + a_1 \left(\frac{p}{q}\right) + a_0 = 0.$$

Multiplying through by q^n, we obtain

$$a_n p^n + a_{n-1} p^{n-1} q + \cdots + a_1 p q^{n-1} + a_0 q^n = 0$$

and

$$a_n p^n = -q(a_{n-1} p^{n-1} + \cdots + a_1 p q^{n-2} + a_0 q^{n-1}).$$

Since all the symbols in this equation are integers, q divides the right side and so $q \mid a_n p^n$. As p and q have no factors in common, q must divide a_n.

Similarly,

$$a_0 q^n = -p(a_n p^{n-1} + a_{n-1} p^{n-2} q + \cdots + a_1 q^{n-1})$$

and it follows that p must divide a_0. $\square$

Example 9.32.   Find all the rational roots of the polynomial

$$g(x) = x^4 + \frac{x^3}{2} - 6x^2 + \frac{3}{2}.$$

Solution.   The roots of $g(x)$ are the same as those of the integer equation

$$2x^4 + x^3 - 12x^2 + 3 = 0.$$

If p/q is a rational root, in its lowest terms, then, by the Rational Roots Theorem, $p \mid 3$ and $q \mid 2$. Hence p can be ± 1 or ± 3, and q can be ± 1 or ± 2. Therefore, the only rational numbers that can possibly be roots are ± 1, ± 3, $\pm 1/2$, or $\pm 3/2$.

Test each of these in turn to see if they are in fact roots.

x	-3	$-3/2$	-1	$-1/2$	$1/2$	1	$3/2$	3
$2g(x)$	30	$-69/4$	-8	0	$1/4$	-6	$-21/2$	84

Hence $x = -1/2$ is the only rational root of $g(x)$. $\square$

Example 9.33. Prove that $\sqrt[5]{3}$ is irrational.

Solution.   The real number $\sqrt[5]{3}$ is a solution of the equation $x^5 - 3 = 0$. If p/q is a rational solution, in its lowest terms, then by the Rational Roots Theorem, $p|(-3)$ and $q|1$; hence $p/q = \pm 1$ or ± 3.

x	-3	-1	1	3
$x^5 - 3$	-246	-4	-2	240

None of these values are roots of $x^5 - 3$, and so $x^5 - 3 = 0$ has no rational solutions. Therefore, $\sqrt[5]{3}$ is irrational.   □

Example 9.34. Is $\sqrt{2} + \sqrt{3}$ rational or irrational?

Solution.   Put $x = \sqrt{2} + \sqrt{3}$, and let us eliminate the root signs. Squaring, we have $x^2 = 2 + 2\sqrt{6} + 3$, and $x^2 - 5 = 2\sqrt{6}$. Squaring again, we have $x^4 - 10x^2 + 25 = 24$, and so x satisfies the equation

$$x^4 - 10x^2 + 1 = 0.$$

By the Rational Roots Theorem, the only possible rational solutions of this equation are ± 1. Clearly, neither of these is a solution, so $\sqrt{2} + \sqrt{3}$ must be irrational.   □

Example 9.35. Find all the real solutions to the equation

$$3x^4 + 13x^3 + 16x^2 + 7x + 1 = 0.$$

Solution.   By the Rational Roots Theorem, the only possible rational solutions are ± 1 or $\pm 1/3$. It is clear that if x is positive, the left side is greater than 1, and so the equation has no real positive solutions. By substitution, we see that both the possible negative rational values, -1 and $-1/3$, are solutions. Therefore, $x + 1$ and $3x + 1$ are factors of the left side. Divide the left side by $(x+1)(3x+1) = 3x^2 + 4x + 1$.

$$
\begin{array}{r}
x^2 + 3x + 1 \\
3x^2 + 4x + 1 \overline{)\, 3x^4 + 13x^3 + 16x^2 + 7x + 1} \\
3x^4 + 4x^3 + x^2 \\
\hline
9x^3 + 15x^2 + 7x \\
9x^3 + 12x^2 + 3x \\
\hline
3x^2 + 4x + 1 \\
3x^2 + 4x + 1 \\
\hline
0
\end{array}
$$

Hence the equation is equivalent to

$$(x + 1)(3x + 1)(x^2 + 3x + 1) = 0.$$

The real solutions are $x = -1, -3$ and, by the Quadratic Formula, $(-3 \pm \sqrt{5})/2$. The last two, of course, are irrational solutions. $\square$

If the leading coefficient and the constant term of an integer polynomial have many factors, then the Rational Roots Theorem will yield a large number of possible rational values to check. For example, if $f(x) = 12x^3 - 5x^2 + 20$, the theorem would tell us that the possible rational roots are p/q, where $p|20$ and $q|12$. That is, $p = \pm 1$, ± 2, ± 4, ± 5, ± 10 or ± 20 and $q = \pm 1$, ± 2, ± 3, ± 4, ± 6 or ± 12. Hence the possible rational roots p/q are

$$\pm 1, \pm 2, \pm 4, \pm 5, \pm 10, \pm 20, \pm \frac{1}{2}, \pm \frac{5}{2}, \pm \frac{1}{3}, \pm \frac{2}{3}, \pm \frac{4}{3}, \pm \frac{5}{3}, \pm \frac{10}{3}, \pm \frac{20}{3}, \pm \frac{1}{4}, \pm \frac{5}{4}, \pm \frac{1}{6}, \pm \frac{5}{6}, \pm \frac{1}{12}, \text{ or } \pm \frac{5}{12},$$

a total of 40 possible numbers to be checked! We obviously would like some methods to reduce this list to one of more reasonable proportions. One of the most effective ways to do this is to use the calculus to sketch the graph of the polynomial in order to find the approximate value of the real roots. We discuss this method in Section 9.4. Another method is outlined in Problems 124, 125, and 126.

9.4   APPROXIMATING REAL ROOTS

There is no algorithm, analogous to that for rational polynomials, for finding the real roots of an arbitrary polynomial with real coefficients. The best we can do is to sketch the graph of the polynomial and try to find the *number* of real roots by determining the number of times the graph intersects the horizontal axis. We can then find the *approximate* values of the real roots to any required degree of accuracy.

Of course, if the real polynomial we are interested in has rational coefficients, we can first apply the Rational Roots Theorem to find all the rational roots. We can then factor out the corresponding linear factors to simplify the polynomial. A prior estimate of the approximate values of all the roots is very useful in eliminating most of the candidates for rational roots.

Any real polynomial $f(x) \in \mathbb{R}[x]$ defines a polynomial function $f : \mathbb{R} \to \mathbb{R}$, and the zero polynomial yields the function that is always zero. It can be shown that the graph of any such function is always continuous; this means that the graph contains no holes or breaks in it. It seems intuitively obvious that if the values of a continuous function change sign in an interval, then the graph must cross the horizontal axis, at least once, in that interval. This fact, known as the Intermediate Value Theorem, will be very useful in approximating real roots. We just state the theorem we need here without proof, as the proof relies on the deeper properties of the real number system.

Intermediate Value Theorem 9.41.   *If $f(x) \in \mathbb{R}[x]$, and $f(a)$ and $f(b)$ are of opposite sign, then $f(x)$ has at least one root between a and b.*

One way to draw the graph of a polynomial function, in order to find its roots, is to plot a large number of points of the graph. However, by using a little calculus, it is possible to reduce this work considerably and at the same time obtain more precise information on the number of real roots. If

$$f(x) \;=\; a_n x^n + a_{n-1} x^{n-1} + \cdots + a_2 x^2 + a_1 x + a_0$$

is a polynomial of degree n, its *derivative*

$$f'(x) \;=\; n a_n x^{n-1} + (n-1) a_{n-1} x^{n-2} + \cdots + 2 a_2 x + a_1$$

is a polynomial of degree $n-1$ whose value at any point x is the slope of the graph of $f(x)$ at the point x. The roots of the derivative $f'(x)$ yield all the stationary values of the function $f(x)$. A stationary value or turning point is either a *local maximum*, a *local minimum*, or a *point of inflection*.

A knowledge of all the turning points and of the behavior of the graph for large positive and negative values of the variable will enable us to find the number of real roots and their approximate values.

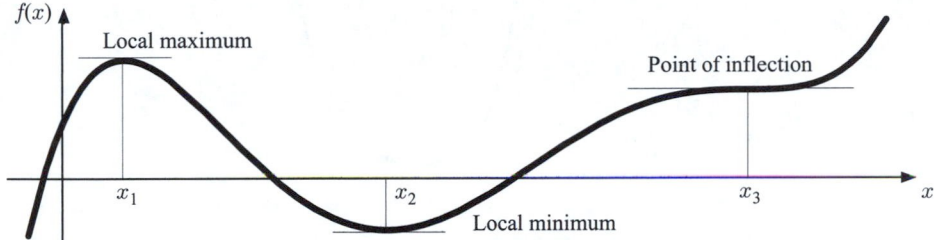

Since any polynomial has only a finite number of roots, its graph must remain on one side of the horizontal axis for sufficiently large positive (and negative) values of the variable. In other words, the sign of the polynomial must remain unchanged for sufficiently large positive (and negative) values of the variable. We show in Proposition 9.45 that the sign of the polynomial $f(x) = a_n x^n + \cdots + a_1 x + a_0$ will be the same as the sign of its leading term, $a_n x^n$, whenever $|x|$ is sufficiently large. For example, if $f(x) = -4x^3 + x^2 - x - 5$, then $f(x)$ is negative for large positive values of x and $f(x)$ is positive for large negative values of x.

Example 9.42.   Determine the rational roots of

$$f(x) \;=\; x^4 - x^3 + x^2 - 12$$

and find the number of real roots.

Solution.   Differentiating

$$f(x) \;=\; x^4 - x^3 + x^2 - 12,$$

we obtain its derivative

$$f'(x) = 4x^3 - 3x^2 + 2x = (4x^2 - 3x + 2)x.$$

This derivative is zero only when $x = 0$, because the quadratic $4x^2 - 3x + 2$ has no real roots. Hence the only turning point of the graph of $f(x)$ occurs at $x = 0$, when $f(x) = -12$. When x is very large, positive or negative, the value of $f(x)$ is the same as its leading term x^4. Hence, when x is large, positive or negative, $f(x)$ is positive. Therefore, it follows from the graph that $f(x)$ has one positive root and one negative root.

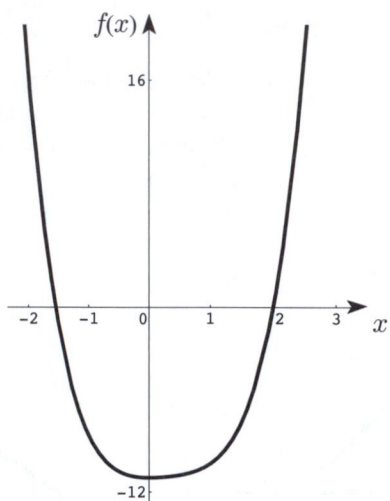

By the Rational Roots Theorem, the possible values of rational roots are ± 1, ± 2, ± 3, ± 4, ± 6, and ± 12.

x	-12	-6	-3	-2	-1	0	1	2	3	4	6	12
$f(x)$				16	-9	-12	-11	0				

The values in the previous table are the only ones we need to calculate. We can see that $x = 2$ is the positive root. As $f(-2)$ and $f(-1)$ have different signs, it follows from the Intermediate Value Theorem that the negative root must lie between -2 and -1; hence it cannot be rational because none of the possible values of the rational roots lie in that range.

Summing up, we see that $f(x)$ has one rational root, namely $x = 2$, and one other real root, an irrational root lying between -2 and -1. $\square$

Example 9.43.   Find the number of real solutions of the equation

$$12x^3 - 5x^2 + 20 = 0$$

and determine whether any of these solutions are rational.

Solution.   Let $f(x) = 12x^3 - 5x^2 + 20$. Differentiating, we have

$$f'(x) = 36x^2 - 10x = 2x(18x - 5).$$

Therefore, the slope of the graph of $f(x)$ is zero when $x = 0$, and when $x = 5/18$. When $x = 0$, $f(x) = 20$, and when $x = 5/18$, $f(x)$ is still positive. For large values of x, $f(x)$ is dominated by $12x^3$. Hence for large positive values of x, $f(x)$ is positive, and for large negative values of x, $f(x)$ is negative. The graph therefore has the form indicated, and we see that $f(x)$ has only one real root, which is negative.

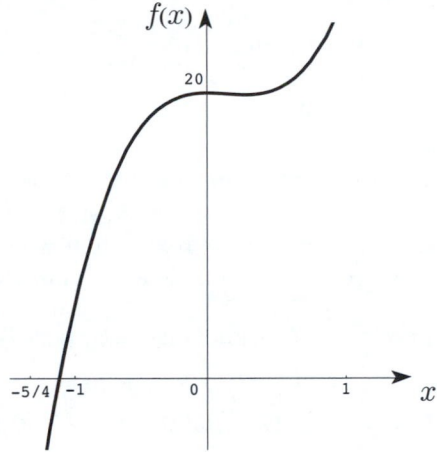

As mentioned at the end of Section 9.3, the possible rational roots of $f(x)$ are

$$\pm 1, \pm 2, \pm 4, \pm 5, \pm 10, \pm 20, \pm \frac{1}{2}, \pm \frac{5}{2}, \pm \frac{1}{3}, \pm \frac{2}{3}, \pm \frac{4}{3}, \pm \frac{5}{3}, \pm \frac{10}{3}, \pm \frac{20}{3}, \pm \frac{1}{4}, \pm \frac{5}{4}, \pm \frac{1}{6}, \pm \frac{5}{6}, \pm \frac{1}{12}, \text{ or } \pm \frac{5}{12}.$$

x	$-5/4$	-1
$f(x)$	$-45/4$	3

Calculating $f(-1)$, we see that it is positive; hence the real root is less than -1. The next possible rational root below -1 is $-5/4$, but $f(-5/4)$ is negative. Hence the only real solution to the equation $f(x) = 0$ lies between -1.25 and -1 and, since there are no other possible rational roots in this range, the solution must be irrational. $\qquad\square$

In the above example, we have located an irrational root lying between -1.25 and -1. We can find the value of this root to any degree of accuracy by successive approximations using the Intermediate Value Theorem. The following example shows one general method.

Example 9.44. Find the real solution to the equation in the previous example,

$$12x^3 - 5x^2 + 20 \ = \ 0$$

correct to two decimal places.

Solution.   We know that the root of $f(x) = 12x^3 - 5x^2 + 20$ lies between -1.25 and -1. Let us now look at the value of $f(-1.1)$. By using the Binomial Theorem, we obtain the approximation $f(-1.1) \approx -12[1 + 3(.1) + 3(.001)] - 5[1.21] + 20 \approx -2.0$.

x	-1.1	-1.0
$f(x)$	-2.0	3.0

The root therefore lies between -1.1 and -1.0.

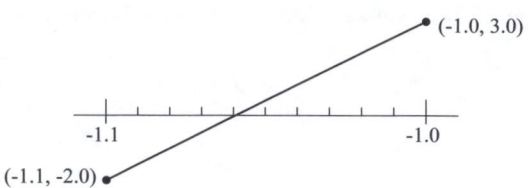

We now wish to approximate the root further to find the value of the second decimal place. By looking at the sketch of the graph between -1.1 and -1.0, it would appear that a second approximation would be -1.06. Again, using the Binomial Theorem, we obtain the following approximations.

$$
\begin{aligned}
f(-1.06) &\approx -12[1 + 3(.06) + 3(.0036)] - 5[1 + 2(.06)] + 20 \\
&\approx -14.28 - 5.60 + 20 = +0.12 \\
f(-1.07) &\approx -12[1 + 3(.07) + 3(.0049)] - 5[1 + 2(.07)] + 20 \\
&\approx -14.64 - 5.70 + 20 = -0.34 \\
f(-1.065) &\approx -12[1 + 3(.065) + 3(.004)] - 5[1 + 2(.065)] + 20 \\
&\approx -14.484 - 5.65 + 20 = -0.134
\end{aligned}
$$

x	-1.07	-1.065	-1.06
$f(x)$	-0.34	-0.134	$+0.12$

Therefore the solution lies between -1.065 and -1.06, and so it is -1.06 correct to two decimal places. $\qquad\square$

One crude general result follows immediately from the continuity of the polynomial functions.

Proposition 9.45. *If $f(x) \in \mathbb{R}[x]$ is a polynomial of odd degree, then $f(x)$ has at least one real root.*

Proof. We first show that the polynomial is dominated by its highest term, and so its graph lies on opposite sides of the x-axis for large positive and large negative values of x. Let

$$
\begin{aligned}
f(x) &= a_n x^n + a_{n-1} x^{n-1} + \cdots + a_1 x + a_0 \\
&= x^n \left(a_n + \frac{a_{n-1}}{x} + \cdots + \frac{a_1}{x^{n-1}} + \frac{a_0}{x^n} \right).
\end{aligned}
$$

Let M be the larger of $(|a_{n-1}| + |a_{n-2}| + \cdots + |a_1| + |a_0|)/|a_n|$ and 1. Then, if $|x| > M$,

$$\left| \frac{a_{n-1}}{x} + \cdots + \frac{a_1}{x^{n-1}} + \frac{a_0}{x^n} \right| \leq \left| \frac{a_{n-1}}{x} \right| + \cdots + \left| \frac{a_1}{x^{n-1}} \right| + \left| \frac{a_0}{x^n} \right|$$

$$\leq \left| \frac{a_{n-1}}{x} \right| + \cdots + \left| \frac{a_1}{x} \right| + \left| \frac{a_0}{x} \right| \leq \frac{M|a_n|}{|x|}$$

$$< |a_n|.$$

Therefore, when $|x| > M$, the function $f(x)$ is dominated by its highest term, and the sign of $f(x)$ is the same as that of $a_n x^n$. If n is odd, then $f(x)$ has different signs when $x > M$ and $x < -M$. Therefore, by the Intermediate Value Theorem 9.41, $f(x)$ has a real root between $-M$ and $+M$. $\qquad \square$

9.5   POLYNOMIAL INEQUALITIES

Now that we have many methods at our disposal for finding real roots of real polynomials, we shall take a look at solving real polynomial inequalities.

There are two basic methods that we can use to solve inequalities of the form $f(x) > 0$, or $f(x) \geq 0$, where $f(x) \in \mathbb{R}[x]$. One method is to sketch the graph of $f(x)$ and determine its roots. The solution to the inequality can then be read off from the graph. The other method is to factor $f(x)$ into irreducible factors in $\mathbb{R}[x]$ as in Theorem 9.26. The solution to the inequality can be found by looking at the signs of all the factors. An irreducible quadratic factor will have the same sign for all $x \in \mathbb{R}$ because it has no real roots.

Example 9.51.   Factor the polynomial

$$f(x) \;=\; 3x^3 - 4x^2 - 2x + 1$$

into irreducible real factors and solve the inequality $f(x) > 0$.

Solution.   Let us first determine whether $f(x)$ has any rational roots. If p/q is a rational root, in its lowest terms, then $p|1$ and $q|3$. Hence the possible rational roots are ± 1 and $\pm 1/3$.

x	-1	$-1/3$	$1/3$	1
$f(x)$	-4	$8/9$	0	-2

Therefore, $1/3$ is a root, and $(3x - 1)$ is a factor of $f(x)$. Divide $f(x)$ by $(3x - 1)$.

$$
\begin{array}{r}
x^2 \;-\; x \;-\; 1 \\[2pt]
3x-1 \,\big)\, \overline{\;3x^3 - 4x^2 - 2x + 1\;} \\[2pt]
3x^3 - \;\; x^2 \\[2pt]
\hline
-\,3x^2 - 2x \\[2pt]
-\,3x^2 + \;\; x \\[2pt]
\hline
-\,3x + 1 \\[2pt]
-\,3x + 1 \\[2pt]
\hline
0
\end{array}
$$

Hence

$$
\begin{aligned}
f(x) \;&=\; (3x-1)(x^2 - x - 1) \\[6pt]
&=\; (3x-1)\left(x - \frac{1}{2} - \frac{\sqrt{5}}{2}\right)\left(x - \frac{1}{2} + \frac{\sqrt{5}}{2}\right).
\end{aligned}
$$

To solve the inequality $f(x) > 0$, consider the following table.

Range	\multicolumn{4}{c}{Sign of}
	$3x-1$	$x - \frac{1}{2} - \frac{\sqrt{5}}{2}$	$x - \frac{1}{2} + \frac{\sqrt{5}}{2}$	$f(x)$
$x < (1-\sqrt{5})/2$	$-$	$-$	$-$	$-$
$(1-\sqrt{5})/2 < x < 1/3$	$-$	$-$	$+$	$+$
$1/3 < x < (1+\sqrt{5})/2$	$+$	$-$	$+$	$-$
$(1+\sqrt{5})/2 < x$	$+$	$+$	$+$	$+$

The solution set to $f(x) > 0$ is

$$
\{x \in \mathbb{R} \mid (1-\sqrt{5})/2 < x < 1/3 \quad\text{or}\quad (1+\sqrt{5})/2 < x\},
$$

which is the union of the open intervals

$$
\left(\frac{1-\sqrt{5}}{2}, \frac{1}{3}\right) \;\cup\; \left(\frac{1+\sqrt{5}}{2}, \infty\right). \qquad\qquad \square
$$

Example 9.52.   Solve the inequality $x^6 + 5x^2 - 6 \le 0$.

Solution.   We shall sketch the graph of $f(x) = x^6 + 5x^2 - 6$. Differentiating, we have $f'(x) = 6x^5 + 10x = 2x(3x^4 + 5)$. Hence the slope of the graph of $f(x)$ is zero only when $x = 0$, in which case $f(x) = -6$.

As $f(x)$ is positive for large positive or negative values of x, it follows from the graph that $f(x)$ has two real roots. The possible rational roots are $\pm 1, \pm 2, \pm 3$, and ± 6. It is seen easily that $f(1) = 0$ and $f(-1) = 0$, so 1 and -1 are the only real roots.

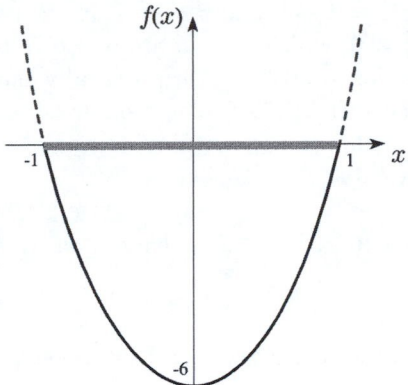

The solution set to $f(x) \le 0$ is therefore

$$\{x \in \mathbb{R} \mid -1 \le x \le 1\},$$

which is the closed interval $[-1, 1]$. $\square$

9.6   CUBIC EQUATIONS

The solution of quadratic equations has been essentially known since Babylonian times, over 3500 years ago. However, it was not until the 1500s that several Italian mathematicians managed to solve cubic and quartic equations. Of course, the polynomials they used in those days had real coefficients, but their methods apply equally well to complex polynomials. If the three roots of a cubic were all real, their methods essentially required the use of complex numbers, though they did not formally define them. In fact, they were uneasy about accepting negative numbers as solutions, and any complex roots were dismissed as imaginary.

These methods are rarely used in practice nowadays (except in computer algebra packages), as irrational roots of cubics and quartics are usually found directly from the equation, by some method of successive approximation. However, the fact that it is possible to solve polynomial equations of degree 4 and less, in terms of nth roots of complex numbers, is of theoretical interest.

The solution of the cubic is a generalization of the method of completing the square used in the proof of the Quadratic Formula, but of course more complicated. We shall just consider monic polynomials, as we can always divide by the coefficient of the highest power of x to obtain such a polynomial.

If we had a monic quadratic equation

$$x^2 + bx + c = 0,$$

recall from the proof of the Quadratic Formula 8.11 that we completed the square in x to obtain

$$\left(x + \frac{b}{2}\right)^2 - \frac{b^2}{4} + c = 0.$$

If we substitute $x = y - \frac{b}{2}$ and $q = \frac{b^2}{4} - c$, then we obtain the simpler quadratic equation $y^2 - q = 0$, which we can always solve in the complex numbers. Notice that the completion of the square essentially eliminates the term involving the first power of the variable. To solve a cubic equation, we start by using a similar substitution to eliminate the term involving the second power of the variable and obtain a simpler cubic.

Cubic Formula 9.61.   *To solve the cubic equation*

$$x^3 + bx^2 + cx + d = 0, \quad \text{where } b,\, c,\, d \in \mathbb{C},$$

first eliminate the second-degree term by the substitution $x = y - \frac{b}{3}$ to obtain the new equation

$$y^3 + py + q = 0,$$

where $p = c - \frac{b^2}{3}$ and $q = \frac{2b^3}{27} - \frac{bc}{3} + d$. The three roots of the original equation are

$$x \;=\; -\frac{b}{3} + u - \frac{p}{3u}, \quad -\frac{b}{3} + \omega u - \frac{\omega^2 p}{3u}, \quad -\frac{b}{3} + \omega^2 u - \frac{\omega p}{3u},$$

where ω is a nonreal cube root of unity, and u is any cube root of $-\frac{q}{2} \pm \sqrt{\frac{q^2}{4} + \frac{p^3}{27}}$.

Proof.   Substitute $x = y - \frac{b}{3}$ in the equation to obtain

$$\left(y - \tfrac{b}{3}\right)^3 + b\left(y - \tfrac{b}{3}\right)^2 + c\left(y - \tfrac{b}{3}\right) + d = 0.$$

After simplifying, the term in y^2 drops out, and we obtain the simpler cubic equation

$$y^3 + py + q = 0, \quad \text{where } p = c - \frac{b^2}{3} \text{ and } q = \frac{2b^3}{27} - \frac{bc}{3} + d.$$

As with the Quadratic Formula, we would like to reduce this further to obtain a cubic we can easily solve, such as

$$(y - t)^3 \;=\; \text{constant}$$

or equivalently, $y^3 - 3ty^2 + 3t^2y = $ another constant. To do this we would need t to satisfy $-3ty + 3t^2 = p$. Therefore, as the next step in simplifying the cubic, we try the substitution

$$y \;=\; t - \frac{p}{3t}.$$

For y to satisfy the cubic $y^3 + py + q = 0$, t must satisfy the sixth-degree equation

$$t^6 + qt^3 - \frac{p^3}{27} \;=\; 0.$$

However, this is possible to solve, since it is a quadratic equation in t^3, and

$$t^3 \;=\; -\frac{q}{2} \pm \sqrt{\frac{q^2}{4} + \frac{p^3}{27}},$$

where $\sqrt{\frac{q^2}{4} + \frac{p^3}{27}}$ denotes either of the complex square roots of $\frac{q^2}{4} + \frac{p^3}{27}$.

Hence a solution to the cubic $y^3 + py + q = 0$ is $y = t - \frac{p}{3t}$, where t is one of the cube roots of $-\frac{q}{2} + \sqrt{\frac{q^2}{4} + \frac{p^3}{27}}$ or $-\frac{q}{2} - \sqrt{\frac{q^2}{4} + \frac{p^3}{27}}$. Let u be a specific cube root of $-\frac{q}{2} + \sqrt{\frac{q^2}{4} + \frac{p^3}{27}}$, so that the other cube roots will be ωu and $\omega^2 u$, where ω is a nonreal cube root of unity. Let v be a specific cube root of $-\frac{q}{2} - \sqrt{\frac{q^2}{4} + \frac{p^3}{27}}$, so that the other cube roots are ωv and $\omega^2 v$. At first sight it appears that our cubic has six solutions, corresponding to the six values of t, namely u, ωu, $\omega^2 u$, v, ωv, and $\omega^2 v$. However, these six values only lead to three roots, which therefore preserves the integrity of Theorem 9.17. The three roots of the cubic $y^3 + py + q = 0$ are

$$
\begin{aligned}
y_1 &= u - \frac{p}{3u} &= v - \frac{p}{3v} \\
y_2 &= \omega u - \frac{p}{3\omega u} &= \omega^2 v - \frac{p}{3\omega^2 v} \\
y_3 &= \omega^2 u - \frac{p}{3\omega^2 u} &= \omega v - \frac{p}{3\omega v}
\end{aligned}
$$

since it is possible to choose v to be the cube root such that $3uv + p = 0$. Using the fact that $\omega^{-1} = \omega^2$, we can write the three roots of the original cubic as

$$
x_1 = -\frac{b}{3} + u - \frac{p}{3u}, \quad x_2 = -\frac{b}{3} + \omega u - \frac{\omega^2 p}{3u}, \quad x_3 = -\frac{b}{3} + \omega^2 u - \frac{\omega p}{3u},
$$

where u is any cube root of one of the values of $-\frac{q}{2} \pm \sqrt{\frac{q^2}{4} + \frac{p^3}{27}}$. ☐

Example 9.62.   Find all the complex roots of $x^3 + 3x^2 + 9x + 5 = 0$.

Solution.   First sketch the graph of

$$
f(x) = x^3 + 3x^2 + 9x + 5.
$$

Differentiating, we have

$$
f'(x) = 3x^2 + 6x + 9,
$$

which is never zero for real values of x.

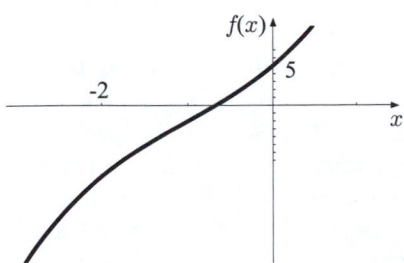

Hence the cubic has no maxima or minima and has only one real root that is clearly negative. By the Rational Roots Theorem 9.31, the only possible negative rational roots are -1 and -5. However, $f(-1) = -2$ and $f(-5) = -90$, so there is one real irrational root between -1 and 0, and a pair of complex conjugate roots.

Following the Cubic Formula 9.61, first put $x = y - 1$ to eliminate the second degree term so that

$$
\begin{aligned}
(y-1)^3 + 3(y-1)^2 + 9(y-1) + 5 &= 0; \\
y^3 + 6y - 2 &= 0.
\end{aligned}
$$

Now put $y = t - \frac{2}{t}$ so that

$$
\begin{aligned}
\left(t - \frac{2}{t}\right)^3 + 6\left(t - \frac{2}{t}\right) - 2 &= 0; \\
t^3 - \frac{8}{t^3} - 2 &= 0; \\
t^6 - 2t^3 - 8 &= 0; \\
(t^3 - 4)(t^3 + 2) &= 0.
\end{aligned}
$$

If we choose u to be the real root $\sqrt[3]{4}$, then the three roots of the original cubic are

$$
\begin{aligned}
x_1 &= -1 + \sqrt[3]{4} - \sqrt[3]{2} &\approx& \ -0.6725 \\
x_2 &= -1 + \sqrt[3]{4}\omega - \sqrt[3]{2}\omega^2 &=& \ -1 + \frac{-\sqrt[3]{4} + \sqrt[3]{2}}{2} + i\sqrt{3}\left(\frac{\sqrt[3]{4} + \sqrt[3]{2}}{2}\right) \\
x_3 &= -1 + \sqrt[3]{4}\omega^2 - \sqrt[3]{2}\omega &=& \ -1 + \frac{-\sqrt[3]{4} + \sqrt[3]{2}}{2} - i\sqrt{3}\left(\frac{\sqrt[3]{4} + \sqrt[3]{2}}{2}\right),
\end{aligned}
$$

where $\omega = \frac{-1 + i\sqrt{3}}{2}$ and $\omega^2 = \frac{-1 - i\sqrt{3}}{2}$. The first solution is the real root between -1 and 0, while the latter two are complex conjugates of each other with approximate values of $-1.1637 \pm 2.4659i$. $\square$

It is possible to solve an equation of degree 4 in a similar way that uses the solution of the cubic equation. Hence it is possible to solve any complex polynomial equation of degree less than 5 if we know how to extract square and cube roots of complex numbers. It is natural to ask whether there are methods for solving polynomials of any degree, assuming that we can extract nth roots of complex numbers. In fact, there is no general formula for solving polynomial equations of degree larger than 4; though of course, certain particular equations can be solved. In 1826, the Norwegian mathematician Niels Henrik Abel proved the impossibility of the solution of quintic equations by roots. Soon after, in 1831, the brilliant French mathematician Evariste Galois developed a general theory of the solvability of polynomial equations, shortly before his tragic death in a duel at the age of 20. This theory, now called *Galois theory*, relies on deep connections between groups and fields. It marked the beginning of a new kind of algebra, called *modern algebra* or *abstract algebra*.

9.7 MULTIPLE ROOTS

If a polynomial contains a multiple root, there is an algorithm that will find the repeated factors. This algorithm involves the derivative of the polynomial and the concept of the greatest common divisor of polynomials. The characterization of the greatest common divisor for integers, in Proposition 2.29, generalizes to polynomials.

Definition. Let $f(x), g(x) \in \mathbb{F}[x]$, where $\mathbb{F}$ is a field. The monic polynomial $d(x) \in \mathbb{F}[x]$ is the **greatest common divisor** of $f(x)$ and $g(x)$ if

(i) $d(x)$ divides both $f(x)$ and $g(x)$.

(ii) Any divisor of both $f(x)$ and $g(x)$ also divides $d(x)$.

If this is the case, we write $d(x) = \gcd(f(x), g(x))$.

For example, if $f(x)$ and $g(x)$ have no common factors of positive degree, then $\gcd(f(x), g(x)) = 1$. Most of the results concerning greatest common divisor of integers carry over to give results about the greatest common divisor of polynomials. The greatest common divisor of two polynomials can be obtained from the factorization of the polynomials into irreducible factors (corresponding to the prime factorization of integers). For example, if $f(x) = (x - 2)^2(x + 1)^3(x + \sqrt{2})(x - \sqrt{2})$ and $g(x) = (x - 2)^3(x + \sqrt{2})$ are two polynomials in $\mathbb{R}[x]$, then $\gcd(f(x), g(x)) = (x - 2)^2(x + \sqrt{2})$. There is even a Euclidean Algorithm for polynomials that can be used to find greatest common divisors. The statement of the result is identical to the Euclidean Algorithm for integers, except that the integers are replaced by polynomials in $\mathbb{F}[x]$.

Polynomials in $\mathbb{F}[x]$, where $\mathbb{F}$ is any field, can be differentiated using the same formulas as for real polynomials. However, for some coefficient fields this differentiation will have no interpretation in terms of limits, or slopes of graphs, and so is called formal differentiation. The *formal derivative* of the polynomial $f(x) = x^n$ is the polynomial $f'(x) = nx^{n-1}$, where the coefficient n is interpreted as the element 1 in the field, added to itself n times. The derivatives of the other polynomials can be obtained from the usual differentiation formulas.

The algorithm for finding the multiple factors of a polynomial $f(x)$ consists of finding the greatest common divisor of $f(x)$ and its derivative $f'(x)$.

Multiple Roots Theorem 9.71. *If $f(x)$ is a polynomial with coefficients in the field $\mathbb{F}$, with formal derivative $f'(x)$, then the multiple roots of $f(x)$ in $\mathbb{F}$ are precisely the roots of $\gcd(f(x), f'(x))$ in $\mathbb{F}$.*

Proof. Suppose that $r \in \mathbb{F}$ is a multiple root of $f(x)$, so that $(x - r)^2$ is a factor of $f(x)$ and we can write

$$f(x) = (x - r)^2 q(x), \quad \text{where } q(x) \in \mathbb{F}[x].$$

By the rule for differentiating a product, we have

$$f'(x) = 2(x - r)q(x) + (x - r)^2 q'(x),$$

so $f'(r) = 0$. Hence $(x - r)$ is a factor of both $f(x)$ and $f'(x)$ and will be a factor of $\gcd(f(x), f'(x))$.

Now suppose that $c \in \mathbb{F}$ is a root of $f(x)$, but is not a multiple root, so that

$$f(x) \;=\; (x - c)p(x), \quad \text{where } p(c) \neq 0.$$

Differentiating, we have $f'(x) \;=\; p(x) + (x - c)p'(x)$, so $f'(c) = p(c) \neq 0$. Hence $(x - c)$ is not a factor of $f'(x)$, or of $\gcd(f(x), f'(x))$. Therefore, the only roots of $f(x)$ that are roots of $\gcd(f(x), f'(x))$ are the multiple roots.

On the other hand, $\gcd(f(x), f'(x))$ is a factor of both $f(x)$ and $f'(x)$, so any root of the greatest common divisor must be a root of $f(x)$ and $f'(x)$ and, by what we have shown, must be a multiple root of $f(x)$ $\square$

Example 9.72.   It is known that the real polynomial

$$f(x) \;=\; 3x^3 - 5\sqrt{2}x^2 + 2x + 2\sqrt{2}$$

has a multiple root. Find all its roots.

Solution.   Differentiating $f(x)$, we have

$$f'(x) \;=\; 9x^2 - 10\sqrt{2}x + 2.$$

We shall not find $\gcd(f(x), f'(x))$ directly but will check to see which roots of $f'(x)$ are also roots of $f(x)$. By the Quadratic Formula, the roots of $f'(x)$ are

$$x \;=\; \frac{10\sqrt{2} \pm \sqrt{200 - 72}}{18} = \frac{5\sqrt{2} \pm \sqrt{32}}{9} = \frac{5\sqrt{2} \pm 4\sqrt{2}}{9} = \sqrt{2} \ \text{ or } \ \frac{\sqrt{2}}{9}.$$

Now $f(\sqrt{2}) = 6\sqrt{2} - 10\sqrt{2} + 2\sqrt{2} + 2\sqrt{2} = 0$ and so by the previous theorem, $\sqrt{2}$ must be a multiple root of $f(x)$. As $f(x)$ is only a cubic, it cannot have two different multiple roots; hence $\sqrt{2}/9$ is not a multiple root.

The polynomial $(x - \sqrt{2})^2 = x^2 - 2\sqrt{2}x + 2$ must be a factor of $f(x)$, and we can use long division to find the remaining factor.

$$
\begin{array}{r}
3x \;+\; \sqrt{2} \\
x^2 - 2\sqrt{2}x + 2 \ \overline{\smash{\big)}\ 3x^3 - 5\sqrt{2}x^2 + 2x + 2\sqrt{2}} \\
\underline{3x^3 - 6\sqrt{2}x^2 + 6x} \\
\sqrt{2}x^2 - 4x + 2\sqrt{2} \\
\underline{\sqrt{2}x^2 - 4x + 2\sqrt{2}} \\
0
\end{array}
$$

Therefore, $f(x) = (x - \sqrt{2})^2(3x + \sqrt{2})$, and its roots are $\sqrt{2}$, $\sqrt{2}$, and $-\sqrt{2}/3$. $\square$

9.8 PARTIAL FRACTIONS

A quotient $f(x)/g(x)$, of a polynomial $f(x)$ by a nonzero polynomial $g(x)$, is called a *rational function* of x. Such rational functions can be manipulated in a similar way to rational numbers. For example,

$$\frac{1}{x-3} - \frac{1}{x+2} = \frac{(x+2)-(x-3)}{(x-3)(x+2)} = \frac{5}{x^2-x-6}.$$

It is useful in integral calculus, and in the theory of differential equations, to be able to express a rational function $f(x)/g(x)$ as a sum of fractional forms, called *partial fractions*, whose denominators are simpler than $g(x)$. Taking the above numerical example, we see that $5/(x^2-x-6)$ can be written in terms of the simpler fractions $1/(x-3)$ and $1/(x+2)$.

If the degree of $f(x)$ is less than the degree of $g(x)$, then $f(x)/g(x)$ is called a *proper fraction*. Just as in rational numbers, any rational function can be expressed as a whole polynomial plus a proper fraction. Let $f(x)/g(x)$ be any rational function and, using the Division Algorithm, write

$$f(x) = q(x)g(x) + r(x), \quad \text{where } r(x) = 0 \text{ or } \deg r(x) < \deg g(x).$$

Then

$$\frac{f(x)}{g(x)} = q(x) + \frac{r(x)}{g(x)}, \quad \text{where } \frac{r(x)}{g(x)} \text{ is a proper fraction.}$$

Recall that any real polynomial can be factored into linear and irreducible quadratic factors in $\mathbb{R}[x]$. This allows us to decompose any fraction into partial fractions as follows.

Partial Fraction Decomposition 9.81. *Any real rational function $f(x)/g(x)$ can be written as a sum of a real polynomial and real partial fractions of the following types.*

(i) *For each nonrepeating linear factor $(ax + b)$ of $g(x)$, the partial fraction decomposition contains a term of the form*

$$\frac{A}{ax+b}, \quad \text{where } A \in \mathbb{R}.$$

(ii) *For each repeating linear factor $(ax+b)^r$ of $g(x)$, the decomposition contains terms of the form*

$$\frac{A_1}{ax+b} + \frac{A_2}{(ax+b)^2} + \cdots + \frac{A_r}{(ax+b)^r}, \quad \text{where each } A_i \in \mathbb{R}.$$

(iii) *For each nonrepeating quadratic factor $(ax^2 + bx + c)$ of $g(x)$, the decomposition contains a term of the form*

$$\frac{Ax+B}{ax^2+bx+c}, \quad \text{where } A, B \in \mathbb{R}.$$

(iv) *For each repeating quadratic factor $(ax^2+bx+c)^r$ of $g(x)$, the decomposition contains terms of the form*

$$\frac{A_1 x + B_1}{ax^2 + bx + c} + \frac{A_2 x + B_2}{(ax^2 + bx + c)^2} + \cdots + \frac{A_r x + B_r}{(ax^2 + bx + c)^r},$$

where each A_i and $B_i \in \mathbb{R}$.

Before we prove this result, we give various examples to show how the constants in the partial fraction decomposition can be found.

Example 9.82.   Decompose $\dfrac{7x+8}{2x^2 - x - 1}$ into partial fractions.

Solution.   By the above result, we can write

$$\frac{7x+8}{2x^2 - x - 1} \;=\; \frac{7x+8}{(x-1)(2x+1)} \;=\; \frac{A}{x-1} + \frac{B}{2x+1}, \quad \text{where } A, B \in \mathbb{R}.$$

Multiply each side by $(x-1)(2x+1)$ to obtain

$$\begin{aligned}
7x + 8 \;&=\; A(2x+1) + B(x-1) \\
&=\; (2A+B)x + (A-B).
\end{aligned}$$

The polynomial $7x+8$ must be identical to the polynomial $(2A+B)x + (A-B)$. Hence they must have the same coefficients, so $7 = 2A + B$ and $8 = A - B$.

Solving these two equations, we obtain $A = 5$ and $B = -3$. Hence

$$\frac{7x+8}{2x^2 - x - 1} \;=\; \frac{5}{x-1} - \frac{3}{2x+1}.$$

Alternative Method.   Start from the relation

$$\frac{7x+8}{2x^2 - x - 1} \;=\; \frac{A}{x-1} + \frac{B}{2x+1}, \quad \text{where } A, B \in \mathbb{R}.$$

Multiply both sides by $(x-1)$ to obtain

$$\frac{7x+8}{2x+1} \;=\; A + \frac{B(x-1)}{2x+1}.$$

The rational functions on each side must be identical. Therefore, the equality holds for all values of x (except when $2x+1 = 0$). In particular, substitute $x = 1$ to obtain $A = 5$. Similarly, by multiplying the partial fraction decomposition by $(2x+1)$ and then putting $x = -1/2$, we obtain $B = -3$. $\square$

Example 9.83. Decompose

$$\frac{x^2 - 10x + 7}{x^3 - 3x^2 + 4}$$

into partial fractions.

Solution.   We first have to factor the denominator $g(x) = x^3 - 3x^2 + 4$. The possible rational roots of $g(x)$ are ± 1, ± 2, and ± 4. It is seen that $g(-1) = 0$, and $g(2) = 0$. Hence $(x + 1)(x - 2) = x^2 - x - 2$ is a factor of $g(x)$.

$$
\begin{array}{r}
x - 2 \\
x^2 - x - 2 \overline{\smash{\big)}\, x^3 - 3x^2 + 0x + 4} \\
\underline{x^3 - x^2 - 2x} \\
-2x^2 + 2x + 4 \\
\underline{-2x^2 + 2x + 4} \\
0
\end{array}
$$

Therefore $g(x) = (x - 2)^2(x + 1)$ and we can write the original expression in terms of partial fractions as

$$\frac{x^2 - 10x + 7}{(x - 2)^2(x + 1)} = \frac{A}{x - 2} + \frac{B}{(x - 2)^2} + \frac{C}{x + 1}.$$

Multiplying each side by one of the partial fraction denominators $(x + 1)$, and then putting $x = -1$, we obtain

$$\frac{1 + 10 + 7}{(-1 - 2)^2} = C$$

and so $C = 2$. Multiplying each side by the highest exponent of the other partial fraction denominator $(x - 2)^2$, and then putting $x = 2$, we obtain

$$\frac{4 - 20 + 7}{3} = B$$

and so $B = -3$. We cannot obtain A by a similar trick. We could find A by multiplying each side by $(x - 2)^2(x + 1)$ and then comparing coefficients. A simpler way is to put $x = 0$ in the expressions as they stand to obtain

$$\frac{7}{4} = \frac{A}{-2} + \frac{B}{4} + C = \frac{A}{-2} - \frac{3}{4} + 2$$

and so $A = -1$.

The partial fraction decomposition is therefore

$$\frac{x^2 - 10x + 7}{x^3 - 3x^2 + 4} = \frac{2}{x + 1} - \frac{1}{x - 2} - \frac{3}{(x - 2)^2}. \qquad \square$$

The following lemmas will help us to prove the existence of the Partial Fraction Decomposition.

Lemma 9.84.   *If* $\gcd(g(x), h(x)) = 1$ *with* $g(x)$ *and* $h(x)$ *nonzero, then, for any real polynomial* $f(x)$, *there exist real polynomials* $s(x)$ *and* $t(x)$ *such that*

$$\frac{f(x)}{g(x)h(x)} = \frac{s(x)}{g(x)} + \frac{t(x)}{h(x)}.$$

Proof.   The Euclidean Algorithm for polynomials allows us to construct real polynomials $p(x)$ and $q(x)$ such that

$$p(x)g(x) + q(x)h(x) = \gcd(g(x), h(x)) = 1.$$

Multiplying this equation by $f(x)$, and dividing by $g(x)h(x)$, we obtain

$$\frac{f(x)p(x)}{h(x)} + \frac{f(x)q(x)}{g(x)} = \frac{f(x)}{g(x)h(x)}.$$

The result now follows with $s(x) = f(x)q(x)$ and $t(x) = f(x)p(x)$. $\square$

Lemma 9.85.   *If* $f(x)$ *and* $g(x)$ *are real polynomials with* $g(x)$ *nonzero, then, for each integer* r, *there exist real polynomials* $p_0(x), p_1(x), \ldots, p_r(x)$ *such that*

$$\frac{f(x)}{[g(x)]^r} = p_0(x) + \frac{p_1(x)}{g(x)} + \frac{p_2(x)}{[g(x)]^2} + \cdots + \frac{p_r(x)}{[g(x)]^r},$$

where $p_1(x), p_2(x), \ldots, p_r(x)$ *are zero or have degrees that are all less than the degree of* $g(x)$.

Proof.   Dividing $f(x)$ by $g(x)$, using the Division Algorithm, we obtain real polynomials $q_r(x)$ and $p_r(x)$ such that

$$f(x) = q_r(x)g(x) + p_r(x),$$

where $\deg p_r(x) < \deg g(x)$ or $p_r(x) = 0$.

By repeated use of the Division Algorithm, we obtain

$$
\begin{aligned}
q_r(x) &= q_{r-1}(x)g(x) + p_{r-1}(x) \\
q_{r-1}(x) &= q_{r-2}(x)g(x) + p_{r-2}(x) \\
&\ \ \vdots \\
q_2(x) &= q_1(x)g(x) + p_1(x),
\end{aligned}
$$

where, for each $k = r-1, r-2, \ldots, 1$, either $\deg p_k(x) < \deg g(x)$ or $p_k(x) = 0$. Putting $q_1(x) = p_0(x)$, and combining the above equations, we obtain

$$
f(x) = p_0(x)[g(x)]^r + p_1(x)[g(x)]^{r-1} + \cdots + p_{r-1}(x)g(x) + p_r(x).
$$

We obtain the required result on dividing by $[g(x)]^r$.   □

Proof of the Partial Fraction Decomposition 9.81.   By Theorem 9.26, the real polynomial $g(x)$ can be factored into irreducible linear and quadratic factors in $\mathbb{R}[x]$. Two distinct irreducible factors have greatest common divisor 1. Therefore, by grouping together all the multiple factors and repeatedly using Lemma 9.84 and then Lemma 9.85, we obtain the result.   □

9.9   EQUATIONS OVER A FINITE FIELD

Let p be a prime number, and let $\mathbb{Z}_p = \{0, 1, 2, \ldots, p-1\}$, the finite field of integers modulo p. Then any solution to an equation in $\mathbb{Z}_p$ must be one of the elements $0, 1, 2, \ldots, p-1$. The brute force method for solving such equations is to try each of the values $0, 1, 2, \ldots, p-1$. If p is small this method is very effective. This is the only general method we mention here, besides those given in Section 3.5 for solving linear equations in $\mathbb{Z}_p$.

Example 9.91.   Solve the equation $x^3 + 4x + 1 = 0$ for $x \in \mathbb{Z}_5$.

Solution.   Let $f(x) = x^3 + 4x + 1 \in \mathbb{Z}_5[x]$. We will evaluate $f(x)$ for each element of the finite field $\mathbb{Z}_5$.

In $\mathbb{Z}_5$ x	0	1	2	3	4
x^3	0	1	3	2	4
$x^3 + 4x + 1$	1	1	2	0	1

Hence $f(x) = 0$ if $x = 3$. Therefore, the only solution is $x = 3 \in \mathbb{Z}_5$.   □

Example 9.92.   Factor the polynomial $f(x) = x^4 + 3x^2 + 5x + 4$ in $\mathbb{Z}_7[x]$.

Solution.   We can find any linear factors of $f(x)$ by looking for the roots of $f(x)$.

In $\mathbb{Z}_7$	x	0	1	2	3	4	5	6
	x^2	0	1	4	2	2	4	1
	x^4	0	1	2	4	4	2	1
$x^4 + 3x^2 + 5x + 4$		4	6	0	1	6	1	3

We see that $x = 2$ is the only root and so $(x - 2)$ must be a factor. Since $-2 = 5$ in $\mathbb{Z}_7$, we can write this factor as $(x+5)$. Divide $f(x)$ by $(x+5)$ using long division.

$$
\boxed{\text{In } \mathbb{Z}_7}
$$

$$
\begin{array}{r}
x^3 + 2x^2 \qquad\qquad + 5 \\
x + 5 \enclose{longdiv}{x^4 + 0x^3 + 3x^2 + 5x + 4} \\
\underline{x^4 + 5x^3} \qquad\qquad\qquad\quad \\
2x^3 + 3x^2 \qquad\quad \\
\underline{2x^3 + 3x^2} \qquad\quad \\
5x + 4 \\
\underline{5x + 4} \\
0
\end{array}
$$

Hence $f(x) = (x + 5)(x^3 + 2x^2 + 5)$.

Does $x^3 + 2x^2 + 5$ factor? Any root of $x^3 + 2x^2 + 5$ must also be a root of $f(x)$. Hence 2 is the only element that is possibly a root. When $x = 2$, $x^3 + 2x^2 + 5 = 1 + 1 + 5 = 0$. Therefore, $(x - 2) = (x + 5)$ is also a factor of $x^3 + 2x^2 + 5$. Divide $x^3 + 2x^2 + 5$ by $x + 5$ using long division.

$$
\boxed{\text{In } \mathbb{Z}_7}
$$

$$
\begin{array}{r}
x^2 + 4x + 1 \\
x + 5 \enclose{longdiv}{x^3 + 2x^2 + 0x + 5} \\
\underline{x^3 + 5x^2} \qquad\qquad\quad \\
4x^2 + 0x \quad\ \\
\underline{4x^2 + 6x} \quad\ \\
x + 5 \\
\underline{x + 5} \\
0
\end{array}
$$

Hence $x^3 + 2x^2 + 5 = (x+5)(x^2 + 4x + 1)$. Again, the only possible root of $x^2 + 4x + 1$ could be 2. When $x = 2$, $x^2 + 4x + 1 = 4 + 1 + 1 = 6 \neq 0$. Therefore, $x^2 + 4x + 1$ has no roots and, by Proposition 9.18, is irreducible.

The complete factorization of $f(x)$ into irreducible factors in $\mathbb{Z}_7[x]$ is

$$f(x) \;=\; (x+5)^2(x^2+4x+1).$$

Check.   This can be checked by multiplying out the factors. $\square$

Example 9.93.   Find all the quadratic irreducible polynomials in $\mathbb{Z}_2[x]$, and determine whether x^4+x^2+1 and x^4+x+1 are irreducible in $\mathbb{Z}_2[x]$.

Solution.   Any quadratic polynomial is of the form $a_2x^2 + a_1x + a_0$, where $a_2 \neq 0$. In $\mathbb{Z}_2[x]$, the coefficients must either be 0 or 1. Hence there are four quadratic polynomials in $\mathbb{Z}_2[x]$, namely, $x^2 + a_1x + a_0$, where $a_1, a_0 \in \mathbb{Z}_2 = \{0, 1\}$.

Any irreducible polynomial must have a nonzero constant term; otherwise it would contain a factor x. Therefore, let us look at the remaining two quadratics, $x^2 + 1$ and $x^2 + x + 1$.

In $\mathbb{Z}_2$	x	0	1
	x^2	0	1
	$x^2 + 1$	1	0

In $\mathbb{Z}_2$	x	0	1
	x^2	0	1
	$x^2 + x + 1$	1	1

We see that $x^2 + 1$ has 1 as a root and is therefore reducible over $\mathbb{Z}_2$. In fact, $x^2 + 1 = (x+1)^2$ in $\mathbb{Z}_2[x]$. However, $x^2 + x + 1$ has no root and therefore, by Proposition 9.18, is irreducible.

Hence the only irreducible quadratic in $\mathbb{Z}_2[x]$ is $x^2 + x + 1$.

Let us now determine whether $x^4 + x^2 + 1$ and $x^4 + x + 1$ have any linear factors in $\mathbb{Z}_2[x]$.

In $\mathbb{Z}_2$	x	0	1
	x^2	0	1
	x^4	0	1
	$x^4 + x^2 + 1$	1	1
	$x^4 + x + 1$	1	1

We see that $x^4 + x^2 + 1$ and $x^4 + x + 1$ have no roots, and therefore no linear factors. However, as they are of degree 4, we cannot apply Proposition 9.18 to conclude that they are irreducible.

If either quartic factors, it must factor as two irreducible quadratics in $\mathbb{Z}_2[x]$. Since $x^2 + x + 1$ is the only irreducible quadratic, the only possible factorization is as $(x^2 + x + 1)^2$. By multiplication in $\mathbb{Z}_2[x]$, we see that

$$(x^2 + x + 1)^2 \;=\; x^4 + x^2 + 1.$$

Hence $x^4 + x^2 + 1$ is reducible, and $x^4 + x + 1$ must be irreducible, in $\mathbb{Z}_2[x]$. $\square$

Exercise Set 9

1–8. Find the sum, difference, and product of each of the following pairs of polynomials with coefficients in the indicated field.

1. $2x^3 + x^2 - 7x + 1$ and $x^2 - 2x - 1$ in $\mathbb{Q}[x]$
2. $x^2 + \frac{1}{2}x + \frac{1}{4}$ and $x^2 - \frac{1}{2}x + 1$ in $\mathbb{Q}[x]$
3. $x^2 - \sqrt{2}x + 1$ and $x^2 + \sqrt{2}x + 1$ in $\mathbb{R}[x]$
4. $ix^3 + (1 + 2i)x - 3$ and $x^2 - ix + 1 - i$ in $\mathbb{C}[x]$
5. $x^3 + 2x + 2$ and $x^4 + x^2 + x + 1$ in $\mathbb{Z}_3[x]$
6. $x^2 + x + 1$ and $x^3 + x^2 + x + 1$ in $\mathbb{Z}_2[x]$
7. $2x^2 + x + 4$ and $3x^2 + 2x + 1$ in $\mathbb{Z}_5[x]$
8. $x^5 + 2x + 4$ and $x^7 + 5x^3 + 4$ in $\mathbb{Z}_7[x]$

9. Do all polynomials of degree 1 with integer coefficients have an integer root? Do all polynomials of degree 1 with rational coefficients have a rational root?
10. Do all polynomials of degree 1 with coefficients in $\mathbb{Z}_5$ have a root in $\mathbb{Z}_5$? Do all polynomials of degree 2 with coefficients in $\mathbb{Z}_5$ have a root in $\mathbb{Z}_5$?
11. List all the polynomials of degree 3 with coefficients in $\mathbb{Z}_2$.
12. Find a polynomial in $\mathbb{Z}_2[x]$ that has no roots in $\mathbb{Z}_2$.

13–16. Find all the roots, with their multiplicities, for the following polynomials in $\mathbb{C}[x]$.

13. $(x - 1)(x + 2)(3x - 5)$
14. $(x + 1)^2(4x + 1)^4(x - i)$
15. $(x^2 - 1)(2x^2 + x - 5)$
16. $x^2 + 2ix - 1$

17–26. Find the quotient and remainder when $f(x)$ is divided by $g(x)$.

17. $f(x) = x^3 + x^2 + 2x - 1$ and $g(x) = x^2 - 2x + 1$ in $\mathbb{Q}[x]$
18. $f(x) = x^2 - 2x + 1$ and $g(x) = x^3 + x^2 + 2x - 1$ in $\mathbb{R}[x]$
19. $f(x) = x^4 + \frac{5}{2}x^3 + \frac{5}{2}x^2 + \frac{3}{2}x + \frac{1}{2}$ and $g(x) = x^2 + 2x + 1$ in $\mathbb{Q}[x]$
20. $f(x) = x^8 + 1$ and $g(x) = x - 1$ in $\mathbb{C}[x]$
21. $f(x) = x^3 + ix^2 + (1 + i)x + 1$ and $g(x) = x + i$ in $\mathbb{C}[x]$
22. $f(x) = x^4 + x + 1$ and $g(x) = ix^2 + x - 2$ in $\mathbb{C}[x]$
23. $f(x) = x^3 + 2x^2 + 2$ and $g(x) = 2x^2 + x + 1$ in $\mathbb{Z}_3[x]$
24. $f(x) = 3x^5 + 3x^4 + 4x^2 + 4x + 3$ and $g(x) = 5x^3 + 2x^2 + 6x + 3$ in $\mathbb{Z}_7[x]$
25. $f(x) = x^{10} + x^8 + x + 1$ and $g(x) = x^3 + x + 1$ in $\mathbb{Z}_2[x]$
26. $f(x) = 10x^5 + 4x^3 + x + 3$ and $g(x) = 3x^5 + 4$ in $\mathbb{Z}_{11}[x]$

27. Let $f(x)$ divide $g(x)$. If c is a root of $f(x)$, show that c is a root of $g(x)$. Does the converse hold?

28–33. *Find the remainder when $f(x)$ is divided by $g(x)$.*

28. $f(x) = 4x^5 + x^4 - 3x^3 + x + 5$ and $g(x) = x - 1$ in $\mathbb{Q}[x]$
29. $f(x) = x^7 + 10x^5 - 4x^3 + x^2$ and $g(x) = x + 2$ in $\mathbb{Q}[x]$
30. $f(x) = ix^6 + (1 - 2i)x^5 + 5ix^4 - x + 4$ and $g(x) = x + i$ in $\mathbb{C}[x]$
31. $f(x) = x^3 + 3x + 1$ and $g(x) = x - 1 + 4i$ in $\mathbb{C}[x]$
32. $f(x) = x^{72} + 2x^{31} - 1$ and $g(x) = x + \frac{1}{2} + \frac{\sqrt{3}i}{2}$ in $\mathbb{C}[x]$
33. $f(x) = x^5 + 2x + 1$ and $g(x) = 3x - 4$ in $\mathbb{Q}[x]$

34. Factor the polynomials $x^2 - 5x + 5$, $x^2 - 5x + 6$ and $x^2 - 5x + 7$ into irreducible factors in $\mathbb{C}[x]$, $\mathbb{R}[x]$, and $\mathbb{Q}[x]$.

35. Is $x^3 + x^2 + 1$ an irreducible polynomial in $\mathbb{Z}_2[x]$?

36. When is $x^3 + px + q$ exactly divisible by $x^2 + mx - 1$?

37. When is $x^4 + px + q$ exactly divisible by $x^2 + mx + 1$?

38–41. *Solve the following polynomial equations over the indicated finite field, stating the multiplicity of any repeated solutions.*

38. $x^3 + 2x + 2 = 0$ in $\mathbb{Z}_5$ **39.** $x^6 + x^5 + x + 1 = 0$ in $\mathbb{Z}_2$
40. $x^5 + x^4 + x^3 + 2x^2 + 2x + 2 = 0$ in $\mathbb{Z}_3$
41. $x^4 + 2x = 4$ in $\mathbb{Z}_7$

42–46. *Find all the rational roots of each of the following rational polynomials.*

42. $x^3 + \frac{x^2}{2} + \frac{x}{2} - \frac{1}{2}$ **43.** $x^4 + x + 1$

44. $x^4 + \frac{5}{6}x^3 - \frac{5x^2}{18} + \frac{1}{9}$ **45.** $x^3 - 6x^2 + 15x - 14$

46. $x^{17} - 2x^{14} + 1$

47. If p is a prime and n is an integer greater than 1, prove that $\sqrt[n]{p}$ is irrational.

48. Factor $2x^4 + x^3 + x^2 + 2x - 6$ in $\mathbb{Q}[x]$ and in $\mathbb{C}[x]$.

49. Find a rational polynomial with $\sqrt{2} + \sqrt{5}$ as a root, and then show that $\sqrt{2} + \sqrt{5}$ is irrational.

50–53. *Find all the real solutions to each of the following equations.*

50. $x^3 + 7x^2 + 8x - 10 = 0$
51. $2x^5 + x^4 - 8x^3 - 4x^2 + 8x + 4 = 0$
52. $4x^4 + x^2 + 3x + 1 = 0$ **53.** $x^3 + \frac{23}{6}x^2 + \frac{29}{6}x + 2 = 0$

54. By means of the substitution $x = iy$, solve the equation $x^3 - 3ix^2 - x - 2i = 0$ in $\mathbb{C}$.

55–56. *Find all the complex roots of the following polynomials.*

55. $z^5 + 4z$ **56.** $z^4 + iz^2 + 2$

57. Solve the equation $x^4 - 8x^3 + 18x^2 - 8x + 1 = 0$, given that $2 - \sqrt{3}$ is one solution.

58. If $1+i$ and $\sqrt{8}-1$ are roots of $f(x) = 2x^6 - 3x^5 - 17x^4 + 63x^3 - 91x^2 + 60x - 14$, find all its roots, and factor $f(x)$ into irreducible polynomials in $\mathbb{C}[x]$, $\mathbb{R}[x]$, and $\mathbb{Q}[x]$.

59. Find a polynomial of lowest degree with integer coefficients that has $4 - \sqrt{2}$ and $3 - \sqrt{2}$ as roots.

60. Find a polynomial of degree 5 in $\mathbb{Q}[x]$ that has $2 - i$ and $\sqrt{2}i$ as roots.

61–64.   Find all the complex roots of each of the following real polynomials. In each case one root $x = c$ is given.

61. $x^4 + 2x^3 + 2x^2 + 6x - 3; \quad c = -1 - \sqrt{2}$
62. $x^4 - 2x^3 - 2x^2 + 10x - 15; \quad c = 1 + \sqrt{2}i$
63. $x^4 - 10x^3 + 23x^2 + 30x - 78; \quad c = 5 - i$
64. $x^4 - (2 + 2\sqrt{2})x^3 + (2 + 4\sqrt{2})x^2 + (2\sqrt{2} - 6)x - 3; \quad c = \sqrt{2} + i$

65. Let $f(x) = x^2 - 5 \in \mathbb{Q}[x]$. Is $f(x)$ irreducible in $\mathbb{Q}[x]$ or in $\mathbb{R}[x]$?

66–69.   In each of the following equations, find the rational solutions, the number of real solutions, and, for each irrational solution, c, find the integer a for which $a < c < a + 1$.

66. $x^3 + x - 5 = 0$ 　　　　　 **67.** $x^4 - 5x^3 - 6x^2 + 20x + 20 = 0$
68. $2x^4 + x^3 + 2x^2 + 3x + 1 = 0$ 　　 **69.** $x^5 + 4x - 7 = 0$

70–73.   Find all the rational solutions and the real solutions, to one decimal place, for each of the following equations.

70. $x^4 - 4x^3 + 23 = 0$ 　　　　 **71.** $4x^3 - 9x^2 - 5x + 3 = 0$
72. $2x^5 + 12x^4 + x^2 + 11x + 30 = 0$ 　 **73.** $3x^3 - 2x^2 + 5x + 1 = 0$

74–77.   Solve the following inequalities for $x \in \mathbb{R}$.

74. $(x - 1)(3x + 2)(x + 6) \leq 0$ 　　 **75.** $(x - 1)^2(x + 4)(x - 7) < 0$
76. $x^3 - 6x - 4 > 0$ 　　　　　 **77.** $4x^4 - 13x^2 - 4x + 5 > 0$

78. Solve the inequality $x^3 - 27x + 5 > 0$, for $x \in \mathbb{R}$, to within one decimal place.

79–82.   By solving the equation $f'(x) = 0$, find the multiple roots, with their multiplicities, of each of the following polynomials.

79. $f(x) = 27x^3 - 54x^2 + 36x - 8$ 　　 **80.** $f(x) = x^3 + 2x - 4$
81. $f(x) = x^4 - 4x^3 + 2x^2 + 4x + 1$
82. $f(x) = \sqrt{3}x^4 - 8x^3 + 6\sqrt{3}x^2 - 3\sqrt{3}$

83. Find a and b, if $ax^{n+1} + bx^n + 1$ is divisible by $(x - 1)^2$.

84–86. *Find the greatest common divisor of each of the following pairs of polynomials.*

84. $f(x) = x^3 + x^2 + x + 1$ and $g(x) = x^2 + 4x + 3$ in $\mathbb{R}[x]$
85. $f(x) = x^3 - x^2 + ix - i$ and $g(x) = x^2 + i$ in $\mathbb{C}[x]$
86. $f(x) = x^5 + x^4 + x^2 + x$ and $g(x) = x^3 + x^2 + x + 1$ in $\mathbb{Z}_2[x]$

87–94. *Decompose the following into real partial fractions.*

87. $\dfrac{3}{4x^2 - x}$

88. $\dfrac{2x^2 + 1}{x^3 + x^2}$

89. $\dfrac{11x + 1}{4x^2 + 3x - 1}$

90. $\dfrac{1}{x^2 + 8x + 15}$

91. $\dfrac{(x + 3)(x - 3)}{(x + 1)^2(x + 5)}$

92. $\dfrac{x}{x^2 - 2}$

93. $\dfrac{2x^2 + x - 3}{x^3 - 1}$

94. $\dfrac{x^2 + 1}{x^3 - x + 6}$

95–96. *Find the roots of each polynomial in the fields, $\mathbb{Q}$, $\mathbb{R}$, $\mathbb{C}$, and $\mathbb{Z}_7$.*

95. $x^3 + 15x^2 + 15x + 14$

96. $2x^3 - 7x^2 + 16x - 15$

97–100. *Factor each polynomial into irreducible factors in $\mathbb{Q}[x]$, $\mathbb{R}[x]$, $\mathbb{C}[x]$, and $\mathbb{Z}_p[x]$, for the given p.*

97. $2x^3 + 2x^2 - 11x - 2$; $p = 3$
98. $3x^3 - x^2 + x + 2$; $p = 2$
99. $2x^4 + 5x^3 + 6x^2 + 4x + 1$; $p = 5$
100. $6x^4 + x^3 + 2x^2 - 4x + 1$; $p = 3$

Problem Set 9

101. Let $f(x)$ and $g(x)$ be two polynomials with coefficients in a field $\mathbb{F}$. If $f(x)$ divides $g(x)$, and $g(x)$ divides $f(x)$, show that $f(x) = cg(x)$ for some $c \in \mathbb{F}$.

102. Use the Factor Theorem to factor $x^3 + y^3 + z^3 - 3xyz$.

103. Let $c = \operatorname{cis}(2\pi/n)$. Show that 1, c, $c^2, \ldots, c^{n-1}$ are distinct solutions of $z^n = 1$, and show that

$$1 + c + c^2 + \cdots + c^{n-1} = 0.$$

104. Prove that $x^{4n} + x^{3n} + x^{2n} + x^n + 1$ is divisible by $x^4 + x^3 + x^2 + x + 1$ whenever n is a positive integer that is not a multiple of 5.

105. *(Wilson's Theorem)* If p is a prime, factor $x^{p-1} - 1$ in $\mathbb{Z}_p[x]$, and prove that $(p - 1)! \equiv -1 \pmod{p}$.

106. If p and q are the roots of $x^2 + bx + c$, show that $p + q = -b$ and $pq = c$.

107. If p, q, r are the roots of $x^3 + bx^2 + cx + d$, show that $p + q + r = -b$, $pq + qr + rp = c$, and $pqr = -d$.

108. If p and q are roots of $x^2 - 10x + 2$, find a quadratic with $3p$ and $3q$ as roots.

109. If p and q are roots of $x^2 - 10x + 2$, find a quadratic with p^2 and q^2 as roots.

110. If p, q, and r are roots of $x^3 + 2x^2 - 3x + 5$, find $p^3 + q^3 + r^3$.

111. If p, q, and r are roots of $x^3 + 2x^2 - 3x + 5$, find a cubic with p^2, q^2, and r^2 as roots.

112. If $f(x)$ and $g(x)$ are two polynomials of degree n that give the same value for $n + 1$ different values of x, show that they have the same coefficients.

113. Let $\mathbb{F}$ be a field. When is the following theorem true? "If two polynomials in $\mathbb{F}[x]$ take the same values for all $x \in \mathbb{F}$, then they have the same coefficients."

114–118. If $f : \mathbb{N} \to \mathbb{R}$ is a function from the nonnegative integers, $\mathbb{N}$, to the real numbers, define a new function

$$\Delta f : \mathbb{N} \longrightarrow \mathbb{R} \quad by \quad \Delta f(n) = f(n + 1) - f(n)$$

and inductively define $\Delta^{r+1} f = \Delta(\Delta^r f)$. The function $f : \mathbb{N} \to \mathbb{R}$ is called a polynomial function of degree r if

$$f(n) \;=\; a_r n^r + \cdots + a_1 n + a_0 \quad \text{for all } n \in \mathbb{N},$$

where $a_1, a_1, \ldots, a_r \in \mathbb{R}$ and $a_r \neq 0$.

114. If $f(n) = n^2 - 3n + 5$, calculate $\Delta f(n)$, $\Delta^2 f(n)$, and $\Delta^3 f(n)$ for $n = 0$, 1, 2, 3, 4, and 5.

115. If $f : \mathbb{N} \to \mathbb{R}$ is any polynomial function of degree r, show that $\Delta^{r+1} f$ is the zero function while $\Delta^r f$ is not the zero function.

116. Certain values of a polynomial function f are given in the following table.

n	12	13	14	15	16
$f(n)$	60	64	67	72	82

What is the smallest possible degree of f? If f does have this smallest degree, calculate $f(17)$ and $f(18)$.

117. **(a)** Calculate Δf when $f(n) = \binom{n}{r}$.

   (b) Show that, for any polynomial function of degree r,

$$f(n) \;=\; c_r \binom{n}{r} + c_{r-1} \binom{n}{r - 1} + \cdots + c_0,$$

   where $c_i = \Delta^i f(0)$.

   (c) If a function f satisfies $\Delta^{r+1} f = 0$, $\Delta^r f \neq 0$, show that f is a polynomial function of degree r.

118. **(a)** Show that $f(n+1) - f(0) = \Delta f(0) + \Delta f(1) + \cdots + \Delta f(n)$.

    (b) Calculate

$$1 + 2 + \cdots + n$$
$$1^2 + 2^2 + \cdots + n^2$$
$$1^3 + 2^3 + \cdots + n^3$$

using Problems 117(c) and 118(a).

119. Find all irreducible polynomials of degree ≤ 4 in $\mathbb{Z}_2[x]$.

120. Find one irreducible quadratic, and one irreducible cubic, in $\mathbb{Z}_3[x]$.

121. **(a)** Show that $f_n(x) = x^n + x + 1 \in \mathbb{Z}_2[x]$ has no roots in $\mathbb{Z}_2$.

    (b) For each degree $n \geq 2$, find polynomials in $\mathbb{Z}_2[x]$ of degree n that have both elements of $\mathbb{Z}_2$ as roots.

122. Check that $x^3 + 5x \in \mathbb{Z}_6[x]$ has roots 0, 1, 2, 3, 4, and 5 in $\mathbb{Z}_6$. Does this contradict Theorem 9.17?

123. Is $\sqrt{3} + \sqrt[3]{3}$ rational or irrational?

124. *(Special Case of Gauss's Lemma)* If p/q is a rational root, in its lowest terms, of a polynomial $f(x)$ with integer coefficients, then show that

$$f(x) = (qx - p)g(x)$$

for some polynomial $g(x)$ with *integer* coefficients.

125. Let $f(x)$ be a polynomial with integer coefficients. Show that the rational number p/q, in its lowest terms, cannot be a root of $f(x)$ unless $|q - p|$ divides $|f(1)|$. This result can be used to shorten the list of possible rational roots of an integer polynomial.

126. Let $f(x)$ be a polynomial with integer coefficients. For each integer m, generalize the result of the previous problem by finding a necessary condition for p/q to be a rational root of $f(x)$, in terms of the divisibility of $f(m)$.

127–130. Find all the rational roots of the following equations.

127. $x^3 + 20x - 144 = 0$

128. $8x^3 + 20x^2 + 14x - 93 = 0$

129. $9x^4 - 2x^2 - 2x + 24 = 0$

130. $20x^5 + 8x^4 + 15x^3 + x^2 - 27x - 10 = 0$

131. Show that an even-degree polynomial, with odd integer coefficients, contains no rational roots.

132. Prove that there is no polynomial of positive degree with integer coefficients that takes prime values for all positive integers.

133. **(a)** Use De Moivre's Theorem and the Binomial Theorem to prove that, for each integer r, $\cos(\frac{2r\pi}{5})$ is a solution to the equation

$$16x^5 - 20x^3 + 5x - 1 = 0.$$

    (b) Factor $16x^5 - 20x^3 + 5x - 1$ into irreducible factors in $\mathbb{R}[x]$, and find the value of $\cos(\frac{2\pi}{5})$.

134. **(a)** Write $\cos 4\theta$ in terms of $\cos\theta$, and find the values of $\cos\theta$ for which

$$\cos 4\theta = \cos\theta.$$

(b) Which values of θ satisfy the equation $\cos 4\theta = \cos\theta$?

135. If the rational polynomial $x^3 + bx^2 + cx + d$ has a rational root, r, and if $c + r^2 = 0$, show that all of the roots are rational.

136. Prove Theorem 9.28 by dividing $f(x)$ by the quadratic

$$(x - a - b\sqrt{c})(x - a + b\sqrt{c}).$$

137. **(a)** Let $f(x)$ be a real polynomial that changes sign between a and b. By factoring $f(x)$ into irreducible real factors, prove that $f(x)$ has a real root between a and b.

(b) Hence prove that an odd-degree real polynomial always has at least one real root. (These proofs of the Intermediate Value Theorem 9.41 and Proposition 9.45 appear to avoid the use of continuity. However, continuity is needed in the proof of the Fundamental Theorem of Algebra, which is used in Theorem 9.26 for the factorization of real polynomials.)

138. The polynomial $f(x) = x^7 - 8x^6 + 17x^5 + 5x^4 - 22x^3 - 37x^2 - 26x + 10$ is known to have $3 - i$ and $2 + \sqrt{3}$ as roots. Find all its roots and factor it into irreducible factors in $\mathbb{C}[x]$, $\mathbb{R}[x]$, and $\mathbb{Q}[x]$.

139. Let

$$f_n(x) = 1 + x + \frac{x^2}{2!} + \cdots + \frac{x^n}{n!}.$$

Show that $f_n(x)$ has no real roots if n is even, and one real root if n is odd.

140–143. *Solve the following inequalities for $x \in \mathbb{R}$, giving your answer, if necessary, to one decimal place.*

140. $\dfrac{x}{2x^2 + x - 1} \geq 4$

141. $\dfrac{1}{(x-2)(x-3)} > \dfrac{1}{(x-1)(x-4)}$

142. $\dfrac{4}{x} \geq x^2 + 3x - 1$

143. $|x^3 - 3x - 1| < 2x$

144. If a cubic polynomial with rational coefficients has a double root, show that all its roots are rational.

145. Show that the remainder, when the real polynomial $f(x)$ is divided by $(x-c)^2$, is $f'(c)(x-c) + f(c)$, where $f'(x)$ is the derivative of $f(x)$.

146. Generalize the previous problem, and find the remainder when the real polynomial $f(x)$ is divided by $(x-c)^r$.

147. If $f(x)$ and $g(x)$ are irreducible polynomials in $\mathbb{F}[x]$, with $f(x) \neq kg(x)$ for any $k \in \mathbb{F}$, show that $\gcd(f(x), g(x)) = 1$.

148. Find the repeated root of the polynomial $x^4 - 6x^3 + 10x^2 - 6x + 9$.

149. By looking at $f(x) + f'(x)$, show that the polynomial

$$f(x) \;=\; x^6 - 6x^5 + 32x^4 - 128x^3 + 388x^2 - 776x + 777$$

has no repeated roots.

150. Show that $f(x) = x^3 + px + q$ has a double root if and only if $\dfrac{4p^3}{27} + q^2 = 0$.

151. Find polynomials $s(x), t(x) \in \mathbb{Z}_3[x]$ such that

$$s(x)(x^4 + 2x^2 + 1) + t(x)(x^4 + x^3 + 2x^2 + x + 1) \;=\; x^2 + 1 \;\; \text{in } \mathbb{Z}_3[x].$$

*152–153. For each polynomial $f(x)$, find all the values of the rational number m for which $f(x)$ has a rational root. For each of these values of m that are **integers**, factor $f(x)$ into irreducible factors in $\mathbb{Q}[x]$, $\mathbb{R}[x]$, and $\mathbb{C}[x]$.*

152. $f(x) = x^3 + 3x^2 + mx + 5$ **153.** $f(x) = x^3 + mx^2 - 5x - 3$

154.   **(a)** Let $f(x) = x^5 + 3x^4 + 2x^3 + 2x^2 + x - 1$. It is known that i is a root of $f(x)$. Factor $f(x)$ completely over the fields $\mathbb{Q}$, $\mathbb{R}$, $\mathbb{C}$, and $\mathbb{Z}_5$.

     (b) Let $g(x) = x^4 + x^3 - 2x^2 + 3x - 1$ in $\mathbb{Q}[x]$. Find $\gcd(f(x), g(x))$, where $f(x) \in \mathbb{Q}[x]$ is the polynomial in part (a).

155.   **(a)** Let $f(x) = x^4 - 6x^3 + 14x^2 - 16x + 8$. Find a repeated root of $f(x)$ in $\mathbb{Q}$, or prove that one does not exist.

     (b) Factor $f(x)$ into a product of irreducible polynomials in $\mathbb{Q}[x]$, $\mathbb{R}[x]$, $\mathbb{C}[x]$, and $\mathbb{Z}_5[x]$.

156. Find $\gcd(f(x), f(x+1))$ when $f(x) = x^4 - 5x^2 + 4$.

157. If a, b, and c are distinct integers, prove that there is no polynomial $p(x)$, with integer coefficients, for which $p(a) = b$, $p(b) = c$, and $p(c) = a$.

158. If $p(x) = 4x^5 + 8x^4 + 15x^3 + 30x^2 - 4x - 8$, then $p(2i) = 0$. Find all the real roots of $p(x)$, and hence factor $p(x)$ into irreducible factors in $\mathbb{Q}[x]$, $\mathbb{R}[x]$, $\mathbb{C}[x]$, and $\mathbb{Z}_7[x]$.

159. The polynomial $f(x) = x^5 + 2x^4 - 3x^2 - 4x - 2$ has $\sqrt{2}$ as one of its roots. Factor $f(x)$ into irreducible polynomials over the rational numbers, the real numbers, and the complex numbers.

160. The polynomial $f(x) = 3x^5 - 2x^4 + x^3 + 24x^2 - 16x + 8$ has $1 + \sqrt{3}i$ as a root. Find the irreducible factors of $f(x)$ over $\mathbb{Q}[x]$, $\mathbb{R}[x]$, and $\mathbb{C}[x]$.

161–164. Use the Cubic Formula 9.61, or any other method, to find all the complex solutions to each equation.

161. $x^3 + 12x + 12 = 0$ **162.** $x^3 + 3x - 2 = 0$
163. $x^3 - 3ix^2 - 4i = 0$ **164.** $27x^3 - 27x^2 + 12x - 1 = 0$

Appendix

TRIGONOMETRY

Mathematicians normally measure angles in radians. The *radian measure* of an angle is the ratio of the arc of a circle, subtended by the angle, to the radius of the circle. Therefore, an angle of θ radians subtends an arc of length θ on a unit circle. One revolution, or $360°$, is equal to 2π radians. An angle is just a real number (modulo 2π).

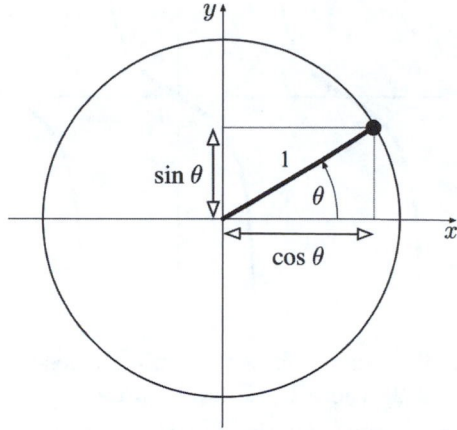

The values of the trigonometric functions **sine** and **cosine** are illustrated in the above diagram. Consider the unit circle in the coordinate plane and a radius vector making an angle θ with the positive x-axis. A positive angle is measured in a counterclockwise direction. The sine of θ, $\sin\theta$, is the projection of the radius vector onto the y-axis and the cosine of θ, $\cos\theta$, is the projection onto the x-axis.

The sine and cosine are real-valued functions whose domain is the set of all real numbers, $\mathbb{R}$, and whose image is the set of real numbers from -1 to 1.

All trigonometric functions are periodic of period 2π so, for example,

$$\sin(\theta + 2k\pi) \;=\; \sin\theta \quad \text{for all } \theta \in \mathbb{R} \text{ and all } k \in \mathbb{Z}.$$

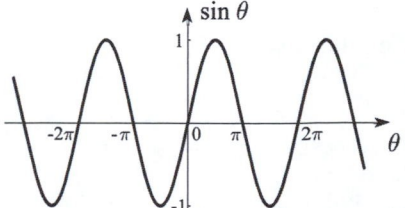 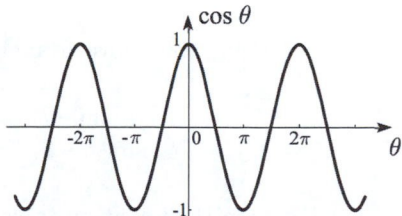

Four other trigonometric functions can be defined in terms of sine and cosine.

The **tangent** of θ, $\tan\theta = \sin\theta/\cos\theta$.

The **secant** of θ, $\sec\theta = 1/\cos\theta$.

The **cosecant** of θ, $\operatorname{cosec}\theta = 1/\sin\theta$.

The **cotangent** of θ, $\cot\theta = 1/\tan\theta$.

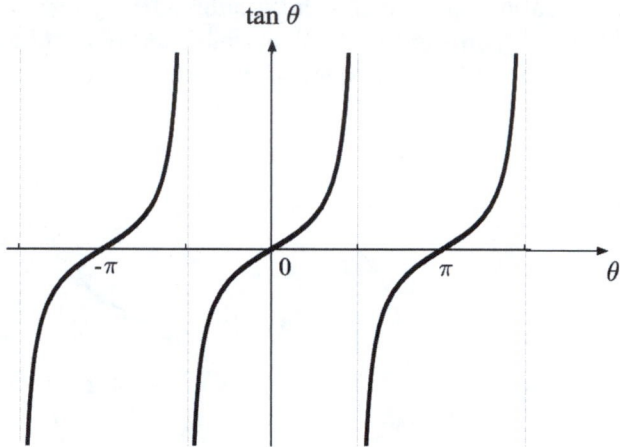

For example, $\tan\theta$ defines a function whose domain is all the real numbers, except $(2k+1)\pi/2$, where $k \in \mathbb{Z}$, and whose image is $\mathbb{R}$.

The trigonometric functions of the angles $\pi/4$, $\pi/6$, and $\pi/3$ should be remembered by means of the following right-angled triangles.

Using the above triangles and the relations

$$\begin{array}{rclcrcl}
\sin(-\theta) & = & -\sin\theta, & & \cos(-\theta) & = & \cos\theta, \\
\sin(\pi-\theta) & = & \sin\theta, & & \cos(\pi-\theta) & = & -\cos\theta,
\end{array}$$

we obtain the following table of values of their trigonometric functions.

θ		$\sin \theta$			$\cos \theta$			$\tan \theta$		
Radians	Degrees									
0	$0°$	0	$=$	0.0	1	$=$	1.0	0	$=$	0.0
$\pi/6$	$30°$	$1/2$	$=$	0.5	$\sqrt{3}/2$	$\approx$	0.886	$1/\sqrt{3}$	$\approx$	0.577
$\pi/4$	$45°$	$1/\sqrt{2}$	$\approx$	0.707	$1/\sqrt{2}$	$\approx$	0.707	1	$=$	1.0
$\pi/3$	$60°$	$\sqrt{3}/2$	$\approx$	0.866	$1/2$	$=$	0.5	$\sqrt{3}$	$\approx$	1.732
$\pi/2$	$90°$	1	$=$	1.0	0	$=$	0.0	∞	$=$	∞
$2\pi/3$	$120°$	$\sqrt{3}/2$	$\approx$	0.866	$-1/2$	$=$	-0.5	$-\sqrt{3}$	$\approx$	-1.732
$3\pi/4$	$135°$	$1/\sqrt{2}$	$\approx$	0.707	$-1/\sqrt{2}$	$\approx$	-0.707	-1	$=$	-1.0
$5\pi/6$	$150°$	$1/2$	$=$	0.5	$-\sqrt{3}/2$	$\approx$	-0.866	$-1/\sqrt{3}$	$\approx$	-0.577
π	$180°$	0	$=$	0.0	-1	$=$	-1.0	0	$=$	0.0
$7\pi/6$	$210°$	$-1/2$	$=$	-0.5	$-\sqrt{3}/2$	$\approx$	-0.866	$1/\sqrt{3}$	$\approx$	0.577
$5\pi/4$	$225°$	$-1/\sqrt{2}$	$\approx$	-0.707	$-1/\sqrt{2}$	$\approx$	-0.707	1	$=$	1.0
$4\pi/3$	$240°$	$-\sqrt{3}/2$	$\approx$	-0.866	$-1/2$	$=$	-0.5	$\sqrt{3}$	$\approx$	1.732
$3\pi/2$	$270°$	-1	$=$	-1.0	0	$=$	0.0	∞	$=$	∞
$5\pi/3$	$300°$	$-\sqrt{3}/2$	$\approx$	-0.866	$1/2$	$=$	0.5	$-\sqrt{3}$	$\approx$	-1.732
$7\pi/4$	$315°$	$-1/\sqrt{2}$	$\approx$	-0.707	$1/\sqrt{2}$	$\approx$	0.707	-1	$=$	-1.0
$11\pi/6$	$330°$	$-1/2$	$=$	-0.5	$\sqrt{3}/2$	$\approx$	0.866	$-1/\sqrt{3}$	$\approx$	-0.577
2π	$360°$	0	$=$	0.0	1	$=$	1.0	0	$=$	0.0

Standard Trigonometric Formulas.

$$\sin^2 \theta + \cos^2 \theta = 1$$

$$\sin\left(\tfrac{\pi}{2} - \theta\right) = \cos \theta$$

$$\cos\left(\tfrac{\pi}{2} - \theta\right) = \sin \theta$$

$$\sin(\theta + \phi) = \sin\theta \cos\phi + \cos\theta \sin\phi$$

$$\sin(\theta - \phi) = \sin\theta \cos\phi - \cos\theta \sin\phi$$

$$\cos(\theta + \phi) = \cos\theta \cos\phi - \sin\theta \sin\phi$$

$$\cos(\theta - \phi) = \cos\theta \cos\phi + \sin\theta \sin\phi$$

$$\sin 2\theta = 2 \sin\theta \cos\theta$$

$$\cos 2\theta = \cos^2 \theta - \sin^2 \theta$$

$$= 2 \cos^2 \theta - 1 = 1 - 2 \sin^2 \theta$$

$$\sin\theta + \sin\phi = 2 \sin\left(\frac{\theta + \phi}{2}\right) \cos\left(\frac{\theta - \phi}{2}\right)$$

$$\sin\theta - \sin\phi = 2 \cos\left(\frac{\theta + \phi}{2}\right) \sin\left(\frac{\theta - \phi}{2}\right)$$

$$\cos\theta + \cos\phi = 2 \cos\left(\frac{\theta + \phi}{2}\right) \cos\left(\frac{\theta - \phi}{2}\right)$$

$$\cos\theta - \cos\phi = -2 \sin\left(\frac{\theta + \phi}{2}\right) \sin\left(\frac{\theta - \phi}{2}\right)$$

INEQUALITIES

It is not always possible to reduce a mathematical problem to an exact equation, though it may be possible to obtain an inequality involving real numbers instead. Any inequality of the form $a > b$ may be rewritten as $a - b > 0$; in other words, an inequality can be reduced to the question of whether a certain expression is positive or not. We state some basic assumptions about the positive real numbers from which the properties of inequalities can be derived.

The *positive real numbers* have the following properties.

(i) If x and y are positive, then $x + y$ and xy are positive.

(ii) For each $x \in \mathbb{R}$, exactly one of the following three relations hold: x is positive, $-x$ is positive, or $x = 0$.

The relation *greater than* may then be defined in terms of the positive numbers by

$$a > b \text{ if and only if } a - b \text{ is positive.}$$

By "$a \geq b$," we mean $a > b$, or $a = b$. We say that a is *negative* if $-a$ is positive. For example, $6 > 4$, $-2 > -3$, $7 \geq 4$ and $7 \geq 7$.

Properties of Inequalities.

(i) If $a > b$ and $b > c$, then $a > c$. *(transitive property)*

(ii) If $a > b$ and $c > d$, then $a + c > b + d$.

(iii) If $a > b$ and $c > 0$, then $ac > bc$.

(iv) If $a > b$ and $c < 0$, then $ac < bc$.

(v) If $a > b$, then $-a < -b$.

(vi) If $ab > 0$, then either a and b are both positive, or a and b are both negative.

(vii) If $ab < 0$, then either a is positive and b negative, or a is negative and b positive.

(viii) If $a \neq 0$, then a^2 is positive.

Property (ii) shows that we may freely add inequalities, but Properties (v) and (iv) show that we have to be more careful when subtracting inequalities or multiplying both sides of an inequality by any expression. If we multiply both sides by a positive number, then the inequality sign remains the same, but if we multiply by a negative number, then the inequality sign is reversed.

The problem of solving an inequality can be reduced to the problem of solving the corresponding equality. If the graph of the function f is known, then the values of x for which $f(x) > 0$ or $f(x) \geq 0$ can be read off. The crucial parts of the graph are the points in which it crosses the x-axis. These are the values of x for which $f(x) = 0$. Therefore, the solution of an inequality depends on the solution of the corresponding equality.

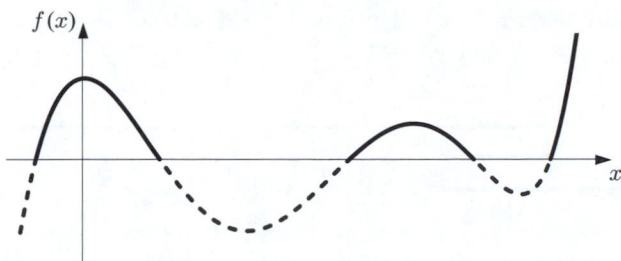

Even if the graph is not known exactly, it is helpful to check a solution to an inequality by making a rough sketch of the graph to see when it is above the axis.

Example.   Solve the inequality $x^2 - 3x - 4 > 0$.

Solution.   The quadratic $x^2 - 3x - 4$ can be factored, and the inequality becomes

$$(x - 4)(x + 1) > 0.$$

By Property (vi), either (i) $x - 4 > 0$ and $x + 1 > 0$, or (ii) $x - 4 < 0$ and $x + 1 < 0$. That is, either (i) $x > 4$ and $x > -1$, or (ii) $x < 4$ and $x < -1$. Hence either (i) $x > 4$, or (ii) $x < -1$.

The solution to the inequality is therefore $x > 4$ or $x < -1$. The solution set can be written as $\{x \in \mathbb{R} \mid x > 4 \text{ or } x < -1\}$, which is the union of the open intervals $(-\infty, -1) \cup (4, \infty)$.

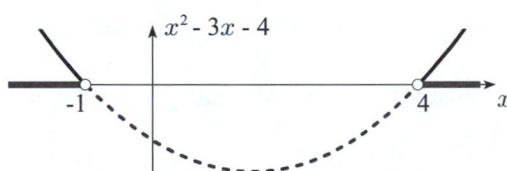

Check.   The graph of $f(x) = x^2 - 3x - 4$ is a parabola that crosses the x-axis at $x = -1$ and $x = 4$. From the graph we see that the solution to $f(x) > 0$ is $x > 4$ or $x < -1$. $\square$

Example.   Solve the inequality $\dfrac{x - 1}{x + 1} \ge 3$.

Solution.   Notice that x cannot be -1 because the left side is not defined there. Rewriting the inequality, we have

$$\frac{x - 1 - 3x - 3}{x + 1} \ge 0 \qquad \text{and so} \qquad \frac{2(x + 2)}{x + 1} \le 0.$$

Since $(x + 1) \ne 0$, we can multiply both sides of the inequality by the positive quantity $(x + 1)^2$ to obtain

$$2(x + 2)(x + 1) \le 0.$$

Therefore, either (i) $x + 2 \geq 0$ and $x + 1 < 0$, or (ii) $x + 2 \leq 0$ and $x + 1 > 0$. In case (i), $x \geq -2$ and $x < -1$ and we see that $-2 \leq x < -1$.

Case (i)   Case (ii)

In case (ii), $x \leq -2$ and $x > -1$, which is impossible.

The solution set to the original inequality is therefore $\{x \in \mathbb{R} \mid -2 \leq x < -1\}$, which is the half-open interval $[-2, -1)$.

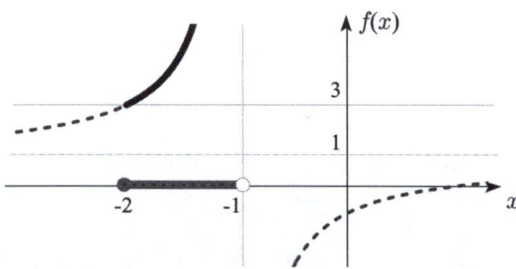

Check.   If we sketch the function

$$f(x) = \frac{x - 1}{x + 1}$$

we obtain a hyperbola and we see that $f(x) \geq 3$ when $-2 \leq x < -1$.   □

Example.   Solve the inequality $|x + 2| > 3x$.

Solution.   Recall that the absolute value $|x+2|$ is defined as $x+2$ if $x+2 \geq 0$ and as $-(x + 2)$ if $x + 2 < 0$. Therefore, whenever an inequality contains an absolute value, we must consider the different ranges of x separately.

Case (i) Let $x \geq -2$. The inequality now becomes

$$x + 2 > 3x$$

or $2 > 2x$. Hence $x < 1$, which, together with condition $x \geq -2$, implies that the inequality is satisfied when $-2 \leq x < 1$.

Case (ii) Let $x < -2$. The inequality, in this range, becomes $-x - 2 > 3x$ or $-2 > 4x$. Hence $x < -1/2$, which, together with the condition $x < -2$, implies that the inequality is satisfied when $x < -2$.

The complete solution set to the original inequality is therefore

$$\{x \in \mathbb{R} \mid x < -2 \text{ or } -2 \leq x < 1\} \;=\; \{x \in \mathbb{R} \mid x < 1\},$$

which is the open interval $(-\infty, 1)$.

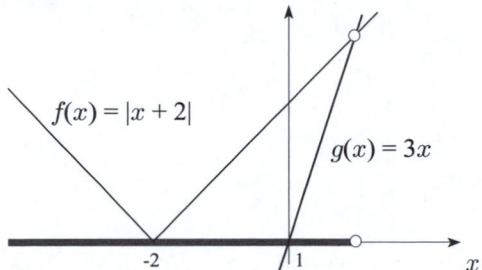

Check.   If we sketch the graphs of $f(x) = |x + 2|$ and $g(x) = 3x$ we see that

$$f(x) > g(x)$$

whenever $x < 1$.   □

One of the most important inequalities in many variables is the connection between the arithmetic and geometric means. Let $a_1, a_2, \ldots, a_n$ be n positive real numbers. Their *arithmetic mean* is the number

$$\text{AM}(a_1, a_2, \ldots, a_n) \quad = \quad \frac{a_1 + a_2 + \cdots + a_n}{n},$$

that is, the ordinary average of the numbers. Their *geometric mean* is the number

$$\text{GM}(a_1, a_2, \ldots, a_n) \quad = \quad \sqrt[n]{a_1 \cdot a_2 \cdots a_n}.$$

Theorem of the Means.   If a_1 and a_2 are positive real numbers, then

$$\text{AM}(a_1, a_2) \geq \text{GM}(a_1, a_2)$$

with equality holding if and only if $a_1 = a_2$.

Proof.   Consider the difference between the two means

$$\begin{aligned}
\text{AM}(a_1, a_2) - \text{GM}(a_1, a_2) \quad &= \quad \frac{a_1 + a_2}{2} - \sqrt{a_1 a_2} \\
&= \quad \frac{a_1 - 2\sqrt{a_1}\sqrt{a_2} + a_2}{2} \\
&= \quad \frac{(\sqrt{a_1} - \sqrt{a_2})^2}{2} \quad \geq 0
\end{aligned}$$

with equality holding if and only if $\sqrt{a_1} = \sqrt{a_2}$; that is, if and only if $a_1 = a_2$.

Hence $\dfrac{a_1 + a_2}{2} \geq \sqrt{a_1 a_2}$, with equality holding if and only if $a_1 = a_2$.   □

This theorem can be generalized to n variables.

Further Reading

Courant, Richard and Herbert Robbins, revised by Ian Stewart *What Is Mathematics?: An Elementary Approach to Ideas and Methods*, Oxford University Press, 1996.

Daepp, Ulrich and Pamela Gorkin *Reading, Writing, and Proving: A Closer Look at Mathematics*, Springer, 2003.

Liebeck, Martin *A Concise Introduction to Pure Mathematics*, CRC Press, 2000.

Schumacher, Carol *Chapter Zero: Fundamental Notions of Abstract Mathematics*, Addison-Wesley, 1996.

Stewart, Ian and David Tall *The Foundations of Mathematics*, Oxford University Press, 1977.

Weiss, Edwin *First Course in Algebra and Number Theory*, Academic Press, 1971.

Answers

To the Odd-Numbered Exercises and Problems

EXERCISE SET 1, PAGE 20

1. True statement

5. Not a statement

3. Not a statement

7.

P	NOT(NOT P)
T	T
F	F

9.

P	Q	R	$P \Longrightarrow$ $(Q$ OR $R)$
T	T	T	T
T	T	F	T
T	F	T	T
T	F	F	F
F	T	T	T
F	T	F	T
F	F	T	T
F	F	F	T

11.

P	Q	R	$(P$ OR NOT $Q)$ $\Longrightarrow R$
T	T	T	T
T	T	F	F
T	F	T	T
T	F	F	F
F	T	T	T
F	T	F	T
F	F	T	T
F	F	F	F

15.

P	Q	P NOR Q
T	T	F
T	F	F
F	T	F
F	F	T

17. $Q \Longrightarrow P$

19. $P \Longrightarrow Q$

21. $P \Longleftrightarrow Q$

25. Equivalent

27. Q AND NOT P

29. If I have broken my leg, then I cannot walk.

31. If I take the bus, then I have broken my leg or I cannot walk.

33. $\exists x \, \forall y \ (0 < x \leq y)$; integers

35. $\forall x \, \exists y \, \exists z \ (x = yz)$; integers

37. $\exists x \, \forall y \ (x^3 + x = y)$; real numbers

39. $\exists x \, \exists y \ (x^2 - 2y^2 = 3)$; integers

41. $\exists x \, (\text{NOT } P(x) \text{ AND NOT } Q(x))$

43. $\forall x \, (P(x) \text{ AND NOT } Q(x))$

45. Every real number is as large as any real number. False

47. There is a smallest real number. False

49. For every real number x there is a real numbers y such that $x^2 + y^2 = 1$. False

51. Equivalent

53. Not equivalent; look for a counterexample where $(\forall x P(x))$ is false.

55. If I don't go to the party, then Tom will not go to the party.
If I go to the party, then Tom will go to the party.

57. If $x^2 \leq 9$, then $x \leq 3$. If $x^2 > 9$, then $x > 3$.

59. If an integer is prime, then it is not divisible by 2.
    If an integer is not prime, then it is divisible by 2.
61. If $x < -3$ or $x > 3$, then $x^2 + y^2 \neq 9$. If $-3 \leq x \leq 3$, then $x^2 + y^2 = 9$.
67. Counterexample when $x = 0$

PROBLEM SET 1, PAGE 22

73. $\exists x \, ((x \in S) \text{ AND } (x \in T) \text{ AND } (x \notin U))$.
75. $\exists \epsilon > 0 \; \forall \delta > 0 \; \exists x \quad (0 < |x - a| < \delta \text{ AND } |f(x) - L| \geq \epsilon)$
79. Yes   **83.** P AND NOT Q

EXERCISE SET 2, PAGE 50

1. $q = 4$, $r = 1$ **3.** $q = 1$, $r = 0$
5. $q = -3$, $r = 0$ **7.** $q = -23$, $r = 7$
11. *Hint:* Let $g = \gcd(a, b)$ so that $\exists \, x, y \in \mathbb{Z}$ with $ax + by = g$.
13. 11   **15.** 1
17. 16
19. $11 = 14 \cdot 484 + (-15)451$ is one form
21. $1 = (-7)17 + 8 \cdot 15$ is one form
23. $5 = (-1)100 + (-3)(-35)$ is one form
25. $17 = 0 \cdot 51 + 1 \cdot 17$ is one form   **29.** $195/782$
31. $x = 2$, $y = -1$ is one solution   **33.** $x = 9$, $y = 0$ is one solution
35. No solutions
37. $x = 4 - 9n$, $y = 7n - 3$, where $n \in \mathbb{Z}$, is one way to write all the solutions
39. $x = 8n - 1$, $y = 5n - 1$, where $n \in \mathbb{Z}$, is one way to write all the solutions
41. No solutions
43. $(x, y) = (2, 108)$, $(11, 94)$, $(20, 80)$, $(29, 66)$, $(38, 52)$, $(47, 38)$, $(56, 24)$, or $(65, 10)$
45. No nonnegative solutions
47. Yes; $1000 = 847 + 153 = 660 + 340 = 473 + 527 = 286 + 714 = 99 + 901$
49. Yes; $120 = 84 + 36$ **51.** Yes, in two ways
53. 311   **55.** 16033
57. $(1628)_9$ **59.** $(1010)_3$
61. $(BAX)_{60}$, where $B = 11$, $A = 10$ and $X = 39$
63. $(3233)_4$ and $(1001110)_4$ **67.** 6
71. $2^5 \cdot 3 \cdot 5 \cdot 11$, $2^3 \cdot 5^2 \cdot 17^2$, $\gcd = 40$, $\text{lcm} = 7629600$

PROBLEM SET 2, PAGE 52

75. No; for example, $a = b = 1$, $c = 3$
77. $c = 13, 18, 21, 23, 26, 67, 70, 72, 75, 80$ and all c in the range $28 \leq c \leq 65$ except $30, 32, 35, 40, 53, 58, 61, 63$.

79. 18 small and 10 large trucks, or 1 small and 24 large trucks

81. No

85. Solve $ax + \gcd(b, c) \cdot t = e$ and $by' + cz' = \gcd(b, c)$ and take $y = ty'$, $z = tz'$.

87. One solution is $x = 10$, $y = -125$, $z = 25$.

91. Base 4 for 1000, and base 4, 5 or 6 for 10^6. *Hint:* To name all numbers less than 10^s in base b, we have to name all numbers less than b^r, where r is $\log_b 10^s$ rounded up to an integer.

93. *Hint:* If p, q are odd primes with $p < q$, then $p < k < q$, where $k = (p + q)/2$ is a composite integer.

95. *Hint:* Look at the powers of 2 in the numerator and denominator.

99. True

101. True; look at the prime factorizations.

103. $n = 41$ yields the number $1681 = 41^2$.

105. (a) $-44, 44$ (b) $(19291)_{-10}$, $(1811)_{-10}$

107. *Hint:* Notice that $3(3b + 2c) - 2(2b + 3c) = 5b$.

EXERCISE SET 3, PAGE 82

1. -12, -4, and 0; -11, 1, and 5; -6, 2, and 10; -9, -1, 3, and 7.

3. 1   **5.** 6

9. $k = 16$ **11.** $k = 9$

13. The sum of the digits is divisible by 3 and the number formed by the last two digits is divisible by 4.

15. The sum of the digits is divisible by 7.

17. 2, 4, 5, 8 and 10   **19.** 2, 3, 4, 6 and 11

21. The calculation is incorrect.

23. Reflexive and transitive   **25.** Reflexive and symmetric

27.

+	[0]	[1]
[0]	[0]	[1]
[1]	[1]	[0]

·	[0]	[1]
[0]	[0]	[0]
[1]	[0]	[1]

$[1]^{-1} = [1]$

29.

+	[0]	[1]	[2]	[3]	[4]	[5]	[6]
[0]	[0]	[1]	[2]	[3]	[4]	[5]	[6]
[1]	[1]	[2]	[3]	[4]	[5]	[6]	[0]
[2]	[2]	[3]	[4]	[5]	[6]	[0]	[1]
[3]	[3]	[4]	[5]	[6]	[0]	[1]	[2]
[4]	[4]	[5]	[6]	[0]	[1]	[2]	[3]
[5]	[5]	[6]	[0]	[1]	[2]	[3]	[4]
[6]	[6]	[0]	[1]	[2]	[3]	[4]	[5]

·	[0]	[1]	[2]	[3]	[4]	[5]	[6]
[0]	[0]	[0]	[0]	[0]	[0]	[0]	[0]
[1]	[0]	[1]	[2]	[3]	[4]	[5]	[6]
[2]	[0]	[2]	[4]	[6]	[1]	[3]	[5]
[3]	[0]	[3]	[6]	[2]	[5]	[1]	[4]
[4]	[0]	[4]	[1]	[5]	[2]	[6]	[3]
[5]	[0]	[5]	[3]	[1]	[6]	[4]	[2]
[6]	[0]	[6]	[5]	[4]	[3]	[2]	[1]

$[1]^{-1} = [1]$, $[2]^{-1} = [4]$, $[3]^{-1} = [5]$, $[4]^{-1} = [2]$, $[5]^{-1} = [3]$, $[6]^{-1} = [6]$

33. $x \equiv 5$ or 12 (mod 14)   **35.** $x \equiv 103$ (mod 128)

37. No solution   **39.** $x \equiv 1$ or 5 (mod 8)

41. $x \equiv 1$ (mod 2)   **43.** [21]

45. $x = [5]$ **47.** $x = [0], [2], [3]$ or $[5]$

49. $x \equiv 19 \pmod{20}$

51. $x \equiv 17 \pmod{42}$

53. $x \equiv 65 \pmod{161}$

55. $x = 124$ or 299

PROBLEM SET 3, PAGE 84

59. Yes

61. $2^{91} \not\equiv 2 \pmod{91}$

63. $k = 1$ only

65. Prove $2^{32} \equiv -1 \pmod{641}$ using $5^5 \cdot 2^3 \equiv 1$ and $5 \cdot 2^7 \equiv -1 \pmod{641}$.

67. (b) $-\pi$ and π, $11\pi/6$ and $23\pi/6$.

71. $x = [1]$, $y = [6]$

73. Yes

75. 4

79. *Hint:* If $a \not\equiv 0 \pmod{p}$ there exists b such that $ab \equiv 1 \pmod{p}$. Hence the nonzero congruence classes in $\mathbb{Z}_p$ can be paired off, $[a]$ with $[b]$, where $[a][b] = [1]$; $[1]$ and $[p-1]$ being paired with themselves.

81. $x \equiv 25, 51$ or $77 \pmod{78}$, or, equivalently, $x \equiv 25 \pmod{26}$

83. 144 days and 720 days

85. No solution

89. 119

91. $x \equiv 55a_1 + 45a_2 \pmod{99}$

97. Use the Division Algorithm.

99. 26

101. 49

103. 75

EXERCISE SET 4, PAGE 104

1. 10

3. 70

7. $n = 7$

9. $\sum_{r=3}^{n}(r^2 - 1)$

19. $(n+1)! - 1$

21. True

23. True

35. $a^5 - 5a^4 + 10a^3 - 10a^2 + 5a - 1$

37. $256x^8 - 768x^6y^3 + 864x^4y^6 - 432x^2y^9 + 81y^{12}$

39. 79.925

41. $26400000x^{22}$

PROBLEM SET 4, PAGE 106

49. *Hint:* In the convex $(k+1)$-gon with consecutive vertices $a_1, a_2, \ldots, a_k, a_{k+1}$, consider the diagonal a_1a_k that divides the $(k+1)$-gon into a triangle with vertices a_1, a_k, a_{k+1}, and a k-gon with vertices $a_1, a_2, \ldots, a_k$.

51. Prove $r! | (n+1)(n+2) \cdots (n+r)$ whenever $n + r = k$, by induction on k.

55. Move the smallest disc to peg B, if n is odd, or to peg C, if n is even.

59. 1, 1, 2, 3, 5, 8, 13, 21, 34, 55, 89, 144, 233, 377, 610

61. The remainders are the previous terms in the sequence.

69. *Hint:* Use as induction hypothesis, "If n guests had a gold star on their back, then all those logicians with a gold star would know when the host asked for the nth time."

71. $(n^2 + n + 2)/2$. *Hint:* Notice that if a straight line is added to a collection of straight lines in the plane and intersects the collection in r points, then the number of regions into which the collection of lines divides the plane is increased by $r + 1$.

79. $x_n x_{n-1} \cdots x_2 x_1$

81. $x_1 - 2x_2 + 4x_3 + \cdots + (-2)^{n-1}x_n$

83. $e(0) = 1$; $e(n) = e(n-1) + \frac{1}{n!}$ for $n > 0$

EXERCISE SET 5, PAGE 121

9. a, c, e

11. $f/2$, $f/2.8$, $f/4$, $f/5.7$, $f/8$, $f/11.3$, $f/16$ approximately

13. 16   **15.** 27

17. $\sqrt{6}$ **19.** 4

21. $3a^{13/4} - a^{3/2}$ **23.** $.08\dot{3}$

25. $.175$ or $.174\dot{9}$ **27.** $421/200$

29. $19/45$ **31.** $487/3700$

PROBLEM SET 5, PAGE 122

33. $\mathbb{Z}$ is the quotient set of $\mathbb{P} \times \mathbb{P}$ under the equivalence relation $\sim$, defined by $(a, b) \sim (c, d)$ if and only if $a + d = b + c$.

35. (a) $\sqrt{2}$ and $1 - \sqrt{2}$ (b) $\sqrt{2}$ and $\sqrt{2}$ (c) Yes; consider $x = \sqrt{2}^{\sqrt{2}}$. If x is rational this provides an example. If x is irrational, then $x^{\sqrt{2}} = (\sqrt{2}^{\sqrt{2}})^{\sqrt{2}} = (\sqrt{2})^2 = 2$ provides an example.

39. $3\sqrt[3]{9} + 3\sqrt[3]{6} + 4\sqrt[3]{4}$. *Hint:* Use the formula $a^3 - b^3 = (a - b)(a^2 + ab + b^2)$.

41. (b) Yes   (c) See Problem 42

43. $(0.13)_6$ **45.** $0.2453\dot{7}\dot{0}$

47. $(0.\dot{0}1\dot{1}0)_3$ **49.** $1315/504$

EXERCISE SET 6, PAGE 153

1. Not a function   **3.** Not a function

5. $X = \mathbb{R}$
$Y = \{y \mid y \geq -1\}$
$f : X \to Y$ has no inverse

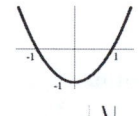

7. $X = \mathbb{R} - \{-1, 0\}$
$Y = \mathbb{R} - \{0, 1\}$
$f : X \to Y$ has an inverse
$f^{-1}(x) = \frac{1-x}{x}$

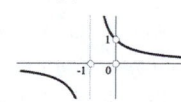

9. $X = \mathbb{R}$
$Y = [-1, 1]$
$f : X \to$ has no inverse

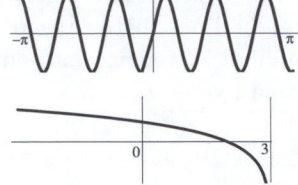

11. $X = \{x \mid x < 3\}$
$Y = \mathbb{R}$
$f : X \to Y$ has an inverse
$f^{-1}(x) = 3 - 10^x$

13. $X = \bigcup_{k \in \mathbb{Z}} [2k\pi, (2k+1)\pi]$, $Y = [0, 1]$, $f : X \to Y$ has no inverse.

15.

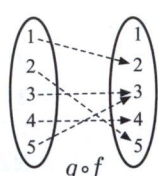

$f \circ g$ $g \circ f$

17.

$g \circ h$ $f \circ (g \circ h)$ $(f \circ g) \circ h$

19. $X = \{x \mid x \le 3\}$, $Y = \{y \mid y \ge 0\}$, $f^{-1}(x) = 3 - x^2$

21. $X = Y = \mathbb{R} - \{1/2\}$, $f^{-1}(x) = f(x)$

23. $X = \{x \mid x \ge 0\}$, $Y = \{y \mid y \ge 1\}$, $f^{-1}(x) = (\log_3 x)^2$

25. Not an injection   **27.** Not an injection

29.

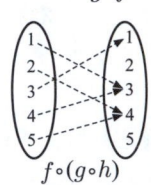

31.

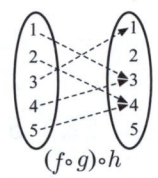

33. $f \circ g(x) = x^2 - 2x$, $g \circ f(x) = x^2 - 2$, $f \circ f(x) = x^4 - 2x^2$, $g \circ f \circ g(x) = x^2 - 2x - 1$

35. $f^{-1}(n) = f(n)$ **39.** Bijective, $f^{-1}(x) = \sqrt[3]{x} + 2$

41. Injective only   **43.** Neither need be bijective

45. *Hint::* Use the distributive law $(A \cup B) \cap C = (A \cap C) \cup (B \cap C)$.

49. The domain is $(-\infty, -1] \cup [1, \infty)$ and the image is $[-\pi, -\pi/2) \cup [0, \pi/2)$ or $[0, \pi/2) \cup (\pi/2, \pi]$; $y = \mathrm{Sec}^{-1}x$ if $x = \sec y$.

51. $x = 1$ **53.** $\mathrm{Sin}^{-1}x + \mathrm{Cos}^{-1}x = \pi/2$

55.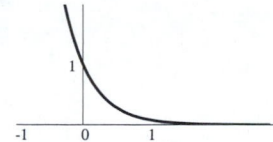

57.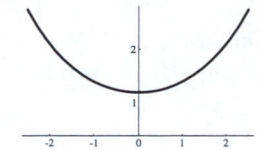

59. $x = 256$

61. $x = \log_e \sqrt{2} = (\log_e 2)/2$

63. $x = 1/\sqrt[3]{5}$

65. $e^{x \log_e x}$

67. About 13%

69. 18 radians or about 3 turns

73. $\begin{pmatrix} 1 & 2 & 3 & 4 \\ 3 & 2 & 1 & 4 \end{pmatrix}$

75. $\begin{pmatrix} 1 & 2 & 3 & 4 \\ 3 & 4 & 2 & 1 \end{pmatrix}$

77. $\begin{pmatrix} 1 & 2 & 3 & 4 \\ 4 & 2 & 1 & 3 \end{pmatrix}$

79. $\begin{pmatrix} 1 & 2 & 3 & 4 \\ 3 & 1 & 2 & 4 \end{pmatrix}$

PROBLEM SET 6, PAGE 157

81. (a) 0, (b) $f(-x) = -f(x)$, (c) np, (d) np, (e) rp

83. $f(x) = 2^x$ **85.** $f(x) = 0$

87. $ad - bc \neq 0$, and $a = -d$ or $a = d$, $b = c = 0$.
   Hint: If $a_2 x^2 + a_1 x + a_0 = 0$ for all $x \in \mathbb{R}$, then $a_2 = a_1 = a_0 = 0$.

89. $(2,1), (3,2), (3,1), (4,3), (4,2), (4,1)$

91.

93. Call the number and see who answers. Then check this name in the telephone directory to see if it corresponds to the given number. Alternatively, check the Web for a Reverse Phone Number Search for your area; in this case a computer has been used to find the inverse function.

95. f does have an inverse; f can be shown to be monotone increasing by either using the calculus, or directly. It is difficult to find an equation for the inverse function.

97. $n!/(n - r)!$

99. *Hint:* To prove f is injective, suppose $f(x_1) = f(x_2)$, and take $T = \{t\}$ with $g(t) = x_1$ and $h(t) = x_2$.

107. $57.04

101. *Hint:* Suppose there is a surjection $g : X \to \mathcal{P}(X)$. Consider the subset of X, $T = \{a \in X \mid a \notin g(a)\}$. Since $T \in \mathcal{P}(X)$, it is the image of an element $b \in X$. Is b an element of T or not?

111. (a) *Hint:* If $1 \leq n < m - 1$ use the fact that
$$\frac{1}{n+1} + \frac{1}{(n+1)(n+2)} + \cdots + \frac{1}{(n+1)(n+2)\cdots(m)} < \frac{1}{n+1} + \frac{1}{(n+1)^2} + \cdots + \frac{1}{(n+1)^m}.$$
   (b) 2.7182

113. (ij)
115. One form is $(13) \circ (12)$.
119. When $\gcd(a, b) = 1$

EXERCISE SET 7, PAGE 180

3. $(25, 323), (265, 323)$
5. $(5, 10379), (8141, 10379)$
7. $(4493, 7663)$
9. $(621, 47083)$
13. 71 and 97
15. 1291 and 2131
17. 115
19. 3884
21. 2
23. 4218
25. 833
27. 2469
29. 2210
31. 24
33. 20
35. 143
37. 65018
39. About 0.03 seconds

PROBLEM SET 7, PAGE 182

47. (b) 7
49. Check that the product of your numbers is n.
51. 1 and 106 (mod 133)

EXERCISE SET 8, PAGE 218

1. 2
3. $3 + 11i$
5. 3
7. $13 + i$
9. $2 + 6i$
11. $(2 - i)/5$
13. $1 - i$
15. i
17. 0
19. $2 - 11i$
21. $i/5$
23. 4 and -7
25. -3 and 0
27. $3 + 2i$ and $\sqrt{13}$
29. i and 1
31. $2 - \frac{i}{\sqrt{2}}$ and $\frac{3\sqrt{2}}{2}$

33.

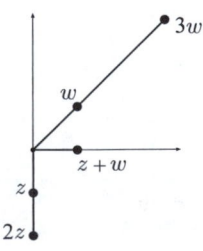

35.

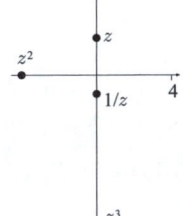

37.

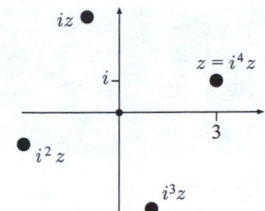

39. $(\sqrt{2}, 7\pi/4)$

41. $(2, 4\pi/3)$

43. $(-\sqrt{6}/2, \sqrt{6}/2)$

45. $(-1/2, -\sqrt{3}/2)$

47. $8(\cos\pi + i\sin\pi)$

49. $2(\cos\frac{3\pi}{2} + i\sin\frac{3\pi}{2})$

51. $\frac{\sqrt{6}}{3}(\cos\frac{\pi}{4} + i\sin\frac{\pi}{4})$

53. 4

55. $-\sqrt{2} - \sqrt{2}i$

57. 3 and $\frac{3\pi}{2}$

59. 2^{15} and $\frac{3\pi}{2}$

61. $0, 1$ or $(-1 \pm \sqrt{3}i)/2$

63. $(\sqrt{3} + i)/2$

65. $(-1 + \sqrt{3}i)/2048$

67. $-7/9$

71. $\pm 1, \pm i$

73. $\frac{\pm 1 \pm i}{\sqrt{2}}$

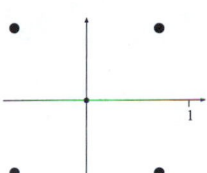

75. $\pm\frac{\sqrt[4]{2}}{2}(1 + \sqrt{3}i)$
$\pm\frac{\sqrt[4]{2}}{2}(\sqrt{3} - i)$

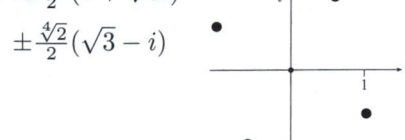

77. $\pm\frac{(1+i)}{\sqrt{2}}$

79. $\frac{1 \pm \sqrt{111}i}{14}$

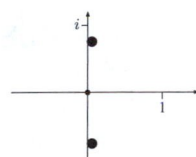

81. $\frac{-i \pm \sqrt{15}}{8}$

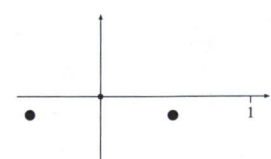

83. $\frac{1+i}{\sqrt[3]{2}}$
$\frac{-1 - i \pm \sqrt{3}(1-i)}{2\sqrt[3]{2}}$

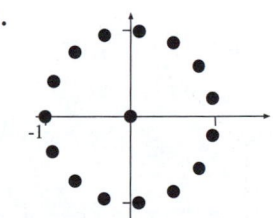

89.

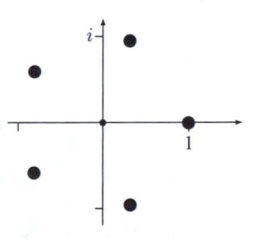

91.

93.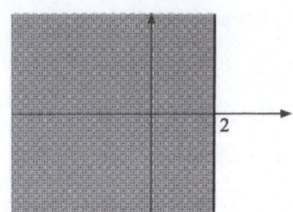

95.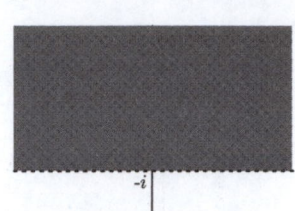

97.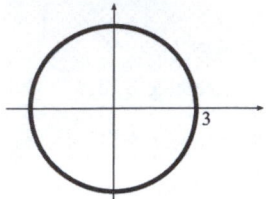

99.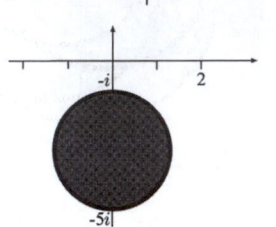

101.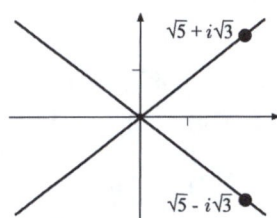

103.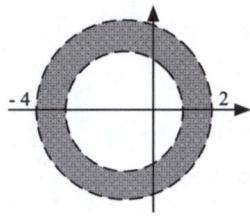

105. $e^{(\log_e 2 + 3i\pi/2)}$

107. $e^{\log_e 3}$

PROBLEM SET 8, PAGE 221

111. $(-\sqrt{3} - 3i)/2\sqrt{2}$

113. The sum of the squares of the diagonals of a parallelogram equals the sum of the squares of its sides.

115.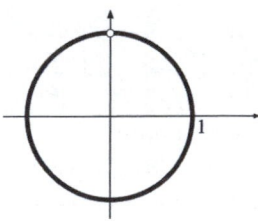

117.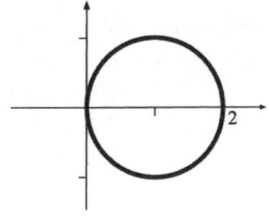

119. 0 and $\operatorname{cis}(2k\pi/n)$ for $k = 0, 1, 2, \ldots, n - 1$ and $\mathbb{R}$, if $n = 2$

121. (b) $\tan 5t = -11\sqrt{2} \approx -15.556$

125. *Hint:* Apply the geometric series and De Moivre's Theorem to $\sum (\operatorname{cis} \theta)^k$.

127. $\pm 4 \pm i$ 　　　　　　**129.** $\pm(1 - i), \pm(1 + i)/\sqrt{2}$

131. $U = \frac{e^{ax}}{a^2 + b^2}(a \cos bx + b \sin bx) + C, \quad V = \frac{e^{ax}}{a^2 + b^2}(a \sin bx - b \cos bx) + C$

133.

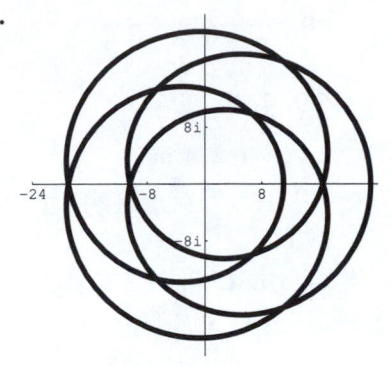

135. $(\sqrt{5}-1)/4$

137. $-2i, -4, 2+i$

139. $(1110)_{-1+i}, (110011)_{-1+i}$

EXERCISE SET 9, PAGE 262

1. $2x^3 + 2x^2 - 9x, 2x^3 - 5x + 2, 2x^5 - 3x^4 - 11x^3 + 14x^2 + 5x - 1$

3. $2x^2 + 2, -2\sqrt{2}x, x^4 + 1$

5. $x^4 + x^3 + x^2, 2x^4 + x^3 + 2x^2 + x + 1, x^7 + x^2 + x + 2$

7. $3x, 4x^2 + 4x + 3, x^4 + 2x^3 + x^2 + 4x + 4$

9. No; yes

11. $x^3, x^3 + 1, x^3 + x, x^3 + x + 1, x^3 + x^2, x^3 + x^2 + 1, x^3 + x^2 + x, x^3 + x^2 + x + 1$

13. $1, -2, 5/3$

15. $1, -1, (-1 + \sqrt{41})/4, (-1 - \sqrt{41})/4$

17. $x + 3$ and $7x - 4$

19. $x^2 + \frac{1}{2}x + \frac{1}{2}$ and 0

21. $x^2 + 1 + i$ and $2 - i$

23. $2x$ and $x + 2$

25. $x^7 + x^4 + x^2 + x + 1$ and x

27. No

29. -412

31. $-43 + 40i$

33. $1915/243$

35. Yes

37. $p = m^3 - 2m, q = m^2 - 1$

39. $1, 1$

41. No solutions

43. No solutions

45. 2

49. $x^4 - 14x^2 + 9$

51. $-1/2, \sqrt{2}, \sqrt{2}, -\sqrt{2}, -\sqrt{2}$

53. $-1, -4/3, -3/2$

55. $0, 1 + i, 1 - i, -1 + i, -1 - i$

57. $2 - \sqrt{3}, 2 - \sqrt{3}, 2 + \sqrt{3}, 2 + \sqrt{3}$

59. $x^4 - 14x^3 + 69x^2 - 140x + 98$

61. $-1 - \sqrt{2}, -1 + \sqrt{2}, \sqrt{3}i, -\sqrt{3}i$

63. $5 + i, 5 - i, \sqrt{3}, -\sqrt{3}$

65. In $\mathbb{Q}[x]$ only

67. Four real solutions; the rational solution -1 and three irrational solutions, one between -2 and -1, one between 3 and 4 and one between 5 and 6

69. One irrational solution between 1 and 2

71. $-3/4, (3 - \sqrt{5})/2 \approx 0.4$ and $(3 + \sqrt{5})/2 \approx 2.6$

73. -0.2

75. $-4 < x < 1$ and $1 < x < 7$

77. $x < (1 - \sqrt{41})/4, -1 < x < 1/2$ and $(1 + \sqrt{41})/4 < x$

79. $2/3$ with multiplicity 3

81. $1 \pm \sqrt{2}$, each of multiplicity 2

83. $a = n, b = -n - 1$

85. $x^2 + i$

87. $\dfrac{12}{4x-1} - \dfrac{3}{x}$

89. $\dfrac{3}{4x-1} + \dfrac{2}{x+1}$

91. $\dfrac{1}{x+5} - \dfrac{2}{(x+1)^2}$

93. $\dfrac{2x+3}{x^2+x+1}$

95. -14 in $\mathbb{Q}$ and $\mathbb{R}$; -14, $(-1 \pm \sqrt{3}i)/2$ in $\mathbb{C}$; 0, 2, 4 in $\mathbb{Z}_7$

97. $(x-2)(2x^2+6x+1)$ in $\mathbb{Q}[x]$; $2(x-2)\left(x+\frac{3-\sqrt{7}}{2}\right)\left(x+\frac{3+\sqrt{7}}{2}\right)$ in $\mathbb{R}[x]$ and $\mathbb{C}[x]$; $2(x+1)^2(x+2)$ in $\mathbb{Z}_3[x]$

99. $(x+1)(2x+1)\left(x+\frac{1+\sqrt{3}i}{2}\right)\left(x+\frac{1-\sqrt{3}i}{2}\right)$ in $\mathbb{C}[x]$; $(x+1)(2x+1)(x^2+x+1)$ in $\mathbb{Q}[x]$ and $\mathbb{R}[x]$; $2(x+1)(x+3)(x^2+x+1)$ in $\mathbb{Z}_5[x]$

PROBLEM SET 9, PAGE 265

109. $x^2 - 96x + 4$

111. $x^3 - 10x^2 - 11x - 25$

113. When $\mathbb{F}$ is infinite

117. (a) $\triangle f(n) = \binom{n}{r-1}$

119. x, $x+1$, x^2+x+1, x^3+x+1, x^3+x^2+1, x^4+x+1, x^4+x^3+1, $x^4+x^3+x^2+x+1$

121. $x^n + x$

123. Irrational

127. 4

129. No rational roots

133. $\cos \frac{2\pi}{5} = \left(\sqrt{5}-1\right)/4$

139. *Hint:* Prove the result by induction, using the facts that $f'_n(x) = f_{n-1}(x)$ and that $f_{2r}(x) = x^{2r}/(2r)! > 0$ whenever $f'_{2r}(x) = 0$.

141. $1 < x < 2$ and $3 < x < 4$

143. $1.3 < x < 2.3$

151. $s(x) = 1+x$, $t(x) = -x$

153. $m = -1$, 7, $5/3$; if $m = -1$, then $f(x) = (x+1)^2(x-3)$ in $\mathbb{Q}[x]$, $\mathbb{R}[x]$, and $\mathbb{C}[x]$; if $m = 7$, then $f(x) = (x-1)(x^2+8x+3)$ in $\mathbb{Q}[x]$ and $f(x) = (x-1)(x+4-\sqrt{13})(x+4+\sqrt{13})$ in $\mathbb{R}[x]$ and $\mathbb{C}[x]$

155. (b) $(x-2)^2(x^2-2x+2)$ in $\mathbb{Q}[x]$ and $\mathbb{R}[x]$; $(x-2)^2(x-1+i)(x-1-i)$ in $\mathbb{C}[x]$; $(x+1)(x+2)(x+3)^2$ in $\mathbb{Z}_5[x]$

159. $(x+1)(x^2-2)(x^2+x+1)$ over the rationals; $(x+1)(x-\sqrt{2})(x+\sqrt{2})(x^2+x+1)$ over the reals and $(x+1)(x-\sqrt{2})(x+\sqrt{2})\left(x+\frac{1+\sqrt{3}i}{2}\right)\left(x+\frac{1-\sqrt{3}i}{2}\right)$ over the complex numbers

161. $2^{2/3} - 2^{4/3}$, $2^{1/3} - 2^{-1/3} \pm i\sqrt{3}\left(2^{1/3} + 2^{-1/3}\right)$

163. $-i$, $2i$, $2i$

List of Symbols

Symbol	Meaning	Page
$\prod$	product	105
$\sum$	sum	92
$\phi(n)$	Euler ϕ-function	78
$\in$	belongs to	7
$\notin$	does not belong to	7
$\subseteq$	is contained in	8
$\supseteq$	contains	8
$\cap$	intersection	8
$\cup$	union	8
$\approx$	approximately equal to	40
$\sim P$	NOT P	3
$\implies$	implies	4, 7
$\iff$	if and only if	6, 7
$\#X$	number of elements or cardinality of the set X	138
$g \circ f$	composition of functions	130
$a\|b$	a divides b	25
$a \equiv b \pmod{m}$	a is congruent to b modulo m	57
aRb	a is related to b	62
S/R	quotient set of S by R	63
$\binom{n}{r}$	n choose r	99
$\begin{pmatrix} 1 & 2 & \dots & n \\ \sigma_1 & \sigma_2 & \dots & \sigma_n \end{pmatrix}$	permutation of $1, 2, \dots, n$	151
$n!$	factorial n	91
$f : X \to Y$	function f with domain X and codomain Y	125
$x \mapsto y$	x maps to y	130
b^x	exponential function	146
a^{-1}	inverse of a	66
$\sqrt[n]{a}$	nth root of a	114
$a^{m/n}$	nth root of a^m	115
f^{-1}	inverse of the function f	133
$f(x)$	image of x under f	125
$f(X)$	image of the set X under f	125
$X \times Y$	Cartesian product of the sets X and Y	8
$\overline{z}$	conjugate of z	194
$\|z\|$	modulus or absolute value of z	128, 192
$\lfloor x \rfloor$	greatest integer less than or equal to x	27, 154
$[a]$	equivalence class, or congruence class containing a	62, 64
$\{x\}$	set containing x	8
$(r_n \dots r_1 r_0)_b$	base b expansion	42
$b.\dot{a}_1\dot{a}_2$	recurrent decimal	118
(x, y)	Cartesian coordinates	199
(r, θ)	polar coordinates	199
$\square$	end of a proof or example (Q.E.D.)	4

Index